Karen E.

P9-CDE-670

ELECTRONICS
MADE SIMPLE

REVISED EDITION

BY

HENRY JACOBOWITZ

MADE SIMPLE BOOKS

DOUBLEDAY & COMPANY, INC.

GARDEN CITY, NEW YORK

ABOUT THIS BOOK

Electronics today profoundly affects our everyday lives. We fly aboard electronically guided airplanes, step through electronically opened doors, operate electronic "brains," are entertained by radio and television, listen to hi-fi music, and rely on a multiplicity of gadgets and processes unthinkable without electronics.

The thoughtful person, who has a general cultural interest in electronics, has the choice of "digging" into a complicated electronics text, or glossing over a "popular" electronics book that scratches only the surface of electronic applications but generally avoids going into essential principles.

This book is intended to steer a middle course between the formal engineering text and the usually superficial popular primer. It *is* a text in the sense that it is a systematic arrangement of the principles and applications of electronics, but it has been "made simple" by using informal language and dispensing with all mathematics, except arithmetic and elementary algebra. In doing this, however, neither rigor nor accuracy in the presentation of fundamental principles has been sacrificed. Therefore, the book is suitable for serious self-study by all persons interested in electronics. This may include, in addition to the general reader, students in vocational schools and technical institutes, technicians interested in broadening their horizon, high-fidelity enthusiasts and electronic hobbyists and even engineers requiring a quick "refresher" in a domain outside of their immediate specialty.

A logical approach to electronics requires a systematic progression from fundamental electronic principles and devices to their application in operating "circuits," which in turn form the building blocks for complex electronic systems, such as radio communication, radar, or television. Following this approach, the book naturally falls into three major sections. The first, consisting of Chapters 1 through 7, explains basic principles, electron tubes and transistors. The second section, embracing Chapters 9 through 14, delves into a variety of electronic circuits including amplifiers, oscillators, power supplies, modulation and detection. The remaining chapters are devoted to some of the major electronic systems of general interest, including radio and television, radar and electronic navigation, and high-fidelity systems. An elementary knowledge of basic electrical principles, such as is contained in ELECTRICITY MADE SIMPLE of this series, has been assumed throughout the discussion.

I would like to thank Steven Hahn for his intelligent help in editing the manuscript.

HENRY JACOBOWITZ

Library of Congress Catalog Card Number 64–20579

Copyright © 1958, 1965 by Doubleday & Company, Inc.

All Rights Reserved

Printed in the United States of America

TABLE OF CONTENTS

CHAPTER ONE
ATOMIC STRUCTURE 5
   Atomic Number and Atomic Weight . . 5
   Ions and Ionization 6
   Free Electrons 6
   Summary 7

CHAPTER TWO
ELECTRON EMISSION 7
   Constants of Thermionic Emitting
   Thermionic Emission 8
     Substances 8
   Methods of Heating 8
   Summary 9

CHAPTER THREE
DIODES 9
   Operation 9
   Space Charge 10
   Diode Characteristics 11
   Child's Law for Diodes 11
   Summary 11

CHAPTER FOUR
TRIODES 12
   Control Grid Action 12
   Triode Characteristic Curves . . . 14
   Triode Characteristics 15
   Tube Constants 16
   Amplification Factor 16
   Plate Resistance 18
   Transconductance 18
   Interelectrode Capacitances . . . 19
   Summary 20

CHAPTER FIVE
MULTIELECTRODE TUBES . . . 21
   Tetrodes 21
   Tetrode Characteristics 21
   Pentodes 23
   Action of Suppressor Grid . . . 23
   Beam-Power Tube 24
   Variable-mu (Remote Cutoff) Tubes . 25
   Summary 25

CHAPTER SIX
SPECIAL ELECTRON TUBES . . . 26
   Gas-Filled Tubes 26
   Cold-Cathode Gas-Filled Diodes . . 27
   Hot-Cathode Gas-Filled Diodes . . 27
   Gas-Filled Triodes and Tetrodes
     (Thyratrons) 27
   Mercury-Pool Tubes (Ignitrons) . . 28
   Phototubes and Photomultipliers . . 28
   Basic Laws 29
   Photoemissive Tubes 29
   Photomultipliers (Electron Multipliers) . 31
   Cathode-Ray Tubes 31
   Basic Action 31
   Magic-Eye Tube 32
   Operation of Cathode-Ray Tube . . 32

Special UHF Tubes 36
Typical UHF Tubes 37
Ceramic Tubes 38
Microwave Tubes 38
Summary 38

CHAPTER SEVEN
TRANSISTORS AND SEMICONDUCTORS 40
   Germanium Crystal Structure . . . 40
   P-N Junction Diodes 41
   Junction Triode Transistors . . . 42
   N-P-N Junction Transistor . . . 44
   Transistor Characteristic Curves . . 44
   Transistor Symbols and Connections . . 45
   Junction Tetrode Transistor . . . 47
   Point-Contact Transistors 47
   Unijunction Transistors 48
   Tunnel Diodes 48
   Zener Diodes 49
   Silicon-Controlled Rectifier 50
   Summary 51

CHAPTER EIGHT
AUDIO AMPLIFIERS 52
   Basic Voltage Amplifier 52
   Grid Circuit 53
   Plate Circuit 53
   Graphical Analysis 54
   Amplification 55
   Phase Relations 55
   Supply Voltage Sources 56
   Amplifier Coupling Methods . . . 56
   Resistance Coupling 56
   Design Considerations and Frequency
     Response 58
   Impedance Coupling 59
   Transformer Coupling 60
   Direct Coupling 60
   Transistor Amplifiers 61
   Summary 62

CHAPTER NINE
WIDEBAND (VIDEO) AMPLIFIERS . . 63
   Video Amplifier Pulse Response . . . 63
   High-Frequency Compensation . . . 65
   Shunt Compensation 65
   Series Compensation 66
   Low-Frequency Compensation . . . 66
   Summary 67

CHAPTER TEN
POWER AMPLIFIERS, DISTORTION
   AND FEEDBACK 67
   Amplifier Classes 68
   Class A Power Amplifiers 68
   Loadline 70
   A-C Power Output 72
   Distortion 72
   Push-Pull Power Amplifier 74
   Phase Inverters 75
   Feedback Amplifiers 76
   Reduction of Distortion and Noise . . 78

Output Distortion and Speaker Damping .   78
Practical Degenerative Feedback Circuits .   79
Summary   80

CHAPTER ELEVEN
OSCILLATORS   81
Mechanical Oscillators   81
Electron Tube Oscillators   83
Essential Parts of Triode Oscillator . . .   83
Oscillations in a Tank Circuit   83
Tickler Feedback Circuit   84
Hartley Oscillator   86
Colpitts Oscillator   86
Tuned-Plate Tuned-Grid Oscillator . . .   87
Electron-Coupled Oscillator   87
Crystal Oscillators   88
Crystal Oscillator Circuit   89
Microwave Oscillators   90
Klystron Oscillator   91
Magnetrons   92
Traveling-Wave Tubes   93
Summary   93

CHAPTER TWELVE
TUNED RADIO FREQUENCY
AMPLIFIERS   94
Tuned (Resonant) Circuits   94
Parallel-Resonant (Tank) Circuit . . .   95
Coupled Circuits   96
Tuned Amplifiers   98
Intermediate-Frequency Amplifiers . . .   99
Summary   99

CHAPTER THIRTEEN
MODULATION AND DETECTION . . .   100
Amplitude Modulation   101
Types of Modulators   103
Frequency Modulation   105
Basic Principles   105
Reactance Tube Modulator   106
Detection   107
AM Detectors   107
FM Detectors   109
Limiters   109
Phase-Shift Discriminator   110
Ratio Detector   111
Summary   112

CHAPTER FOURTEEN
POWER SUPPLIES   113
Rectifiers   113
Half-Wave Rectifier   113
Full-Wave Rectifier   114
Full-Wave Bridge Rectifier   115
Voltage Doubler   116
Filter Circuits   117
Choke-Input Filter   117
Capacitor-Input Filter   118
Complete Power Supply   118
D-C to A-C Inverters   119
Summary   120

CHAPTER FIFTEEN
RADIO COMMUNICATION   121
Basic Elements   121

Radio Transmitters   121
Continuous-Wave (C-W) Transmitters . .   122
Amplitude-Modulated Transmitters . . .   123
Frequency-Modulated Transmitters . . .   126
Antennas   127
Basic Action of Antenna   127
Wave Propagation   131
The Radio Spectrum   132
Types of Wave Propagation   132
Radio Receivers   134
Amplitude-Modulated (AM) Receivers .   135
Elementary Crystal Receiver   135
Tuned Radio-Frequency Receiver . . .   136
Superheterodyne Receivers   139
Frequency-Modulated (FM) Receivers .   143
C-W Superheterodyne Receiver   144
Summary   144

CHAPTER SIXTEEN
HIGH FIDELITY AND STEREO . . .   145
Fundamentals of Sound   146
Basic Components   148
The Radio Tuner   148
Phono Pickups and Record Players . .   149
Preamplifiers   152
Main Amplifier   154
Loudspeakers and Enclosures   154
Tape Recorders   158
Stereophonic Reproduction   160
Summary   162

CHAPTER SEVENTEEN
TELEVISION   163
Physical Basis   163
Complete Television System   164
Television Cameras   165
Scanning and Synchronizing   166
Synchronization   168
Sweep Oscillator   171
Picture Tube and Associated Circuits . .   173
Automatic Frequency Control . . .   174
Television Bandwidth and Channels . .   175
Color Television   176
Summary   176

CHAPTER EIGHTEEN
RADAR AND NAVIGATIONAL AIDS .   176
Simple Radar System   177
Azimuth and Range Measurement . . .   178
Types of Radar   179
Instrument Landing System (ILS) . . .   181
Radio Ranges   183
Low Frequency Range   183
Visual VHF Omnirange (VOR) . . .   185
VOR-DME Navigation System . . .   185
TACAN Navigation System   186
Summary   186

CHAPTER NINETEEN
NEW HORIZONS IN ELECTRONICS .   187
Microelectronics and Integrated Circuits .   187
The Laser—A New Communication
Medium   189

Chapter One

ATOMIC STRUCTURE

Since the time of the Greeks all matter was thought to be made up of atoms ("atom" is the Greek word for "indivisible"), although the Greeks had rather unscientific ideas about the exact nature of these tiny "indivisible" particles. It was not till 1802 that Dalton suggested that all matter could be broken down into fundamental constituents, or **elements,** the tiniest particles of which he called **atoms.** Since there are 92 elements in nature, there are 92 different types of atoms. Through the work of the scientists Niels Bohr, Rutherford, and others it was revealed that atoms actually had a complex structure, resembling somewhat a miniature solar system. According to Bohr's theory, an atom consists of a central **nucleus** of positive charge around which tiny, negatively charged particles, called **electrons,** revolve in fixed orbits, just as the planets revolve around the sun. In each type of atom, the negative charge of all the orbital electrons just balances the positive charge of the nucleus, thus making the combination electrically neutral.

The positively charged nucleus, in turn, reveals a complex structure, but for the purpose of understanding electronics a vastly simplified picture is adequate. According to this simplified picture, the nucleus of the atom is made up of two fundamental particles, known as the **proton** and the **neutron.** The proton is a relatively heavy particle (1840 times heavier than the electron) with a positive (+) charge, while the neutron has about the same mass as the proton, but has *no charge* at all.

The positive charge on each proton is equal to the negative charge residing on each electron. Since atoms are ordinarily electrically neutral, the number of positive charges equals the number of negative charges, that is, the number of protons in the nucleus is equal to the number of electrons revolving around the nucleus.

Practically the entire weight of the atom is made up by the protons and neutrons in the nucleus, the weight of the orbital electrons around the nucleus being negligible in comparison. Lest you should think that substantial weights are involved, let us hasten to say that the mass of an electron is only about 9.11×10^{-28} grams (a number with 27 zeros *after* the decimal point), while that of the proton is only about 1840 times as much, which is still phantastically little. The proton, on the other hand, is a little smaller than the electron, having a radius of

about 10^{-13} cm, while the electron is more than twice as big. To give you an idea how small this really is, you might consider that an electron is about as small compared to a standard ping-pong ball, as a ping-pong ball is compared to the orbit of the earth, which is 186,000,000 miles in diameter.

ATOMIC NUMBER AND ATOMIC WEIGHT

We have seen that all atoms are made up of protons, neutrons, and electrons. The difference between various types of elements is the number and arrangement of the protons, neutrons, and electrons within their atoms. The elements are arranged according to their **atomic number,** which is *equal* to the number of electrons revolving around the nucleus (or to the number of protons within the nucleus). Thus, an atom of hydrogen (atomic number 1) has a single electron spinning around its nucleus, while an atom of uranium (atomic number 92) has 92 electrons spinning about the nucleus. The orbits of these electrons are arranged in "shells" about the nucleus, each shell having definite maximum capacity of electrons. The capacity of successive shells from the nucleus out is 2, 8, 18, 32, 18, 18, and 2 electrons; however, the **outermost shell contains never more than eight electrons.** It is this outermost shell which determines the chemical valence of an atom and its principal physical characteristics. The outermost shell is also the most important for electronics, since it is the only one from which electrons are relatively easily dislodged to become "free" electrons capable of carrying a current in a tube or conductor. The electrons in the inner shells cannot be easily forced out from their orbits and, hence, are said to be "bound" to the atom.

The weight of an atom, called atomic weight, is determined almost entirely by the sum of the number of protons and neutrons within its nucleus. Atomic weights are relative, that is, they do not state the number of ounces each atom weighs, but compare the weight of one atom with that of another. Thus, the atomic weight of oxygen is 16, that of helium is 4, while that of hydrogen is 1. Hence, an oxygen atom is 16 times as heavy as a hydrogen atom and four times as heavy as a helium atom. You can determine the number of neutrons in an atom by **subtracting** the atomic number (equals the number of protons) from the atomic weight (equals the number of protons + neutrons).

Let us look at a couple of examples of our atomic model. (See Fig. 1.) The top illustration shows an

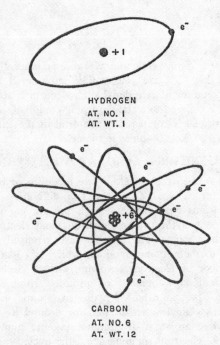

HYDROGEN
AT. NO. I
AT. WT. I

CARBON
AT. NO. 6
AT. WT. 12

Fig. 1. *Structure of Hydrogen and Carbon Atom*

atom of hydrogen, while the bottom illustration shows the more complex structure of an atom of carbon. The hydrogen atom has only one proton in its nucleus (charge +1), surrounded by a solitary orbital electron (symbolized e−). It has no neutrons at all. Consequently, its atomic number is 1 and its atomic weight is also 1. It is the simplest atom of all the elements. The carbon atom, which looks a little more like a miniature solar system, has a nucleus containing six neutrons and six protons (total charge +6). Its atomic weight, therefore, is 12 (number of protons + number of neutrons). To balance the positive charge of the nucleus, the carbon atom is surrounded by six orbital electrons (6 e−), two of which are in the innermost shell, while the other four are in the outer shell.

The most complex atom in nature is uranium, which has an atomic weight of 238 and atomic number 92. As you can figure out for yourself, the uranium atom must have 92 orbital electrons and a nucleus containing 92 protons and 146 neutrons. It is a little difficult to draw a picture of it. Still more complex atoms have been created artificially in the cyclotron and nuclear reactors, but these are generally unstable and break down into lighter elements.

While atoms are the smallest bits of matter in each element, it may be well to keep in mind that most materials in the world are **compounds** of various elements, formed by combinations of different atoms. These smallest combinations are called **molecules.**

IONS AND IONIZATION

An ion is an atom (or molecule) that has become electrically unbalanced by the loss or gain of one or more electrons. An atom that has lost an electron is called a **positive ion,** while an atom that has gained an electron is a **negative ion.** The reason is clear. When an atom loses an electron, its remaining orbital electrons no longer balance the positive charge of the nucleus, and the atom acquires a charge of +1. Similarly, when an atom gains an electron in some way, it acquires an excess negative charge of −1. The process of producing ions is called **ionization.**

Ionization does not change the chemical properties of an atom, but it does produce an electrical change. It can be brought about in a number of ways. As we have seen the electrons in the outermost shell of an atom are rather loosely held and they can be dislodged entirely by collision with other electrons or atoms, or by exposure to X-rays.

Ionization is important in electron tubes. Although tubes are evacuated, there is always a trace of gas left. An electron moving rapidly through a tube may collide with a gas atom, resulting in several possible effects. The electron may become attached to the gas atom, thus changing it into a **negative** ion. The electron may simply bounce off the atom. Finally, the impact of a rapidly moving electron may knock out one or more of the outer-shell electrons in the atom, thus producing a **positive** ion. If one electron is knocked off, the gas is said to be **singly** ionized, if two are dislodged, it is **doubly** ionized, and so on. Positive ions may adversely affect the operation of the tube.

FREE ELECTRONS

Electrons that have become dislodged from the outer shell of an atom are known as **free electrons.** These electrons can exist by themselves outside of the atom, and it is these free electrons which are responsible for most electrical and electronic phenomena. Free electrons carry the current in ordinary conductors (wires), as well as in all types of electron tubes. The motion of free electrons in antennas gives rise to electromagnetic radiations (radio waves). Free electrons also constitute the so-called **cathode rays** and **beta rays,** with which we shall become familiar.

Most substances normally contain a number of free electrons, which are capable of moving freely from atom to atom. Materials, such as silver, copper, or aluminum, which contain relatively many free electrons capable of carrying an electric current, are called **conductors;** materials that contain relatively few free electrons are called **insulators.** Materials that have an intermediate number of available free electrons are classed as **semiconductors.** Actually, there are no perfect conductors and no perfect insulators. The more free electrons a material contains, the better it will conduct. All substances can be arranged in a series according to their conductivity, that is, in accordance with their relative number of available free electrons.

The free electrons in a conductor are ordinarily in a state of chaotic motion in all possible directions. When an electric force (battery) is connected across a conductor, however, the free electrons will move in an orderly fashion, atom to atom, from the negative terminal of the battery, through the wire, and to the positive terminal of the battery. This orderly drifting motion of free electrons is said to constitute an **electric current.** Although the motion is rather slow, the impulse is transmitted almost at the speed of light.

When free electrons are introduced in an **evacuated tube** across which a voltage (electric force) has been applied, they will move very rapidly towards the positive terminal of the voltage. The drifting motion is speeded up, since an evacuated tube has very few atoms to impede the progress of the electrons. The greater the voltage across the tube, the greater is the attraction of the (negatively charged) electrons to the positive terminal and, hence, the greater is the speed of the electron motion.

In the following chapter we shall learn how free electrons may be introduced into such an evacuated tube.

SUMMARY

There are **92 elements** in nature, corresponding to 92 different types of atoms.

An **atom** is the smallest particle of an element that shows its properties.

Atoms resemble **miniature solar systems,** consisting of a **central nucleus of positive charge,** around which tiny, **negatively charged particles,** called **electrons,** revolve in **fixed orbits.** The **negative charge** of all orbital electrons just balances the **positive charge** of the nucleus.

The **nucleus** of the atom is made up of two fundamental particles, called the **proton** and the **neutron.** The **proton** is relatively heavy (1840 times heavier than an electron) and has a **positive** charge, while the **neutron** has about the same mass as the proton, but has *no charge* at all.

The electron orbits are arranged in **shells** about the nucleus, each having a definite capacity of electrons (shells of 2, 8, 18, 32, 18, 18, and 2 from the nucleus out.) The **outermost shell** never contains more than **eight** electrons.

The **atomic number** of an atom is the sum of the electrons in the shells surrounding the nucleus.

The **atomic weight** of an atom is the sum of the number of protons and neutrons in the nucleus. **Number of neutrons** is equal to the difference between atomic weight and atomic number.

An atom that has lost an electron is called a **positive ion,** while one that has gained an electron is a **negative ion. Ionization** may be produced by **collision** with other atoms or electrons.

Free electrons are electrons dislodged from the outer shell of an atom; they may exist by themselves (cathode rays, beta rays).

Conductors contain many free electrons, **insulators** contain few.

Chapter Two

ELECTRON EMISSION

The electron tube depends for its action on a stream of electrons that act as current carriers. To produce this stream of electrons a special metal electrode **(emitter)** is present in every tube. But at ordinary room temperatures the "free" electrons in the metallic emitter cannot leave its surface because of certain restraining forces that act as a barrier. These attractive surface forces tend to keep the electrons within the emitter substance, except for a small portion that happens to have sufficient kinetic energy (energy of motion) to break through the barrier.

The majority of electrons move too slowly for this to happen.

To escape from the surface of the emitter the electrons must perform a certain amount of work to overcome the restraining surface forces. To do this work the electrons must have sufficient energy imparted to them from some **external source of energy,** since their own kinetic energy is inadequate. This external energy may come from a variety of sources, such as **heat energy, light energy,** energy stored in **electric** or **magnetic fields,** or the **kinetic** energy of electric charges bombarding the metal surface. Accordingly, we can classify the four principal methods of obtaining electron emission from the surface of a metallic conductor, as follows:

1. Thermionic (primary) emission. In this method the emitter metal is heated, resulting in increased thermal (or kinetic) energy of the unbound electrons. Thus, a greater number of electrons will attain sufficient speed and energy to escape from the surface of the emitter. The number of electrons released depends on the temperature.

2. Photoelectric emission. In this process the energy of the light radiation (called **"quanta"**) falling upon the metal surface is transferred to the free electrons within the metal and speeds them up sufficiently to enable them to leave the surface. The number of electrons emitted depends upon the intensity (brightness) of the light beam falling upon the emitter surface.

3. Field emission (cold-cathode emission). The application of a strong electric field (i.e., a high *positive* voltage) outside the emitter surface will literally "yank" the electrons out of the emitter surface, because of the attraction of the positive field. The stronger the field, the greater the field emission from the cold emitter surface.

4. Secondary emission. When high-speed electrons suddenly strike a metallic surface they give up their kinetic energy to the electrons and atoms which they strike. Some of the bombarding electrons collide directly with free electrons on the metal surface and may knock them out from the surface, similar to a billiard-ball collision. The electrons freed in this way are known as **secondary emission electrons,** since primary electrons from some other emitter must be available to bombard the secondary electron-emitting surface.

Of these four methods of emission, **thermionic emission** is the most important and the one most commonly used in electron tubes. We shall discuss it a little further in the following.

THERMIONIC EMISSION

Electron emission from a heated metal surface is very similar to the evaporation of a liquid from its surface. When a liquid is heated, an increasing number of molecules acquire sufficient energy to overcome the restraining forces of the liquid surface and are evaporated. The number of molecules evaporated increases rapidly as the temperature is raised. Similarly, when a metallic body is heated, a progressively larger number of electrons overcome the restraining surface barrier and are literally "boiled out" from the metal, like steam from a kettle.

The number of electrons "evaporated" per unit area of an emitting surface is related to the absolute temperature T (Kelvins) of the emitter and a quantity b that is a measure of the work an electron must perform when escaping the emitter surface, according to an equation derived by O. W. Richardson almost 50 years ago:

emission current $I = A T^2 e^{-b/T}$ amps per square cm. where e = 2.7183 (base of natural logs)

and A is a constant which has a value of about 60 for pure metals, such as tungsten or tantalum, but varies widely for other practical emitters.

The combination of the squared and exponential terms in the emission equation makes it extremely sensitive to small changes in temperature. Doubling the temperature of an emitter may increase electron emission by more than 10,000,000 times. For example, the emission from pure tungsten metal is about a millionth ampere per sq.cm at 1600° Kelvin (15,541° F), but rises to the enormous value of about a 100 amps per sq.cm when the temperature is raised to 3200° Kelvin. Halving the **work function** of the material (related to b above) will have about the same effect. Table I gives the constants A and b in the emission equation for some common emitter materials.

TABLE I

CONSTANTS OF THERMIONIC EMITTING SUBSTANCES

Emitter Substance	A amp./cm²/deg²	b °K
Tungsten (W)	60	52,400
Thoriated tungsten (Th-W)	3	30,500
Barium-strontium-oxide (BaO-SrO)	0.01	12,000
Molybdenum (Mo)	55	48,100
Nickel (Ni)	1380	58,300
Tantalum (Ta)	60	47,200

The most commonly used emitters in electron tubes are **tungsten,** heated to temperatures between 2200 and 3000° K, **thoriated tungsten** (thorium and carbon added to tungsten), operated at about 1900° K (1440° F) and **oxide-coated cathodes** (barium-strontium oxide), operated at about 1000 to 1150° K (540 to 690° F).

METHODS OF HEATING

Electron emitters are heated electrically, either *directly* or *indirectly*. In the direct method, the electric current is applied directly to a wire, called **filament,** that also serves as electron emitter. (See Fig. 2.) In the indirect method, the electric current is applied to a separate **heater element,** located inside a cylindrical sleeve (**cathode**) that is coated with the emitting material. (See Fig. 3.) The cathode is thus heated indirectly through heat transfer from the heater element. Alternating or direct current can be used for either method of heating. All the emitters previously described (tungsten, thoriated tungsten, or oxide-coated emitters) can be used for directly heated filaments, but indirectly heated cathodes always use oxide-coated emitters. Most receiving-type tubes are indirectly heated.

SUMMARY

Electron tubes depend for their action on a stream of electrons that act as current carriers. Free electrons may be produced in a tube from a metal electrode, called **emitter.** The process is called **emission.** Emission does not occur at room temperature because of **restraining forces** at the surface of the emitter.

To escape from the surface of an emitter, electrons must be given sufficient **kinetic energy** from an **external** energy source to overcome the **surface barrier.**

When electrons are emitted from an emitter by supplying heat energy (heating it), the process is called **thermionic emission.**

In **photoelectric emission** light energy falling upon the emitter is transferred to free electrons and ejects them from the surface.

In **field** or **cold-cathode emission** the electric field set up by a high positive voltage "yanks" free electrons from the emitter.

When primary electrons strike a metal surface at high speed, some will collide with electrons on the metal surface and project them outward like billiard balls. This is called **secondary emission.**

Thermionic emission from a metal surface is similar to the **evaporation** of liquid molecules produced by boiling a liquid.

The number of electrons "boiled out" from the emitter, called **emission current,** increases extremely fast (exponentially) when the emitter temperature is increased.

The **work function** is a measure of the work an electron must perform to escape from the emitter surface. Emission *increases* rapidly, as the work function *decreases* (for different emitters).

In **direct heating,** electric current is applied to a **filament** wire that serves as emitter. In **indirect heating,** electric current is applied to a separate **heater** element, located inside a cylindrical **cathode** that serves as emitter.

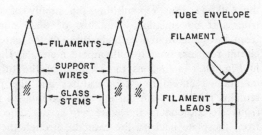

Fig. 2. *Directly Heated Filaments and Schematic Symbol*

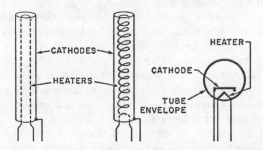

Fig. 3. *Indirectly Heated Cathodes and Schematic Symbol*

Chapter Three

DIODES

The simplest combination of elements constituting an electron tube is the **diode** ("di-ode" means "two electrodes"). It consists of a **cathode,** which serves as an emitter of electrons, and a **plate** or **anode** surrounding the cathode, which acts as a collector of electrons. (See Fig. 4.) Both electrodes are enclosed in a highly evacuated envelope of glass or metal. The emitter may be either directly or indirectly heated. The size of diode tubes varies from tiny metal tubes to large-sized glass-envelope rectifiers. The plate is generally a hollow metallic cylinder made of nickel, molybdenum, graphite, tantalum, monel, or iron.

OPERATION

A basic law of electricity states that **like charges repel each other and unlike charges attract each other.** Electrons emitted from the cathode of an electron tube are negative electric charges. These charges may be either attracted to or repelled from

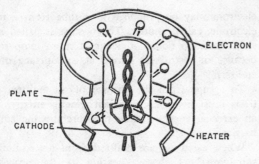

Fig. 4. *Diode Elements*

the plate of a diode tube, depending on whether the plate is positively or negatively charged. Actually, by applying a potential difference (voltage) from a battery or other source between the plate and cathode of a diode, an **electric field** is established within the tube. The **lines of force** of this field always extend from the negatively charged element to the positively charged element. Electrons, being negative electric charges, follow the direction of the lines of force in an electric field. By establishing an electric field of the correct polarity between cathode and plate and "shaping" the lines of force of this field in certain paths, the motion of the electron can be controlled as desired.

Fig. 5 is a simplified schematic of a diode circuit to illustrate its basic action. A battery has been connected between plate and cathode of a diode, so as to make the plate negative with respect to the cathode. The lines of force of the field established within the tube thus extend *from the plate to the cathode*. When a voltage is now applied to the heater element (*H*), the cathode will emit a copious flow of electrons. However, these electrons are strongly repelled from the negatively charged plate and tend to fill the **interelectrode space** between cathode and plate. (Some of the electrons actually fall back into the cathode.) Since no electrons actually reach the plate, the tube acts like an open circuit, and the milliampere meter connected externally between plate and cathode indicates no current flow.

The battery connection has been reversed so as to make the plate positive with respect to the cath-

ode. The lines of force of the electric field now extend in a direction **from the cathode to the plate.** Again, applying a **heater** voltage, results in copious emission of electrons from the cathode. Now, however, the electrons follow the lines of force to the positive plate and strike it at high speed. Since moving charges comprise an electric current, the stream of electrons to the plate *is* an electric current, called the **plate current.** Upon reaching the plate the electron current continues to flow through the external circuit made up of the connecting wires, milliampere meter, and the battery. The arriving electrons are absorbed into the positive terminal of the battery and an equal number of electrons flow out from the negative battery terminal and return to the cathode, thus replenishing the supply of electrons lost by emission. The flow of plate current from the plate to the cathode through the external circuit registers on the milliampere meter, as shown in Fig. 5b. As long as the cathode of the tube is maintained at emitting temperatures and the plate remains positive, plate current will continue to flow from the cathode to the plate *within* the tube and from the plate back to the cathode through the external circuit.

The following conclusions may be drawn from this simplified picture of diode tube operation:

1. Electron current (plate current) flows in the diode only when the plate is made **positive** with respect to the cathode. No current can flow when the plate is negative with respect to the cathode.

2. Current flow *within* a diode takes place only from the cathode to the plate, never from the plate to the cathode. This is known as **unidirectional** or **unilateral** conduction.

3. Because of its unidirectional characteristic a diode can be made to act like a switch or valve, automatically starting or stopping the plate current, depending on whether the plate is positive or negative with respect to the cathode. This ability permits diodes to change alternating current to direct current, or **rectify** it. We shall become acquainted with the operation of rectifiers in a later chapter (Power Supplies).

SPACE CHARGE

The total number of electrons emitted by the cathode of a diode is always the same at a given operating temperature and is determined by the Richardson emission equation, as we have seen. The plate voltage (voltage between plate and cathode) has no effect, therefore, on the *amount* of electrons emitted from the cathode. Whether or not these electrons actually reach the plate, however, is determined by the plate-to-cathode voltage, as well as by a phenomenon known as **space charge.** The term **space charge** is applied to the cloud of electrons

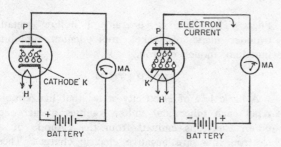

Fig. 5. *Action of a Diode*

that is formed in the interelectrode space between cathode and plate. Since it is made up of (negatively charged) electrons, this cloud constitutes a **negative charge** in the interelectrode space that has a **repelling effect** on the electrons being emitted from the cathode. The effect of this negative space charge *alone,* therefore, is to force a considerable portion of the emitted electrons back into the cathode and prevent others from reaching the plate.

The space charge, however, does not act alone. It is counteracted by the electric field from the positive plate, which reaches through the space charge to attract electrons and thus partially overcomes its effects. At low positive plate voltages only electrons nearest to the plate are attracted to it and constitute a small plate current. The space charge then has a strong effect on limiting the number of electrons reaching the plate. As the plate voltage is increased, a greater number of electrons are attracted to the plate through the negative space charge and correspondingly fewer are repelled back to the cathode. If the plate voltage is made sufficiently high, a point is reached eventually, where all the electrons emitted from the cathode are attracted to the plate and the effect of the space charge is completely overcome. Further increases in the plate voltage cannot increase the plate current through the tube, and the emission from the cathode limits the maximum current flow.

DIODE CHARACTERISTICS

The relation between the plate current in a diode and the plate-to-cathode voltage just discussed can be represented by a **characteristic curve,** obtained by plotting the plate-current (I_b) values for different values of the applied plate voltage (E_b). The **diode characteristics** for a typical diode tube and various cathode operating temperatures are shown in Fig. 6.

It is seen from Fig. 6 that all the curves are the same at low plate voltages, where the negative space

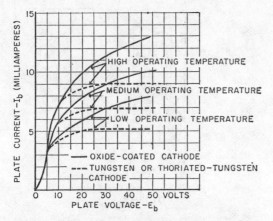

Fig. 6. *Diode Characteristic Plate Current-Plate Voltage Curves*

charge is most effective in limiting the flow of electrons. The plate current in the low plate-voltage region is completely controlled by the voltage at the plate and is independent of the emitter (cathode) temperature. Under these conditions the plate current is said to be **space-charge limited.**

As the plate voltage is made progressively higher, an increasingly greater portion of the total supply of emitted electrons are attracted to the plate and the effect of the space charge is eventually completely overcome. This is seen by the flattening of the characteristic curves, as the plate voltage is increased. When the entire supply of emitted electrons (at a given cathode temperature) is attracted to the plate, the plate current becomes *independent* of the plate voltage and reaches a constant value equal to the total emission current. **Emission saturation** takes place and the plate current is said to be **emission-limited** in the high plate-voltage region. This is seen best by the flattening of the dotted graphs in Fig. 6, which represent tungsten or thoriated tungsten cathode emitters. For each constant operating temperature, the plate current at high plate voltages reaches its specific saturation value, equal to the total emission current determined from Richardson's (emission) equation. Oxide-coated emitters, represented by the solid lines in Fig. 6, do not have such a specific emission saturation value, and the plate current—though tapering off at high plate voltages—never becomes completely independent of the plate voltage. At very high plate voltages oxide-coated cathodes may become damaged because of the abnormally large emission. In general, **electron tubes are operated in the space-charge limited (low E_b) region.**

CHILD'S LAW FOR DIODES

An interesting relation holds for electron tubes that are operated in the plate-voltage region where space charge limits the value of the plate current. (See Fig. 6.) It has been found that the plate current (I_b) in this region is approximately proportional to the 3/2 power of the voltage between the plate and the cathode (E_b). This may be expressed mathematically by the so-called **Child's Law:**

$$\text{Plate current } I_b = K\ E_b^{3/2}$$

where K is a constant that depends on the shape of the electrodes and geometry of the tube. Although this law is a guide, it is not too accurate in practice.

SUMMARY

A **diode** is a two-element electron tube that consists of a **cathode,** serving as electron emitter, and an **anode** or **plate** acting as an electron collector.

By applying a **positive voltage** to the plate of a diode, an **electric field** is established between cathode and plate that *attracts* electrons emitted from the cathode to the plate.

Plate current is the flow of electrons from cathode to plate and their return to the cathode through the **external circuit.**

Plate current flows in a diode when the plate is made **positive** with respect to the cathode. **No current flows when the plate is negative** with respect to the cathode.

Current within a diode flows from **cathode to plate,** *never* from plate to cathode. This is called **unidirectional conduction.**

The cloud of electrons formed in the space between cathode and plate is called **space charge.** The space charge is **negative** and hence has a **repelling effect** on electrons emitted from the cathode.

The **amount of plate current depends on the space charge and the relative strength of the electric field set up by the positive plate voltage.**

At **low plate voltages** the negative space charge limits the flow of electrons and the plate current is completely controlled by the plate voltage and is independent of emitter (cathode) temperature. The plate current thus is **space-charge limited.**

At **high plate voltages** the space charge is drawn off and the plate current reaches **saturation** at a value equal to the total emission current. It is then **independent** of the plate voltage and is said to be **emission-limited** for a specific cathode temperature.

The plate current in the space-charge limited region is approximately **proportional to the 3/2 power** of the plate voltage.

Chapter Four

TRIODES

In 1907 Lee De Forest added a third element—the **control grid**—between the cathode and plate of a diode and so provided the resulting **triode** tube (he called it an "audion") with the ability to **amplify** tiny radio signals. This led to the sensational development of radio communication, broadcasting and electronics in general with which we are all familiar.

The construction of a typical triode is shown in Fig. 7. The control grid in this tube is a circular helix (spiral) of a number of turns of fine wire that completely surround the cathode. Because of its open construction the grid does not directly hinder the flow of electrons to the plate, but when a voltage is placed on the grid it has a profound effect on the electric field between cathode and plate and, hence, on the total electron flow.

Grid structures take many other forms besides a circular helix, such as flat or elliptical helix, ladder types, etc. Different sizes and spacings of the control grid wires are employed, depending on the desired field configuration and design of the tube. Metals used for grids can be nichrome, molybdenum, iron, nickel, tungsten, tantalum, and various alloys. Triode tubes differ widely in size and electrode spacing, depending on power rating and desired function.

CONTROL GRID ACTION

Since the control grid is nearer to the cathode than the plate, a potential placed on the grid has a much larger effect on the electric field within the tube—and hence upon the plate current—than the same potential placed on the plate. The grid thus

has a controlling effect on the flow of plate current in the tube. A triode requires three operating voltages, one on each electrode, to operate correctly. The plate (or anode) of the tube is normally connected to a high positive voltage (called "B+") to attract the stream of electrons. A relatively low "A" voltage (a.c. or d.c.) is connected to the filament or heater to bring the cathode to its proper emitting temperature and thus make available a supply of electrons. Finally, a voltage is placed on the control grid to govern the flow of plate current. This voltage generally consists of two components. One is a fixed

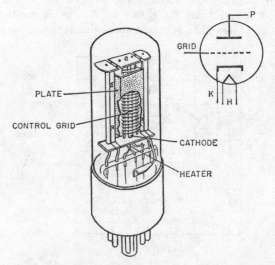

Fig. 7. *Triode Construction and Schematic Symbol*

d-c voltage, called the **bias** (C—) which is normally a few volts *negative* with respect to the cathode. Its purpose is to operate (or "bias") the tube on a definite point on its characteristic curve (remember the diode characteristic) so that a certain amount of plate current is always flowing. Superimposed upon the bias voltage is a varying or alternating voltage, usually called the **signal voltage.** The purpose of this voltage is to vary the flow of plate current through the tube in strict accordance with the signal variations, so as to make the plate current an **amplified replica** of the signal voltage. Amplification takes place, since a small variation of the signal voltage on the grid results in a large variation of the plate current through the tube.

Fig. 8 presents a simplified picture of the action of the grid in a triode tube. Instead of applying a bias *and* a signal voltage to the grid, we have varied the value of the bias voltage in each part of the figure, thus simulating the effect of a varying or signal voltage. The d-c supply voltages (obtained from batteries here) are measured with respect to the **cathode** of the tube and a milliampere meter has been inserted into the plate-to-cathode return circuit to measure the amount of plate current.

Cutoff Bias. Fig. 8a shows a cross section of the cathode, grid wires and plate in a simplified schematic manner. A voltage (from the "A-battery") has

been applied to the heater and the cathode is emitting a normal supply of electrons. The plate is at a high positive potential and would normally attract a large number of electrons from the space charge, if it were not for the large **negative** bias voltage applied to the grid from the "C-battery." Because of this large negative potential, the electrostatic field normally existing between plate and cathode cannot penetrate to the cathode and actually terminates on the grid wires. This is shown by the lines of force extending from the grid to the plate. Under these conditions the grid entirely neutralizes the electrostatic field and, hence, the attraction of the plate. Since there is no electrostatic field near the cathode to draw away the electrons, the plate current through the tube is zero (as indicated by the zero reading of the milliampere meter) and a large space charge accumulates in the region between cathode and grid. The smallest negative voltage between grid and cathode that is just capable of cutting off the plate current is called the **cutoff bias.** Bias voltages that are more negative than this cutoff value have no effect on the action of the tube.

Less Than Cutoff Bias. In (b) of Fig. 8 everything has been left unchanged, except that the negative bias voltage has been reduced to a value less than cutoff. The grid is now no longer capable of neutralizing the field between plate and cathode

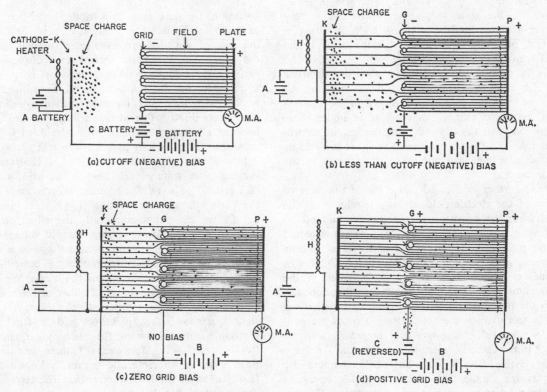

Fig. 8. *Action of Control Grid in a Triode*

completely and some of the lines of force penetrate between the grid wires to the cathode, as shown. Consequently, some electrons are attracted away from the space charge and move between the grid wires toward the positive plate. This results in a moderate flow of plate current, as indicated on the milliampere meter. As the negative grid voltage is further reduced (that is, made *less* negative), progressively more electrons are able to pass between the grid wires to the plate and the plate current continues to increase. Note, however, that electrons are not attracted to the grid itself, as long as it is maintained at a negative bias voltage with respect to the cathode.

Zero Grid Bias. When the C-battery is removed and the grid voltage is zero (Fig. 8c), the positive voltage on the plate produces a substantial electric field at the cathode and large numbers of electrons are attracted through the grid wires to the plate, resulting in a fairly large plate current. The action is similar to that of a diode, except that the grid still has some retarding effect on the electrons because of its shielding action, and hence the plate current is somewhat less than it would be with the grid removed entirely. Again, electrons are not attracted to the grid itself, since it is at zero potential with respect to the cathode.

Positive Grid Bias. In Fig. 8d, the C-battery has been reversed in polarity, thus making the grid **positive** with respect to the cathode. The grid potential now aids the plate voltage and produces a very strong electrostatic field at the cathode, resulting in a large plate-current flow through the tube. If the grid is made sufficiently positive with respect to the cathode, a point will be reached when the electrons are attracted to the plate as fast as they can be emitted from the cathode. As shown in the illustration, no space charge can accumulate under these conditions and the plate current reaches its **saturation value.** Still further increases in either the grid or the plate voltage can not cause an increase in the plate current.

Note further in Fig. 8d that some of the lines of force of the electric field actually terminate on the grid wires itself because of the positive and potential. As a result a part of the electrons are attracted to the positive grid and cause a **grid current** to flow between grid and cathode (through the C-battery). Under these conditions power is dissipated in the grid circuit. To avoid this power consumption and also the large saturation plate current, which eventually can damage the tube, electron tubes are generally operated at **negative** grid potentials with respect to the cathode. We shall see later that the shape and relative linearity of the tube's characteristic curves are another important reason for operating electron tubes at negative grid potentials. (The big

exception are the so-called **pulse circuits,** to be discussed in a later chapter.)

Amplification. As we have seen, the plate current in a triode is determined by the electrostatic field resulting from the combined action of the grid and plate potentials on the space charge near the cathode. Since the grid is placed closer to the space charge than the plate, the grid voltage has a greater effect on the amount of plate current than the plate voltage. Thus, if the grid voltage is made more negative by a fixed number of volts, the plate current is reduced far more than the same decrease in plate voltage would produce. When a resistance load is placed in series with the plate circuit, the voltage drop produced across this resistance is a function of the plate current and, hence, is controlled by the grid voltage. Thus a tiny change in the grid (or signal) voltage can cause a large change in the plate current and in the resulting voltage across the load resistance. In other words, the signal voltage appearing at the grid is **amplified** in the plate circuit of the tube. This amplification takes place without any grid current or power consumption in the grid circuit, as long as the grid voltage is negative with respect to the cathode. We shall see later how the amplification of a triode can be defined more precisely in quantitative terms.

TRIODE CHARACTERISTIC CURVES

The relationships between the plate voltage, grid voltage and plate current in a triode, which we have explored in the last few paragraphs, can be (as in the case of the diode) conveniently summarized in the triode's characteristic performance curves. Actually, a three-dimensional surface model is required to represent the relation between all three quantities at the same time, but for convenience two-dimensional cuts through this surface will give the relation between any two quantities, while the third is held constant. Thus, we can plot a curve that shows the values of the plate current (I_b) as a function of varying plate voltages (E_b), when the grid voltage (E_c) is held at some fixed value. This is known as the plate current-plate voltage ($I_b - E_b$) characteristic. Or we can show graphically the effect on the plate current caused by varying the grid voltage (i.e. the bias), while holding the plate voltage at a constant value. This is called the plate current-grid voltage ($I_b - E_c$) characteristic of the triode. We can, of course, obtain a whole set of either of these characteristics by assuming different values for the constant quantity (either plate or grid voltage) and plotting a curve between the remaining quantities ($I_b - E_b$, or $I_b - E_c$) for each of these conditions. Such a set of characteristic curves is known as a **family of static** triode characteristics. The term **static** denotes that the characteristics are obtained when various **steady** voltages are applied to the tube's electrodes. Later in the book we shall become acquainted with the **dynamic characteristics** of the triode, that is, the characteristics obtained under actual operating conditions, with a signal voltage applied and a load resistance inserted into the plate circuit to extract power from the tube.

We have neglected to mention the filament or heater voltage, which—you remember—also substantially affect the plate current. Sets of curves can also be plotted showing the effect of varying heater voltages (or emission) upon the plate current for different plate voltages and grid voltages. However, the cathode voltage is usually fixed at a value to provide sufficient emission for normal operation and, hence, we need not concern ourselves with these tube characteristics.

TRIODE CHARACTERISTICS

A circuit for obtaining the static characteristics of a triode is illustrated in Fig. 9. Variable voltage dividers (potentiometers) are connected across the plate voltage and grid voltage supplies (denoted by E_{bb} and E_{cc}, respectively) to permit ascertaining the effect of varying either voltage on the tube's plate current, while the remaining voltage is held constant. The potentials at the electrodes and the plate and grid currents resulting are then measured by suitable voltmeters and current meters, inserted into the grid and plate circuits. Note, that resistors cannot be inserted in series with either grid or plate to obtain the **static** characteristics of the triode. There are no voltage drops, therefore, external to the tube and thus the grid-to-cathode (E_c) equals the grid supply voltage (E_{cc}) and the plate-to-cathode voltage (E_b) equals the plate supply voltage (E_{bb}). This is not true for an actually operating (dynamic) amplifier circuit, as we shall see later on.

A grid family of characteristic curves for a type 6J5 triode is shown in Fig. 10. (This is also known as the **static transfer characteristic** of the tube.) In these curves the plate current (I_b) has been plotted as a function of the grid voltage (E_c) for various constant values of the plate voltage (E_b). The shape of the curves is typical of most receiving triodes. Note that each curve intersects the grid-voltage axis at a specific point that indicates the value of the negative grid voltage required to stop the plate current, at the fixed value of the applied plate potential. This is the so-called **cutoff bias.** As the plate voltage is increased, it may be seen that the negative bias required to cut off the plate current also increases.

It is also evident that each of the graphs in Fig. 10 is quite curved in the lower portion, near cutoff, while it is almost a straight line in the central and upper portions. Triodes are almost always operated in the straight-line, **linear** portions of their characteristic and rarely in the curved or **non-linear** portion. This is so, because increases in the grid (signal) voltage do not result in *proportional* increases of the plate current in the non-linear portion of the characteristic. For example, in Fig. 10, a change in grid voltage from −18 to −15 volts on the 300-

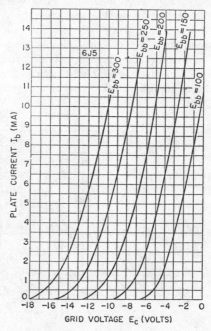

Fig. 10. *Plate Current-Grid Voltage Static Characteristics of 6J5*

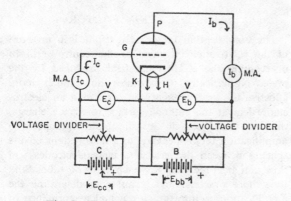

Fig. 9. *Circuit for Determining Triode Characteristics*

volt plate voltage curve (E_{bb} = 300) results in an increase in plate current from 0 to about 1½ ma. If the grid voltage is reduced again by 3 volts, from −15 to −12 volts, the plate current increases from 1½ to about 5¾ ma, or an increase of 4¼ ma. Thus equal (3-volt) grid voltage changes in the curved portion of the characteristic have caused highly unequal plate current changes (1½ ma as against 4¼ ma). On the other hand, a grid-voltage change in the linear portion of the 250-volt plate voltage curve ($E_{bb} = 250$) from −11 to −10 volts results in a plate-current increase from 3 to about 4¾ ma, or a change of 1¾ ma. Decreasing the grid voltage again by 1 volt, from −10 to −9 volts, causes a plate-current increase from 4¾ to 6½ ma, or again a change of 1¾ ma. Thus, **in the linear portion of the characteristic, equal changes in grid voltage cause equal changes in the plate current.** As we shall see later on, this is highly desirable to obtain undistorted, linear amplification of a signal voltage at the grid of the tube.

Plate Current-Plate Voltage Characteristics. We can use the circuit of Fig. 9 to determine the family of static plate current-plate voltage characteristics for the 6J5 triode. The results are shown in Fig. 11. Here again, the curves have a similar shape, being curved in the lower portion near plate-current cutoff and fairly linear in the upper portions. Each curve shows the effect upon the plate current when the plate voltage is varied over a certain range, while holding the grid voltage fixed at a definite value. A new curve results each time the grid voltage is changed to another fixed value, and when this is done over a representative range of constant grid voltages, the entire family of curves is obtained. Note that no new information can be obtained from the plate family of static characteristics; it offers the same data as the grid family, but in a slightly different form.

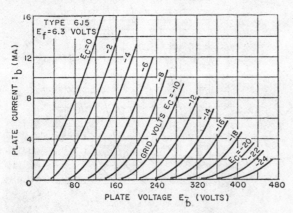

Fig. 11. *Plate Current-Plate Voltage Static Characteristics of 6J5 Triode*

Fig. 12 illustrates another plate family of static characteristics for a type 6J6 twin triode (two triodes in one envelope). The effect of positive grid voltages on the I_b−E_b curves may be seen clearly. Even for low positive values of the grid voltage the plate current is seen to increase very rapidly with small increases in plate voltage. In addition, the grid circuit of the tube draws a grid current (I_c) for positive grid voltages. This grid current evidently *decreases* as the plate voltage is made large relative to the grid voltage, and the electrons are rapidly attracted to the plate.

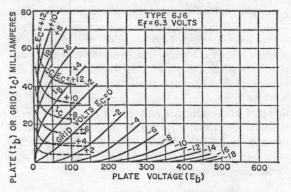

Fig. 12. *Plate Family of Static Characteristics of 6J6 Twin Triode*

TUBE CONSTANTS

The families of tube characteristics, which show the characteristic performance of each type of tube, are not a result of accident. Rather they represent the outcome of purposeful design to make each tube behave in a certain manner. Tube design takes into account the geometric configuration of the tube's electrodes and the spacing between them, the power dissipation capabilities and many other factors. It is these factors that determine the maximum voltages applied to the electrodes, the maximum permissible plate current through the tube, plate-current cutoff, the tube's amplification, etc. The design factors of the tube are summarized by a series of numbers, called the **tube constants.** All major tube manufacturers publish tube manuals in which these constants are listed along with other ratings and the characteristic curves for each type of tube. The three most important tube constants are the **amplification factor,** the **a-c plate resistance** and the **transconductance.**

AMPLIFICATION FACTOR

The **amplification factor of a triode is a measure of the relative effectiveness of the control grid in overcoming the electrostatic field produced by the plate.** Numerically it expresses how much more effective the grid voltage is in producing an electrostatic field at the cathode surface (or space charge) than the field set up by the plate voltage. The amplification factor depends to a large extent on the spacing between the grid and plate electrodes and thus can be predicted for each type of tube. Still, it is more practical and accurate to determine the amplification factor from the actual performance of the tube, as expressed in the characteristic curves.

The characteristic curves reflect the effectiveness of the control-grid voltage in determining the plate current, compared with that of the plate voltage—which is another way of defining the amplification factor. All that is necessary, therefore, to determine the amplification factor is to change the plate voltage by a certain amount, record the change in plate current, and then change the grid voltage (in the opposite direction) by an amount just sufficient to restore the previous plate-current value. By comparing the plate-voltage change to the grid-voltage change for the same change in plate current (essentially holding it constant), we can determine their relative effectiveness, which is the amplification factor. Accordingly, we simply define the amplification factor by the formula:

$$\text{amplification factor} = \frac{\text{small change in plate voltage}}{\text{small change in grid voltage}}$$

to **produce the same change in plate current.** The changes have to be small, since the non-linear tube characteristics lead to errors for large changes.

We can abbreviate this formula by the following symbols:

$$\text{amplification factor } \mu = \frac{\Delta E_b}{\Delta E_c} \text{ (with } I_b \text{ constant)}$$

where ΔE_b = a small change in plate voltage and ΔE_c = a small change in grid voltage

Thus, if the amplification factor, μ, of a tube is 20, a change in grid (or signal) voltage will be 20 times as effective in changing the plate current as the same change in the plate voltage. To illustrate further, suppose a plate-current change of 1 ma is produced by a plate-voltage change of 50 volts and that a grid-voltage change of only 2 volts results in the same 1-ma plate-current change. The amplification factor then is:

$$\mu = \frac{\Delta E_b}{\Delta E_c} = \frac{50V}{2V} = 25$$

In practice we need not perform any measurements to ascertain the amplification factor, but we simply make a paper experiment on the tubes's characteristics. As an example, let us determine μ from the 6J5 plate family of static characteristics, shown in Fig. 11. Figure 13 shows the same family, somewhat enlarged for greater accuracy, and provided with a simple construction for determining μ. Suppose, you want to find the amplification factor near point A in the illustration. (μ changes slightly depending on the chosen point of operation.) The grid voltage at the operating point A is -8 volts, the plate current is 7 ma and the plate voltage is about 235 volts. First move down on the -8 volt curve

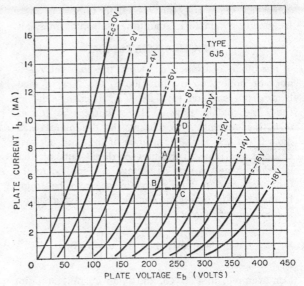

Fig. 13. *Determining the Amplification Factor and Plate Resistance from the 6J5 Plate Family of Characteristics*

to a convenient point B. Draw a horizontal line along a constant plate-current value of 5 ma to the next grid-voltage curve (-10 volts) until it intersects at point C. Now draw a vertical line upward for a constant plate voltage of 255 volts, until it intersects the -8 volt grid-voltage curve again at point D.

This construction has the following significance: Moving from point B to point D along a constant grid voltage of -8 volts, the plate current has increased from 5 ma to about 9.6 ma, or by 4.6 ma. To produce this increase the plate voltage has to be increased from about 215 volts (at B) to about 255 volts (at D), or an increase of 40 volts. A plate-voltage change of 40 volts, thus, is necessary to produce a plate-current increase of 4.6 ma. We now reduce the plate current to its previous value of 5 ma, by moving vertically down from point D to point C, along a constant plate voltage of 255 volts. To reduce the plate current to its former value the grid voltage must be increased from -8 volts (at D) to -10 volts (at C), or an increase of 2 volts. A change of 2 volts in the negative grid voltage thus has produced the same plate-current change as a 40-volt change in the plate voltage. The amplification factor is, therefore:

$$\mu = \frac{\Delta E_b}{\Delta E_c} = \frac{40V}{2V} = 20$$

This, incidentally, is the same value as stated by the manufacturer in the tube manual for a 6J5. The amplification factor could have been equally well determined from the grid family of the 6J5 characteristics (Fig. 10), but the plate family is more frequently available in manufacturer's tube manuals.

PLATE RESISTANCE

The plate resistance describes the **internal resistance** or opposition of the tube to the flow of electrons from cathode to plate. This resistance is *not* the same for the flow of direct current as for alternating current. Accordingly, the **d-c plate resistance** is defined as the resistance of the path between cathode and plate to the flow of direct current, that is, when **steady** values of voltage are applied to the tube's electrodes. This d-c resistance can be found from the plate family of characteristics by a simple application of Ohm's law. Thus,

$$\text{d-c plate resistance, } R_p = \frac{E_b}{I_b} \text{ ohms}$$

For example, in Fig. 13, the plate current at point *A* is 7 ma (.007 amp) and the plate voltage is 235 volts. Hence, the d-c plate resistance

$$R_p = \frac{235V}{.007A} = 33,600 \text{ ohms}$$

More significant for the tube's performance as an amplifier is the **a-c plate resistance**, which is the tube's opposition to the flow of (alternating) plate current, when varying or a-c voltages are applied to the electrodes. The a-c plate resistance is defined as the **ratio of a small change in plate voltage to the change in plate current produced thereby, when the grid voltage is kept at a constant value.** Expressing this definition in equation form, as before, the

$$\text{a-c plate resistance, } r_p = \frac{\Delta E_b}{\Delta I_b} \text{ (with } E_c \text{ constant)}$$

The plate resistance is usually determined from the linear part of the tube's plate characteristics. Returning to our example (Fig. 13), let us determine the plate resistance of the 6J5 near operating point *A*. We have previously found that a change in plate voltage from 215 volts (point *B*) to 255 volts (point *D*) on the −8 volt grid-voltage curve produced a change in plate current from 5 ma (at *B*) to 9.6 ma (at *D*). Thus, the a-c plate resistance near point *A* is

$$r_p = \frac{\Delta E_b}{\Delta I_b} = \frac{(255-215)V}{(9.6-5) \text{ ma}} = \frac{40V}{.0046A} = 8,700 \text{ ohms}$$

The tube manual lists a value of 7700 ohms for the a-c plate resistance, but the discrepancy is not too serious and must be expected, since the a-c plate resistance is a function of the operating point along any particular grid-voltage characteristic. We might also point out that the triangle B-C-D in Fig. 13 could have been made smaller (it is not necessary to draw a line to the next grid-voltage curve, as in the case of μ), with probably greater accuracy resulting.

TRANSCONDUCTANCE

A third constant used in describing the properties of electron tubes is the control grid-to-plate **transconductance** (sometimes called **mutual conductance**) designated by the symbol g_m. The transconductance is the most important of the three tube constants, since it reveals the effectiveness of the control grid in securing changes in the plate current and, hence, in the signal output of the tube. **It is defined as the ratio of a small change in plate current to the small change in control-grid voltage producing it, when the plate voltage is held constant.** Expressing this in equation form,

$$\text{transconductance, } g_m = \frac{\Delta I_b}{\Delta E_c} \text{ (with } E_b \text{ constant)}$$

Since g_m is the ratio of a current to a voltage, it has the form of a conductance and is expressed in units of *mhos* (ohm spelled backward). This unit is, however, too large for electron tube usage and hence one-millionth part of a mho, the micromho (μmho) is generally used. (To obtain μmhos multiply mhos by 1,000,000.) Thus, an electron tube in which a 1-volt change in grid voltage produces a 2-ma (.002 amp) change in plate current has a transconductance of

$$\frac{.002}{1} \times 10^6 = 2,000 \text{ micromhos}$$

Transconductance is easily found from the plate family of tube characteristics. Again returning to our example of the 6J5 triode, we found from Fig. 13 that a change in grid voltage from −10 to −8 volts (from point *C* to point *D*) produced a change in plate current from 5 ma to 9.6 ma. Hence the transconductance

$$g_m = \frac{\Delta I_b}{\Delta E_c} = \frac{(9.6-5) \text{ ma}}{(10-8)V} = \frac{.0046A}{2V} = 2,300 \ \mu\text{mhos}$$

The tube manual lists the transconductance near this operating point as 2600 micromhos, which is in fair agreement with the above.

Relation Between μ, R_p, and g_m. The three tube constants are interrelated, in accordance with their definition, as you can easily see by dividing the relation defining the amplification factor by that defining the plate resistance. Thus, checking back,

$$\frac{\mu}{r_p} = \frac{\Delta E_b/\Delta E_c}{\Delta E_b/\Delta I_b} = \frac{\Delta I_b}{\Delta E_c} = g_m$$

Hence $g_m = \dfrac{\mu}{r_p}$ and equivalently $\mu = g_m \times r_p$

$$\text{and } r_p = \frac{\mu}{g_m}$$

If two of the tube's constants are known, therefore, the third may be obtained from the relations above. The transconductance is frequently used as a **figure of merit** for comparing different tubes in the same general classification. You can easily see why. Transconductance expresses the **ratio** of the amplification factor to the plate resistance of the tube. A **high** amplification factor is desired to obtain large output voltages from small input signals. A **low** plate resistance, on the other hand, permits the flow of a large plate current, resulting in large output **power.** Transconductance, thus, is a measure of the voltage and power amplification possible in any specific tube. But it is not easy for the designer to raise the value of the transconductance, since the plate resistance increases as the amplification factor is made larger. Besides serving as a figure of merit, the transconductance of a tube is the factor most frequently measured by tube testers to compare the performance of the tube with its original ratings.

Although we have stated that the tube constants depend to a large degree on the spacing and geometrical configuration of the original tube design, we have also seen from the examples that these so-called "constants" vary to some degree depending on the choice of operating voltages on the electrodes. To show this variation with operating conditions, we have plotted the 6J5 tube "constants" for a constant plate voltage of 250 volts in Fig. 14. (To keep the plate voltage constant, we had to *decrease* the negative grid voltage, as the plate current increases.) Each of the constants has its own vertical scale, but all have the same horizontal scale, i.e., the plate current in ma. Note that the amplification factor, μ, is essentially constant, while the plate resistance *decreases* with *increasing* plate currents and the transconductance, g_m, increases correspondingly—as expected.

INTERELECTRODE CAPACITANCES

Among the important design factors are the so-called **interelectrode capacitances.** You will remember that various electrostatic fields exist between the charged electrodes of a triode, such as the field between plate and cathode, between plate and grid, and between grid and cathode. From elementary electricity you will also recall that an electrostatic field between any two charged metal plates is the equivalent of an electric capacitor. Capacitance, thus, exists between any two pieces of metal separated by a dielectric. The amount of capacitance depends on the size of the metal pieces, the distance between them, and the type of dielectric. Accordingly, the capacitances existing between the electrodes of a tube depends on their size and spacing, and on the dielectric (usually a vacuum).

The interelectrode capacitances of a triode are shown in Fig. 15. The grid-to-cathode capacitance is symbolized by C_{gk}, the grid-to-plate capacitance by C_{gp} and the plate-to-cathode capacitance by C_{pk}. Although these capacitances are generally small (from 2 to 10 micromicrofarads), their reactance at the higher radio frequencies becomes low and leads to undesirable coupling effects. The grid-to-plate capacitance, C_{gp}, especially has the property of feeding back energy from the plate (output) circuit to the grid (input) circuit, which may lead to instability and oscillations. This property is generally undesirable, although it is sometimes utilized to produce oscillations in suitable circuits, as we shall see later on.

A reduction of the interelectrode capacitances

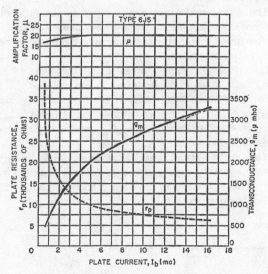

Fig. 14. *Variation in Tube Constants μ, g_m and r_p for 6J5*

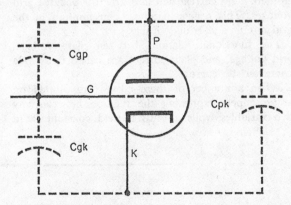

Fig. 15. *Triode Interelectrode Capacitances*

can be achieved by additional shielding electrodes and leads to the design of **multielectrode tubes,** which shall be described in the next Chapter. However, at the ultra-high-frequencies (UHF) used in radar and similar circuitry, interelectrode capacitances become so objectionable that ordinary vacuum tubes can no longer be used. Special UHF tubes of very tiny physical dimensions and with closely spaced electrodes have been designed and are discussed in a later chapter.

SUMMARY

A **triode** is a three-element electron tube, consisting of a cathode (emitter), a plate (collector) and a **control grid** (spiral of fine wire), placed between cathode and plate.

The voltage placed on the control grid usually is made up of a fixed, **negative,** d-c voltage, called **bias,** and a varying or alternating (a-c) voltage, called the **signal.**

The plate current in a triode is determined by the **combined effect** of the electric fields set up by the plate and control-grid voltages. The control grid, being nearer to the cathode than the plate, has a *greater* effect on the plate current than the plate voltage. This permits **amplification** of an input signal.

When a large negative bias voltage is placed on the control grid, the grid neutralizes the electrostatic field of the plate, and no plate current flows. The smallest negative grid voltage capable of cutting off the plate current is called the **cut-off bias.** Larger bias values have no effect.

As the negative grid bias is reduced the grid is no longer capable of neutralizing the electric field of the plate completely, and the plate current increases in proportion to the reduction of the negative grid bias. When the bias is made *positive,* the field of the grid aids that of the plate and a point will be reached where the plate current reaches its **saturation value,** equal to the total supply of electrons emitted. **Grid current** will also flow for **positive** grid voltages. This results in **power consumption** in the grid circuit and is undesirable.

The functional relations between plate voltage, grid voltage, and plate current are called the **triode characteristic curves.**

When a number of these curves are plotted on graph paper, expressing the relation between any two variables while a third is held constant (at a different value for each curve), a **family** of triode characteristic curves results.

Static characteristics are obtained when steady (d.c.) voltages are applied to the tube's electrodes; **dynamic characteristics** are obtained under **actual operating conditions,** that is, with a varying **signal voltage** applied to the grid and a **load resistance** inserted into the plate circuit.

In the **linear portion** of each characteristic equal changes in grid or plate voltage cause equal changes in the plate current. This is necessary for **undistorted amplification.**

The most important design factors of a tube, or **tube constants,** are the **amplification factor,** the a-c **plate resistance,** the **mutual conductance** or **transconductance,** and the **interelectrode capacitances.**

The **amplification factor** is a measure of the relative effectiveness of the control grid in overcoming the electrostatic field of the plate. It is defined as the **ratio of a small change in plate voltage to the small change in grid voltage required to produce the same change in the plate current.**

The a-c or **variational plate resistance** is the internal opposition of the tube to the flow of plate current, when a-c voltages are applied to the electrodes. It is defined as the **ratio of a small change in plate voltage to the corresponding change in plate current produced thereby, when the grid voltage is kept constant.**

The **transconductance** or mutual conductance, is the most important tube constant, since it reveals the effectiveness of the control grid in securing changes in the plate current and, hence, signal output of the tube. It is defined as the **ratio of a small change in plate current to the small change in control-grid voltage producing it, when the plate voltage is held constant.**

The transconductance is also **equal to the ratio of the amplification factor to the a-c plate resistance of the tube.** It sometimes serves as a **figure of merit** for comparing tubes.

Interelectrode capacitances exist between cathode, plate and grid, because of the electric field present between these charged electrodes. They are designated as the **grid-to-cathode** capacitance, C_{gk}, the **grid-to-plate** capacitance, C_{gp}, and the **plate-to-cathode** capacitance, C_{pk}.

The grid-to-plate capacitance **feeds back energy** from the plate (output) circuit to the grid (input) circuit. This may lead to instability and oscillations at radio frequencies.

Chapter Five

MULTIELECTRODE TUBES

Multielectrode tubes are tubes having more than one grid, which permits attaining many desirable characteristics not possible with the triode. Among the most common multielectrode tubes are the **tetrodes,** which have **four** electrodes (two grids), and the **pentodes,** which have **five** electrodes (three grids). Let us first turn to tetrodes.

TETRODES

The development of the tetrode is a direct outcome of the undesirable grid-to-plate capacitance in triodes, which we have previously discussed. This capacitance leads to coupling effects and instability in radio-frequency amplifiers that cannot be eliminated, except by costly and tricky "neutralization circuits." The most effective answer to the feedback problem caused by the grid-to-plate capacitance is to insert an additional shielding electrode, called the **screen grid,** between the control grid and plate of a triode. This additional grid almost completely encloses the plate and thus acts as an effective electrostatic shield between plate and control grid. The grid-to-plate capacitance, which usually runs from 2 to 5 micromicrofarads in a triode, is thereby reduced to values as low as 0.01 micromicrofarad in a tetrode. This effectively cancels out the feedback action and resulting instability in radio-frequency amplifiers. The tetrode produces some undesirable distortion of the tube characteristics, however, as we shall see shortly. Its successor, the pentode, overcomes this disadvantage while retaining the tetrode's desirable characteristics and has

for this reason almost replaced the tetrode. Nevertheless, it is instructive to study the basic action of the tetrode, since the pentode's action is very similar.

The physical arrangement of the four electrodes in a tetrode is illustrated in Fig. 16, along with the tetrode schematic symbol. The cathode, plate and control grid are substantially the same as in a triode. The plate is separated from the control grid either by a single wire screen on one side, or the plate may be completely enclosed by the screen grid. The latter is similar to the control grid, except for a somewhat coarser mesh. The connection to the control grid is usually made through a metal cap at the top of the tube envelope.

TETRODE CHARACTERISTICS

The tetrode is operated similarly to the triode, with the cathode near ground potential, the control grid at a small negative (bias) voltage, and the plate at fairly high positive voltage of several hundred volts. The screen grid is also placed at a high positive potential with respect to the cathode, but somewhat lower than the plate potential. The positive voltage on the screen grid helps to accelerate the electrons on their way from the cathode to the plate and thus aids the electrostatic field of the plate. Some of the electrons strike the screen grid and produce a **screen current,** which is not useful, but most of the electrons pass through the open mesh of the screen toward the plate to produce the plate current.

Because of the shielding effect of the screen grid the electrostatic field of the plate has little effect on the electron space charge near the cathode. As a consequence, **variations in plate voltage have little effect on the plate current** and a much greater change in the plate voltage is required to produce the same change in plate current than would be necessary in a triode. Since the a-c plate resistance is defined as the ratio of a change in plate voltage to the corresponding change produced in the plate current, the **plate resistance of a tetrode is far greater than that of a triode.** (The tetrode plate resistance is in the order of 0.5 to 1 megohm.)

The control grid of a tetrode has about the same effect in governing the flow of plate current as in a triode, since it is not shielded from the space charge by the screen grid. Thus, a small change in control-grid voltage will produce a large change in

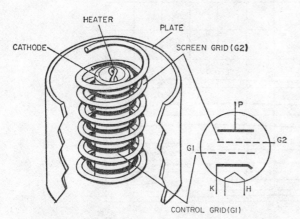

Fig. 16. *Simplified Construction of Tetrode and Schematic Symbol*

the plate current, just as in a triode. On the other hand, the tetrode requires a very large change in **plate voltage** to produce a small change in plate current, as we have just seen. Since the amplification factor (μ) is defined as the ratio of a small change in plate voltage to the small change in grid voltage that will produce the same change in plate current, the **amplification factor of a tetrode, clearly, will be far higher than that of a comparable triode.** Amplification factors for tetrodes run from about 400 to 800, compared to values of about 5 to 50 for triodes.

The price we have to pay for the high amplification factor of a tetrode is a correspondingly high plate resistance, as we have seen. Since the transconductance (g_m) of a tube is the **ratio** of the amplification factor to the plate resistance, we would expect this ratio to be about the same for tetrodes as for triodes—and so it is. The transconductance of tetrodes averages about 1000 to 1500 micromhos, which is about the same as for triodes. There are, however, some special types of power tetrodes, with g_m values as high as 4500 micromhos.

A typical family of plate characteristic curves for a type 32 tetrode is illustrated in Fig. 17. These curves show the relation between the plate current (I_b) and the plate voltage (E_b) for various values of the control-grid voltage (E_{c1}). The screen voltage (E_{c2}) is fixed at 67.5 volts. One curve, showing the variation in the screen-grid current (I_{c2}) with plate voltage, for a control-grid voltage of −3 volts, is also shown.

The distorted shape of the characteristics in the region of low plate voltages may be explained by considering a single I_b − E_b curve, such as the one for a control grid voltage of −3 volts. (See Fig. 17.) Since a positive potential of +67.5 volts is applied to the screen, an electrostatic field exists between screen and cathode, even when the plate voltage is zero. Consequently, the electrons are attracted toward the screen (through the slightly negative control grid) and a small screen current (I_{c2}) of about 2 ma flows, when the plate voltage is zero. As the plate voltage is increased, some of the electrons passing through the screen are diverted to the plate, with a consequent increase in plate current and a decrease in the screen current. The plate current continues to increase and the screen current decreases until the plate voltage is about +12.5 volts. The total **space current** (sum of plate plus screen currents) also is seen to increase slightly.

As the plate voltage is further increased, however, the plate current suddenly begins to fall off, while the screen current increases correspondingly. The reason for this strange behavior is the phenomenon of **secondary emission,** which we mentioned in an earlier chapter. What happens is this. When the plate voltage is raised above a few volts, the electrons are speeded up sufficiently to dislodge loosely held electrons within the plate material and project them as **secondary electrons** into the region between plate and screen grid. These additional, secondary electrons are immediately collected by the screen grid because its potential is *more positive* than that of the plate. Secondary emission also takes place in the triode, but since the plate in a triode is the only positive electrode, the secondary electrons are recollected by the plate. In the tetrode, however, these secondary electrons are diverted from the plate to the screen and are effectively *subtracted* from the plate current. (The flow of secondary electrons to the screen is in the opposite direction to the normal plate-current flow.) The *net* plate current then is the number of **primary electrons** from the cathode received by the plate *minus* the

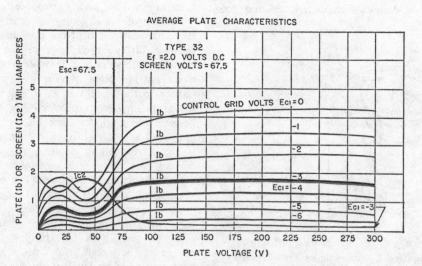

Fig. 17. *Plate Family of Characteristic Curves for Type 32 Tetrode*

number of **secondary electrons** lost to the screen grid. When each primary electron striking the plate produces on the average more than one secondary electron, the screen will receive more electrons from the plate than the plate from the cathode, and the **net plate current will then be negative.** This is shown by the **negative slope** of the $I_b - E_b$ characteristic between plate-voltage values of 12.5 and 50 volts. The screen current (I_{c2}) in the same region is seen to increase in accordance with the plate-current decrease. (The total space current remains about the same in this region, but its division between plate and screen continually changes.)

As the plate voltage is still further increased and begins to approach the value of the screen voltage, the force of attraction exerted by the plate on the secondary electrons becomes greater than that exerted by the screen grid and the secondary electrons return to the plate. Consequently, the plate current begins to rise sharply and the screen current decreases correspondingly. When the plate voltage is made substantially higher than the screen voltage, the plate collects practically all the electrons from the cathode and, in addition, a small number of **secondary electrons from the screen grid.** The screen grid collects only a few electrons whose paths happen to be intercepted by it. The plate current now stabilizes at a high value, almost equal to the total space current, while the screen current drops to a low, constant value. Because of the shielding effect of the screen, further increases in plate voltage have little effect and the plate current becomes almost independent of the plate voltage, as is evident from the flattening out of the curves.

Negative Resistance. The **negative** slope of the plate current-plate voltage characteristic in the low plate voltage region signifies that the plate current *decreases* for increasing plate voltages, *contrary to Ohm's law.* This negative current characteristic is termed **negative resistance** and the tube in this region may be interpreted as a **source of power** rather than a consumer of it. As we shall see in a later chapter (on Oscillators) this behavior leads to instability and oscillations of the plate current under certain conditions. For this reason, the plate voltage in tetrodes must be made sufficiently high so that it never drops to the negative resistance region for any signal voltage on the grid. This disadvantage is overcome in pentodes, to be discussed next. Incidentally, the negative resistance characteristic of tetrodes is utilized in special oscillator circuits.

PENTODES

The insertion of an additional grid, called the **suppressor grid,** between the plate and screen grid of a tetrode overcomes the undesirable effects of secondary emission and the resulting negative-resist-

ance characteristic at low plate voltages. By adding this electrode the tube becomes a five-electrode unit, or **pentode.** While eliminating the distortion of the plate family of characteristics, pentodes retain all the advantages of tetrodes, such as low grid-to-plate capacitance, high amplification factor and high power output.

Fig. 18 illustrates a typical pentode tube, along with its schematic symbol. The pentode contains an emitter (cathode), three grids and a plate. The grid closest to the cathode, G1, is the control grid, next is the screen grid, G2, and the third is the suppressor grid, G3, located between screen grid and

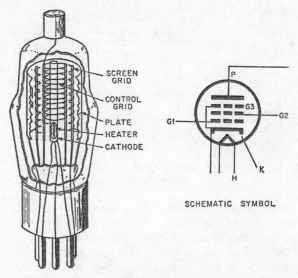

Fig. 18. *Pentode Construction* (*metal-type*) *and Schematic Symbol*

plate. Pentodes resemble tetrodes in external appearance. The connections to all electrodes are generally made from pins in the base of the tube, but some of the older types have top grid caps.

ACTION OF SUPPRESSOR GRID

The suppressor grid is usually connected directly to the cathode and is thus at a substantial **negative** potential with respect to the plate. As you would suspect, the secondary electrons emitted by the plate are **repelled** by the negative suppressor grid and are driven back to the plate. Thus, while *not preventing* secondary emission by the plate, the suppressor grid **does eliminate the effects of secondary emission.** As a result, the plate current rises smoothly with increasing plate voltage, from zero up to the maximum value for each control-grid voltage.

The presence of the suppressor grid further **increases the shielding action** between plate and control grid and thus further reduces the grid-to-plate capacitance. For the same reason, the plate current is even more independent of the plate voltage than in the tetrode, as demonstrated by the further "flattening" (low slope) of the plate characteristics. (See Fig. 19.) With the plate current little affected by the plate voltage because of the high screening, both the amplification factor and the plate resistances of the pentode become extremely high. The amplification factor in pentodes may run as high as 1000 to 1500 and the plate resistance is usually in the order of about one megohm (one million ohms). As we shall see later (in the Chapter on

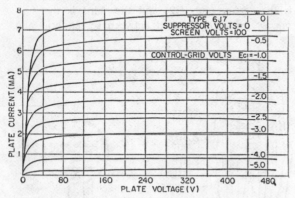

Fig. 19. *Plate Family of Characteristic Curves for 6J7 Pentode*

Audio Amplifiers) the high plate resistance makes it impossible to use more than about one-tenth of the pentode amplification factor. The undistorted plate characteristics of pentodes, however, permit large variations in the input signal voltage at the grid, resulting in large available power output. As is the case for the tetrode, the combination of a high amplification factor and high plate resistance results in only average values of the transconductance.

BEAM-POWER TUBE

A useful hybrid between a tetrode and a pentode is the so-called **beam-power tube.** It has two grids, like a tetrode, but yet is capable of suppressing the effects of secondary emission and thus operate as a pentode. The effect of the pentode suppressor grid is obtained by special electrodes and by arranging the tube elements in such a way as to produce a negative "space charge" *near the plate*. The action of this space charge repels the secondary electrons back to the plate, just as the suppressor grid in a pentode.

The action of the beam-power tube is attained through its structural features, which are illustrated in Fig. 20. As illustrated, the tube produces a dense beam of electrons between screen grid and plate. This beam is formed by a combination of two features. First, the control grid and screen grid is of the same pitch and the grid wires are so aligned that the screen grid lies in the "shadow" of the control grid. The "shadowed" screen grid intercepts few electrons, resulting in low screen current and high plate-current density. Secondly, special **beam-confining** electrodes, electrically connected to the cathode, further assist in producing dense electron beams in the plate-screen region. This bunching up of electrons leads to a **potential minimum** and a resulting **space-charge effect** in the plate-screen region, which effectively repels the secondary electrons from the plate. Moreover, the distance between screen and plate is made larger than in a tetrode (or pentode) to make sure that there will be a large supply of electrons in the region, which further assists in increasing the space-charge effect.

As a matter of fact, the suppressor action of the space charge in a beam-power tube is superior to

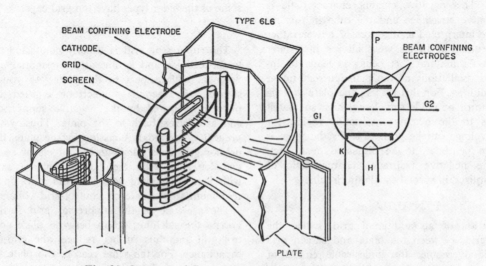

Fig. 20. *Structure of Beam-Power Tube and Schematic Symbol*

that of a pentode, resulting in plate-current characteristics that are even less distorted and have a sharper "knee" (at low plate voltages). Because of this low distortion, low screen current, and large effective cathode and plate areas, the beam-power tube has the advantages of exceptionally high power output for small input signals and a high efficiency. It is often preferred, therefore, in the output stages of amplifiers.

VARIABLE-MU (REMOTE CUTOFF) TUBES

The amplification factor (μ) of electron tubes depends to a large extent on the **spacing** of the control-grid wires. In conventional tubes the turns are **uniformly spaced** throughout the length of the control grid, resulting in a constant amplification factor for most of the plate-current and grid-voltage range of values. Furthermore, when making the grid voltage more negative, all parts of the control-grid structure begin to cut off the plate current at the same time, producing a **sharp cutoff** characteristic. (See Fig. 21.)

It is sometimes desirable to produce a more gradual, or **remote cutoff** characteristic to accommodate large-amplitude signals without the distortion accompanying plate-current cutoff. This is accomplished by spacing the turns of the control-grid wire in tetrodes or pentodes **non-uniformly**, winding them closer near the ends of the structure than at the center (Fig. 21). This effectively gives the tube a

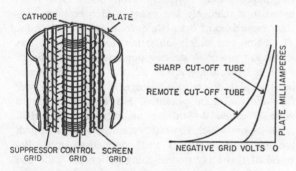

Fig. 21. *Construction of Remote Cutoff Pentode and Comparison of Plate-Current Cutoff Characteristics Between Sharp and Remote Cutoff Pentode*

variable mu (μ), low at the center and high at the ends.

For low values of the grid bias and high plate current values the tube operates normally. As the control grid is made more negative, however, the closely spaced (high-mu) regions at the ends of the grid reach plate-current cutoff first, while the center (low-mu) portions of the grid still permit plate current to flow. As operation is progessively forced into the more central (low-mu) portions of the grid, the

amplification factor decreases considerably and the negative control-grid voltage must be made very high to achieve complete plate-current cutoff. Because of these features, such tubes are called either **variable-mu** or **remote cutoff** tubes. Since the amplification of these tubes depends on the value of the control-grid bias, they are very useful in **automatic-volume-control** (AVC) circuits, where it is required to vary the tube's amplification automatically by changing the grid bias, obtained from the signal amplitude.

SUMMARY

Multielectrode tubes have more than one grid. The **tetrode** has **four electrodes (two grids)** and the **pentode** has **five electrodes (three grids)**.

The **tetrode eliminates the feedback of energy,** caused by the grid-to-plate capacitance, by the insertion of a **screen grid,** between control grid and plate of the tube.

The plate current-plate voltage characteristics of the tetrode are **distorted in the region of low plate voltages,** resulting in a **negative resistance** characteristic and instability in this region.

Negative resistance and distortion at low plate voltages is caused by **secondary emission** from the tetrode plate. The secondary electrons are attracted to the highly positive screen and subtract from the plate current. The **net** plate current will be **negative,** when the number of secondary electrons flowing to the screen exceeds the number of primary electrons arriving at the plate.

At plate voltages higher than the screen voltage, the plate collects all primary and secondary electrons and, because of the shielding action of the screen, the plate current becomes practically **independent** of the plate voltage.

Since variations in plate voltage have little effect on the plate current (at high plate voltages), both the a-c plate resistance and the amplification factor of the tetrode are *far greater* than those of the triode. The transconductance, which is the ratio of the amplification factor to the plate resistance, is about the *same* for tetrodes as for triodes.

In the **pentode, a suppressor grid** is inserted between the plate and screen grid. When connected to the cathode, the suppressor grid is **highly negative** with respect to the plate and is able to overcome the **effects** of secondary emission and thus the distortion of the plate characteristics.

The suppressor grid also **increases the shielding action** between plate and control grid. As a result, both the plate resistance and the amplification factor of pentodes are **extremely high.** The transconductance is about the same as that of triodes or tetrodes.

The pentode permits **higher power output** and **efficiency** than either the triode or the tetrode.

A **beam-power tube** is a hybrid between a tetrode

and a pentode. It has two grids, like a tetrode, but is also equipped with special **beam-confining electrodes,** capable of overcoming the effects of secondary emission, as does a pentode.

In a beam-power tube **dense electron beams** produce an electron space charge in the plate-screen region that repels secondary electrons and returns them to the plate.

A **variable-mu** or **remote cutoff** tube is a pentode that has a **gradual plate-current cutoff** when the negative bias is increased. This is achieved by nonuniform spacing of the control-grid wires, resulting in an amplification factor that is **low at the center and high at the ends** of the grid structure. Thus, the amplification of the tube may be changed by varying the grid bias.

Chapter Six

SPECIAL ELECTRON TUBES

Standard diodes, triodes, tetrodes and pentodes are the heart of practically all electronic circuits, with which we shall become intimately acquainted later on. Now let us turn to a group of specialized electron tubes which are capable of doing various jobs that ordinary tubes cannot perform and which are especially useful in industrial, control, and ultra-high-frequency (UHF) circuits. Although the number of special-purpose tubes that have sprung up are legion, we shall consider here only four of the most important generic types. These are **gas-filled tubes, phototubes and photomultipliers, cathode-ray tubes, and special UHF tubes.**

GAS-FILLED TUBES

In the manufacture of ordinary vacuum tubes as much air as possible is removed from the tube envelope to prevent significant ionization of residual gases and the resulting large, uncontrolled currents. You will remember that the secret of the triod is the fine control of free electrons within the tube by the electrostatic fields of the grid and plate. Ionized gas molecules would interfere with this control and make the tube useless as an amplifier. When fine control is of less importance, but rather the efficient handling and turning on and off of heavy currents in industrial applications, gases such as nitrogen, helium, neon, argon, or mercury vapor may be purposely introduced into the tube envelope.

As we learned in Chapter 1, **ionization** within a gas-filled tube consists of the removal of one or more electrons from a normal gas molecule, leaving the resulting ion with a **positive charge.** Ionization of molecules may occur when electrons traveling from cathode to plate **collide** with gas molecules, or when gas molecules collide with other gas molecules. In either case additional electrons are freed which join the original electron stream and may be capable of liberating still more electrons by colliding with other gas molecules. This process is **cumulative,** as in an avalanche, and results in a sudden, sharp increase in the electron current flowing toward the plate. At the same time the heavier, positively charged ions (ionized gas molecules) slowly drift toward the negative cathode; during the journey they attract electrons from the cathode and recombine with them to form gas molecules.

The energy required to produce ionization of gas molecules by collision with high-speed electrons is supplied by means of the **plate-to-cathode voltage.** A definite voltage value exists for each type of gas at which ionization suddenly begins. Before this point is reached the plate current is about the same as for a vacuum tube at the same plate potential. When ionization occurs, however, the **plate current increases** dramatically to large values and the **plate to-cathode voltage** (voltage drop across the tube) **drops to a relatively low value, which remains constant regardless of the plate current through the tube.** The voltage at which ionization occurs is known as **ionization potential, striking potential,** or **firing point.**

Once ionization has started, the action maintains itself at plate-to-cathode voltages considerably lower than the ionization potential. However, a minimum voltage, called **de-ionizing** or **extinction potential,** exists below which ionization cannot be maintained; the gas then de-ionizes and conduction stops. Because of this relay or switching action the tube can be used as an **electronic switch,** which closes at the striking potential, permitting a large current to flow, and opens at the de-ionization voltage, blocking the current flow.

Because of their **unidirectional characteristics** (current will only flow from cathode to plate), gas tubes are useful for rectification of heavy alternating currents used in industry. However, caution must be exercised when placing gas tubes in a-c circuits, where the polarity at the plate continuously reverses. It takes a certain time, known as **de-ionization time,** for the gas ions to recombine with free electrons and thus stop the ionization current. If the plate voltage is made negative before de-ionization is completed, the gas ions will flow to the negative plate, constituting an inverse current flow, or **arc-back.** The tube thus carries a heavy current on both a-c half-cycles, which might destroy it. Arc-back becomes more severe as the operating frequency is increased and less time is available for de-ionization to be completed. It is important, therefore, to know the **inverse-voltage rating** of the tube at the operating frequency.

Gas-filled tubes may be classified according to the type of electron emission employed into **cold-cathode** types (usually diodes), **hot-cathode (thermionic)** types, available as diodes, triodes, and tetrodes, and mercury-pool tubes, generally triodes known as **ignitrons.** Cold-cathode types obtain **field emission** of electrons from an unheated cathode (see Chapter 2). Gas-filled thermionic tubes have oxide-coated, heated cathodes, just as conventional vacuum tubes, while mercury-pool tubes obtain electron emission from a pool of liquid mercury.

The construction of gas-filled tubes is similar to that of high-vacuum tubes, except that the cathodes, grids, and plates are usually larger to carry the heavy currents through the tube. As we shall see later, grid control in gas-filled tubes is limited to *starting* conduction and cannot be used to stop it, as in high-vacuum tubes.

Schematic symbols for four types of gas-filled tubes are shown in Fig. 22. A small dot within the circle generally indicates that a tube is gas-filled.

COLD-CATHODE GAS-FILLED DIODES

Usually filled with neon gas in combination with other gases, cold-cathode diodes utilize field emission to obtain ionization of the gas. Since there are no emitted electrons to help the process, the striking potential for cold-cathode tubes is higher than for the hot-cathode types and it is also somewhat erratic. Two types of neon tubes may be distinguished. The cathode may have the same shape and size as the plate, such as the neon-glow lamp shown in Fig. 22-4, in which case the tube can conduct in *either direction*, depending on the values of the applied electrode potentials. Since the *negative* electrode (cathode) is surrounded by a characteristic glow (usually orange), the tube is useful to indicate the polarity of a d-c voltage. When an alternating voltage is applied, both electrodes are surrounded by a glow discharge. A strong radio-frequency (r-f) field is capable of ionizing the gas without direct connection to the tube and neon-glow lamps are, therefore, frequently used to indicate the presence of an r-f field.

More frequently the cathode of a cold-cathode tube is *larger* than the plate, in which case the tube permits conduction in one direction only (See Fig. 22-3). This makes the tube useful as a rectifier, to

handle large currents at a relatively low tube voltage drop. The tubes are also employed as **voltage regulators,** since the voltage drop across the tube remains nearly constant over a wide range of current values through it. Various types are available with different voltage and current ratings.

HOT-CATHODE GAS-FILLED DIODES

Designed for use as rectifier, the chief type of hot-cathode gas diodes is the so-called **mercury-vapor** tube, which consists of a thermionic (hot) cathode, a plate and a small amount of liquid mercury that vaporizes when the cathode is heated (Fig. 22-1). When the plate voltage is applied, the vapor ionizes and sustains a heavy current (several times as high as that of an equivalent high-vacuum type) at a low, constant tube voltage drop of about 15 volts. This means that most of the available supply voltage will appear as rectified output voltage and little will be wasted in an internal tube drop. The cathodes of mercury-vapor tubes must be **preheated** for one to two minutes before the plate voltage is applied to permit the mercury to be *completely* vaporized. Only then is the tube capable of carrying its rated plate current.

GAS-FILLED TRIODES AND TETRODES (THYRATRONS)

A triode or tetrode to which a small amount of gas (argon, neon, or mercury vapor) has been added is known as a **thyratron.** (See Fig. 22-2.) Although the electrode structure is basically the same as in conventional vacuum tubes, the tube characteristics of gas triodes and tetrodes are entirely different.

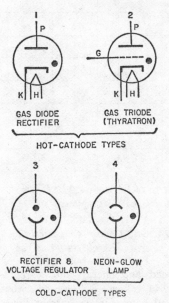

Fig. 22. *Schematic Symbols for Four Types of Gas-Filled Tubes*

The grid in a thyratron is only used to *start conduction* of the plate current by ionization, but cannot be employed to control the amount of plate current or stop it. The action of a thyratron, thus, is essentially that of a **trigger** which starts the plate current. To stop it, the **plate voltage must be removed** from the tube.

The construction of two types of thyratrons is illustrated in Fig. 23; the grid voltage necessary to start the plate current for different plate voltages, is shown in Fig. 24.

When the grid voltage is made sufficiently negative the electrons emitted from the cathode do not acquire the necessary velocity to ionize the gas, and the plate current is substantially zero. As the negative grid potential is reduced, the electrons acquire

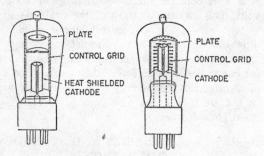

Fig. 23. *Cross Section of Gas-Filled Triodes* (*Thyratrons*)

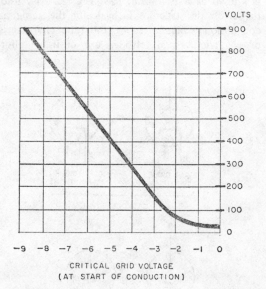

Fig. 24. *Grid Control Characteristics of Typical Thyratron*

more speed and energy and a point—called the **critical grid voltage**—is reached, where ionization occurs and a large plate current flows. After conduction has started, the grid has no further control

over the plate current. As shown in Fig. 24, there is a different value of critical grid voltage for each value of the plate voltage. Points that lie to the right of the curve (in Fig. 24) represent conduction, and points to the left of the curve represent nonconduction of the tube.

Since the grid voltage has no control over the magnitude of the plate current, thyratrons cannot be used as amplifiers like vacuum triodes. Because of their trigger characteristic, however, the tubes are useful in switching and relay applications, where it is desired to start conduction at a certain instant by control of the grid voltage. Thyratrons are also used in motor-control circuits and in so-called **sawtooth sweep generators** for TV and radar applications, as we shall see later on. Some thyratrons are constructed for **cold-cathode** operation, in which case the critical grid voltage is a rather large *positive* value.

MERCURY-POOL TUBES (IGNITRONS)

Used primarily for heavy-duty, industrial rectifier service, mercury-pool tubes are one of the oldest types of gas-filled tubes. In its modern version, called **ignitron,** a pointed electrode (called **ignitor**) dipped into a mercury pool is utilized to trigger the mercury-vapor discharge in the tube at the desired instant. The mercury pool is used itself as a **cathode** and requires no heating power. In practice, a current is passed through the ignitor, which generates sufficient heat to produce a "cathode spot" at its tip. Electrons are emitted from this bright spot and are immediately ionized. Ignitrons are capable of carrying plate currents from 10 amps to 5000 amps.

PHOTOTUBES AND PHOTOMULTIPLIERS

Phototubes are the exciting gadgets that account for a multiplicity of the miraculous control applications we encounter every day. Their operation is based on the principle of **photoemission,** which we have discussed in Chapter 2. When a light beam falls upon a **photoemissive cathode,** part of the light energy or "quanta" are transferred to the electrons within the material and kick them out from the emitter surface. If a positive voltage is placed on the plate of the tube, the electrons are attracted toward it and a plate current results, just as in a thermionic tube. The plate current may be made to operate a relay, whenever the beam of light falls on the tube. This is used for such applications as operating a garage door from a car's headlights, automatically dimming headlights when two cars encounter each other, telegraph and telephone transmission by means of light beams, and many others. On the other hand, an object passing between the phototube and the light source will interrupt the lightbeam and produce a shadow on the photoemissive cathode, which reduces or cuts off the plate current. Again, this effect can open or close a relay, which may operate a mechanical register to count the number of objects passing the tube, open a door, start an escalator, or what-have-you.

There are really **three** types of photosensitive or photoelectric devices, which may be classified according to their function as **photoemissive (or photoelectric), photovoltaic,** and **photoconductive.** Photoemissive or photoelectric tubes make use of the principle of photoelectric emission from a photosensitive cathode surface. Photovoltaic cells convert light energy falling upon them directly into electric current. Finally, photoconductive cells contain a semi-conducting solid material whose resistance decreases as the amount of light falling on it increases. Only photoelectric tubes are truly electronic, since current flow takes place in a vacuum or gas, while in the other two types current flows in a solid or liquid. All three devices depend on the same basic fact—namely, the liberation of electrons from a photosensitive surface when light strikes it.

BASIC LAWS

Regardless of type, all photosensitive surfaces act in accordance with the following empirical laws, when light energy falls upon them:

1. The number of electrons emitted per second from a photosensitive surface is **proportional to the intensity of the incident light.**

2. The maximum kinetic energy of each released electron is **independent of the light intensity,** but is **directly proportional to the frequency** (number of vibrations per second) of the light. This is equivalent to saying that the **initial velocity** of the emitted electrons is proportional to the **square root** of the frequency of the incident light. You will remember from basic physics that light frequencies are measured in **Angstroms** (one $A = 10^{-8}$ cm) and that the entire visible light **spectrum** (rainbow colors) extends from about 4000 Å (.00004 cm) for violet to about 7400 Å for red.

3. From the second law it is also apparent that there must be some frequency, called the **threshold frequency,** below which the kinetic energy of the electrons is insufficient to liberate them from the photoemissive surface. This is indeed so, and each type of material has its characteristic threshold frequency.

PHOTOEMISSIVE TUBES

Photoemissive tubes (or phototubes, for short) consist of cathodes and anodes, similar to conventional diodes. The cathode (emitter) is usually large in area and arranged so that it exposes as much surface area to incident light as possible. The anode, or collector of electrons, is frequently a metal

rod that is surrounded by the cylindrical cathode. Fig. 25 illustrates three typical forms of phototubes.

Cathodes and anodes are enclosed in a sealed glass envelope, which is either highly evacuated or filled with an inert gas at low pressure. The gas-filled tubes are more sensitive than the vacuum types, but are somewhat erratic in operation; their response is non-linear and falls off somewhat at higher audio frequencies above 5000 cycles. Vac-

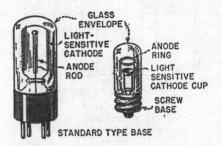
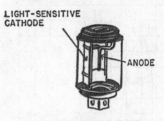
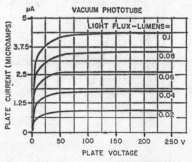

Fig. 25. *Three Typical Forms of Phototubes*

uum types are generally used for high precision laboratory, light-measuring, and critical relay control applications, while the gas types are suitable for sound reproduction and many relay applications.

Plate Characteristics. A typical plate voltage-plate current characteristic family at various light intensities for a vacuum phototube is shown in Fig. 26. As is apparent from the curves, the plate current **saturates** for plate voltages above about 20 volts to a constant value, determined by the total cathode emission for the particular value of light illumina-

Fig. 26. *Plate Characteristic Family of Vacuum Phototube at Various Light Intensities*

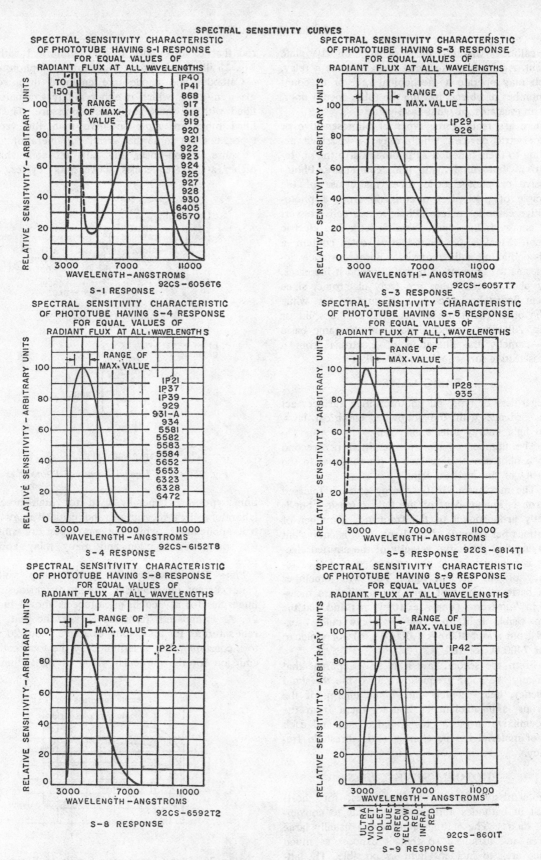

Fig. 27. *Spectral Sensitivity Curves of Commercial Types of Photosensitive Cathode Materials. (Courtesy RCA Electron Tube Division, Harrison, N.J.)*

tion. There is little point, therefore, in increasing the plate potential beyond that required to overcome the space charge near the cathode.

Spectral Sensitivity Characteristics. The cathode of the phototube determines its two most important characteristics, the **luminous sensitivity** and the **spectral response.** The luminous sensitivity, expressed in **microamperes per lumen,** is a measure of the amount of current emitted from the cathode for a given amount of light falling on it. The spectral response of a photocathode expresses the relative amount of photoelectric current produced by light of **different colors.**

Phototube cathodes utilize a light-sensitive surface coating, consisting of some form of an **alkali metal** or **earth.** Among these are cesium, lithium, potassium, rubidium, sodium, barium, calcium, and strontium. The tube is usually designated by the name of the material that serves as light-sensitive surface. Among the most important is the **cesium-silver oxide-silver tube,** which has a luminous sensitivity of about 30 microamps per lumen. This tube is often simply referred to as a "cesium tube" and is prominent in TV and sound reproduction.

The spectral response of a phototube is influenced not only by the type of photosensitive cathode material, but also by the technique used to prepare the surface and by the type of glass employed for the envelope. The spectral response of most phototubes generally resembles that of the human eye. As shown in Fig. 27, the spectral response of all phototubes reaches a peak at some specific wavelength. In some tubes, the peak response is made to coincide with that of the human eye, at about 5500 Angstroms. In other tubes the peak is set at some other value for special applications, and the spectral response may be extended into the ultraviolet and near-infrared regions, which are both invisible to the human eye.

PHOTOMULTIPLIERS (ELECTRON MULTIPLIERS)

The drawback of the conventional phototube is its tiny plate current for relatively large values of the illumination and, hence, the need for additional amplification of the current. **Photomultipliers** (sometimes called **electron multipliers**) overcome this disadvantage by utilizing the principle of secondary emission to multiply the electron current by a tremendous value. A photomultiplier tube consists of a light-sensitive photocathode, a system of nine or ten secondary-emission electrodes, called **dynodes,** and a collector anode. A cross-section view perpendicular to the elements of a partition-type photomultiplier tube is shown in Fig. 28.

When light falls on the **photocathode,** electrons

are liberated and are attracted through a hole, *H,* in a mica shield to the first positive electrode,

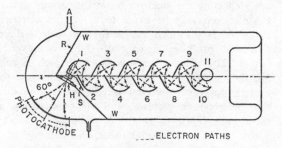

Fig. 28. *Cross Section of Partition-Type Photomultiplier*

dynode 1. An electrostatic shield is attached to this electrode, which has a voltage of about $+150$ volts. The photoelectrons striking the concave side of dynode 1 at high velocity give rise to secondary electrons (by secondary emission) which are attracted to the second dynode, electrode 2, which is again 150 volts more positive than dynode 1. Again, secondary electrons are added to the initial stream of primary electrons. The secondary electrons from dynode 2 give rise to still more secondary electrons upon striking dynode 3, which is again 150 volts positive with respect to dynode 2. Thus, each successive dynode is set at a fixed higher potential relative to its predecessor and each electron striking it gives rise to several secondary electrons. Since the action, clearly, is cumulative, the electron multiplication may be as much as 2.5 million after nine stages, for 150 volts on each dynode. In general, if each dynode gives off r secondary electrons per electron striking it, and the number of dynodes is n, the output current will be amplified r^{n-1} over the initial photoelectric current from the cathode.

CATHODE-RAY TUBES

We are all familiar with one type of **cathode-ray tube,** namely the television picture tube (kinescope), which converts the complex "video signal" into a visual representation of a distant scene. But cathode-ray tubes come in many types, sizes and shapes and about the only thing they have in common is that they all **convert an electrical signal (voltage) into a visual one.** Let's see how this is done.

BASIC ACTION

The electron tubes that we have mentioned till now make use of the **amount** and **intensity** of the electron stream flowing from cathode to plate to produce a useful plate current. Electrons may have been emitted in various ways, but the payoff was always the production of the largest possible plate current. The cathode-ray tube is in a class by itself, since it is less concerned with the intensity of the electron stream than with its *direction*. By giving the tube a certain geometrical configuration and deflecting the electron beam by various means, the electrons may be made to act as an electrical pencil of light, which moves in any desired direction and produces a spot of light, wherever it strikes. The principal applications of such a tube are the measurement of voltages and currents, the visual display of electrical waveforms, the production of television pictures, and the presentation of radar indications.

Just as in conventional tubes, the cathode of a cathode-ray tube makes available a plentiful supply of electrons. These electrons are then formed by various electrodes into a high-velocity beam of electrons (called a **cathode ray),** which is projected into the evacuated space of the tube until it strikes a screen. A fluorescent material is coated upon the screen to produce a spot of light, wherever electrons strike it. The electron beam may be deflected on its journey in any direction by suitable **electrostatic** or **magnetic fields.**

MAGIC-EYE TUBE

Let us first look at a very simple type of cathode-ray tube, called the **electron-ray** or **magic-eye** indicator. This is really a combination of two sets of elements, one being an ordinary triode amplifier tube, while the other is a cathode-ray indicator (see Fig. 29). The plate of the triode section is internally connected to the **ray-control electrode,** so that the voltage on this electrode varies in the same manner as the plate voltage with the applied signal voltage. The ray-control electrode is a metal cylinder (Fig. 29), so placed between the cathode and the fluorescent-coated anode (or **target)** that it deflects a portion of the electrons emitted from the cathode.

All electrons that strike the target cause it to **fluoresce** with a greenish glow. The part of the target that lies in the "shadow" of the ray-control electrode is not struck by electrons and, hence, remains dark. The size or angle of this wedge-shaped shadow

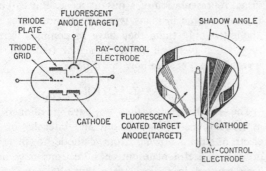

Fig. 29. *Schematic Symbol and Construction of Magic-Eye Tube*

depends on the electric field produced by the ray-control electrode, that is, the voltage placed on it. When the ray-control electrode is at the same potential as the fluorescent anode, the shadow disappears entirely. As the ray-control electrode is made negative (less positive) with respect to the anode, a shadow appears. The width of the shadow is proportional to the voltage on the ray-control electrode and, thus, can be used for rough voltage measurements. The principal uses of magic-eye tubes are as **tuning indicators** in radio receivers and as **balance indicators** in electrical bridge circuits (Wheatstone bridges).

OPERATION OF CATHODE-RAY TUBE

Now let us turn to those cathode-ray tubes (abbreviated CRT) that provide a visual representation of voltage and current waveforms. These tubes are usually meant when the term "cathode-ray tube" is used, rather than the magic-eye tube discussed above. Cathode-ray tubes consist of three basic components:

1. the **electron gun,** which produces, accelerates and focuses the emitted electrons into a narrow beam.

2. a **deflection system,** which deflects the electron beam either electrostatically or magnetically in accordance with the phenomenon (voltage or current waveform) to be displayed.

3. a **fluorescent screen,** upon which the beam of electrons impinges to produce a spot of visible light.

These essential parts of a cathode-ray tube are mounted inside a highly evacuated, funnel-shaped glass envelope, the large end of which has the fluorescent screen coated upon the inside surface. The construction of an electrostatic-deflection type cathode-ray tube is illustrated in Fig. 30. The electron gun consists of the group of electrodes to the left of the vertical deflecting plates. Commercial tubes range from 10 to 20 inches in length and have fluorescent screens varying from one inch diameter for radio servicing to giant 24-inch diameter television picture tubes. Beam-accelerating potentials vary from 800 volts for small tubes to about 12,000 volts for large ones.

The Electron Gun. The electron gun has the job of producing and focusing the electrons into a narrow beam so that it makes a tiny spot, when impinging on the fluorescent screen. A broad beam is of little use for accurate display of waveforms or TV pictures. An electron gun of the electrostatic type (Fig. 30) consists of an indirectly heated cathode, a control grid, an accelerating electrode or grid, a focusing anode, and a final accelerating anode. These electrodes are in the form of cylinders surrounding the cathode. Connections to the electrodes are brought to pins in the base.

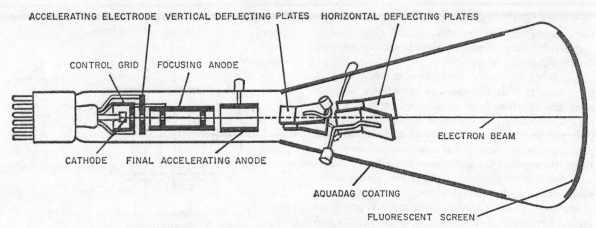

Fig. 30. *Structure of Electrostatic-Deflection Type Cathode-Ray Tube*

The cathode, control grid, and accelerating electrode perform essentially the same function as the cathode, control grid and **screen grid** in a tetrode or pentode. The cathode emits the electrons, the control grid determines the amount of electron flow by means of the negative bias placed on it, and the accelerating electrode—being highly positive with respect to the cathode—speeds up the electrons passing through it.

Electrons passing through the openings in the accelerating electrode are focused into a sharp electron beam by the combined effect of the focusing anode and the final accelerating anode, which are in effect a **multiple electronic lens system.** Both the accelerating and focusing anodes are at a positive potential with respect to the cathode, but the voltage on the focusing anode is always considerably lower (by several hundred to thousand volts) than that on the final accelerating anode.

To understand the focusing action that takes place between the focusing and (final) accelerating anodes, refer to Fig. 31. As shown in the figure a

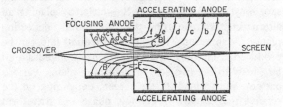

Fig. 31. *Electrostatic Focusing Action*

strong electrostatic field exists between the focusing and accelerating anode cylinders. The field lines (lines of force) are semicircular or radial in shape because of the geometry of the two anodes and they extend **from the focusing anode to the accelerating anode,** because the latter is at a high positive voltage with respect to the focusing anode. Electrons enter this "lens system" from a point near the con-

trol grid, called the **crossover,** where they have been brought to an initial focus by the combined effect of the control-grid aperture and the negative grid bias. From this first crossover point, the electrons diverge rapidly outward and would eventually strike the sides of the tube, as shown by the dotted rays in Fig. 31, if it were not for the action of the radial field between the two anodes.

You will remember that electrons tend to **follow the lines of force** of an electric field. Accordingly, the electrons entering the field between the focusing and accelerating anodes, tend to follow the field lines (shown by arrows) and would normally fall into the accelerating-anode cylinder, if they were **moving slowly enough.** However, because of the rapid **forward acceleration** imparted to the electrons by the attraction of the accelerating electrodes, electrons are not sufficiently long in the space between the anodes to be completely pulled off course. Rather, they move **tangentially** along the lines of force and then out of the field, **converging into a narrow spot on the screen.**

Thus, the radial field lines just have sufficient strength to guide the electrons (along their tangents) to a smoothly converging beam, but they are not strong enough to pull them away from their forward movement toward the sides of the anode cylinder. We might mention in passing that a similar focusing action can be attained by means of a **magnetic field produced** by focus coils around the neck of the tube.

The Fluorescent Screen. If the electron beam leaving the electron gun were not deflected, it would produce a luminous spot at the *center* of the fluorescent screen. The intensity of this spot can be controlled by adjusting the control-grid bias, but it is also dependent on the fluorescent material, called **phosphor,** coated on the inside of the tube. Among commonly used phosphors are **Willemite** (zinc silicate), which gives off predominantly **green** light;

zinc oxide, which gives off a blue color; zinc beryllium silicate or zinc sulphide, which glow yellow; and combination phosphors, which can be selected to give off nearly white light.

Another important consideration in the choice of phosphors is their afterglow or persistence of glow after the electrons have ceased to bombard a spot on the screen. If the afterglow is less than 0.1 second, the screen is said to have short persistence; if it is 1 second or more, it has a long persistence. (Between these limits the persistence is medium.) Willemite has a short persistence. Short persistence is desirable for rapidly changing images, such as those displayed in television. Long persistence is occasionally of advantage, such as in radar presentations and waveform comparisons.

The Aquadag Coating. You can readily see that some means must be provided for removing the electrons from the screen and returning them to the cathode. Otherwise the negative charge on the screen would build up to a point where it would repel arriving electrons, and no more could reach it. The method used for removing the electrons is to place a conducting coating of carbon particles, called Aquadag, along the side walls of the tube (not the screen), and to connect the coating to the cathode or the accelerating anode. (The accelerating anode is usually grounded and the other electrodes are made negative with respect to it. Connecting the Aquadag coating to the accelerating anode thus provides a ground return to the cathode.) Although the coating is not directly connected to the screen, the electrons are removed by means of secondary emission from the screen and no pile-up occurs. The secondary electrons are collected by the coating and then returned to the cathode.

Electrostatic Deflection. We have mentioned that the electron beam may be deflected either electrostatically or magnetically. Let us first consider electrostatic deflection. An electrostatic field can be produced between two metal plates (electrodes) simply by applying a potential difference between them. The lines of force of this field are straight from the negative to the positive plate, except near the edges, where they bulge outward, as shown in Fig. 32. Electrons entering the field between the deflecting plates will be bent in the direction of the lines of force, that is, they will be attracted toward the positive plate. The resultant path of the beam will be the net effect (vector sum) between its forward velocity and—in this case—its upward deflection, or transverse velocity. This path is parabolic between the deflecting plates (inside the field) and becomes a straight line after the beam leaves the field. As a result the electron spot is deflected upward (in this case) from its central position on the screen.

The spot may be deflected downward by reversing the polarity of the voltage on the plates.

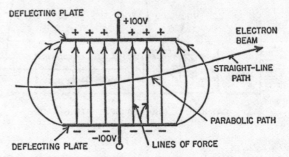

Fig. 32. *Deflection of Electron Beam by Electrostatic Field*

Note in Fig. 30 that two pairs of deflecting plates at right angles to each other are set into the path of the electron beam. The vertical deflecting plates move the beam vertically up and down, while the horizontal deflecting plates move the beam sideways, to the left or right of center. The terms "vertical" and "horizontal" thus refer to the direction in which the beam is deflected and *not* to the position of the plates within the tube. If a d-c voltage is applied to the vertical plates, the electron spot on the screen will be displaced up or down from center, depending on the polarity of the deflection voltage, and by an amount that depends on the relative magnitude of the voltage. The same is true for sideways (left or right) deflection by the horizontal plates.

Suppose now that an alternating (a-c) voltage is applied to the vertical plates. The spot will now move up and down on the screen at the frequency of the a-c voltage. Because of the persistence of the screen and that of the eye, the moving spot will actually appear as a continuous, luminous vertical line. (At very low frequencies the movement of the spot may actually be seen.) Similarly, by placing an alternating voltage on the horizontal deflecting plates, a continuous, luminous horizontal line can be produced. Finally, if a-c voltages are applied simultaneously to the two sets of deflecting plates, various patterns are formed on the screen, depending on the relative magnitudes, frequency, phase and waveform of the deflecting voltages. These patterns are called Lissajous figures.

Fig. 33 illustrates some typical, simple Lissajous figures. Part (a) of the figure illustrates the 45-degree inclined line formed, when a-c voltages of equal magnitude, phase and frequency are applied to the two sets of deflecting plates. In (b) the voltages are equal in magnitude and frequency, but 90 degrees out of phase with each other; as a result a circle appears. In (c), the magnitudes of the deflecting voltages are equal, but the frequency of the vertical deflecting voltage is **twice** that of the horizontal deflecting voltage. In (d) the vertical and horizontal deflecting voltages have been interchanged. Finally, in (e), the frequency of the vertical voltage is **three times** that of the horizontal deflecting voltage. In each case the Lissajous pattern that appears on the screen represents the **vector sum** of the horizontal and vertical deflecting voltages, and thus may be predicted by plotting it on graph paper.

Deflection Sensitivity. The deflection sensitivity of

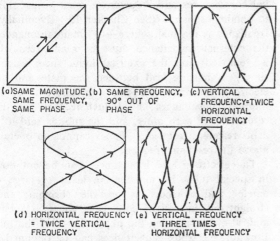

(a) SAME MAGNITUDE, (b) SAME FREQUENCY, SAME FREQUENCY, 90° OUT OF SAME PHASE PHASE (c) VERTICAL FREQUENCY=TWICE HORIZONTAL FREQUENCY

(d) HORIZONTAL FREQUENCY = TWICE VERTICAL FREQUENCY (e) VERTICAL FREQUENCY = THREE TIMES HORIZONTAL FREQUENCY

Fig. 33. *Typical Lissajous Figures on Screen of Cathode-Ray Tube*

an electrostatic-deflection cathode-ray tube is the **amount of spot deflection** on the screen (measured in inches) **per volt applied** to the deflecting plates. An average value is 50 volts per inch, which means that a positive voltage of 50 volts applied to the upper vertical deflection plate will deflect the electron spot upward by one inch on the screen. If an a-c voltage with a peak amplitude of 50 volts is applied to the vertical plates the spot will trace out a vertical line **two inches high,** one inch up from center and one inch down from center.

Note that the deflection sensitivity is **inversely proportional** to the forward velocity of the electrons to which they are accelerated by the **final anode voltage.** It is also **directly** proportional to the **length of the deflecting plates** and the **length of the beam** from the center of the deflecting plates to the screen and **inversely** proportional to the **distance between the deflecting plates** (at right angles to the beam).

Magnetic Deflection. Magnetic deflection of the electron beam is employed wherever it is inconvenient or impossible to obtain suitable electrostatic deflection voltages. Radar indicators and television picture tubes use it. A big advantage of magnetic deflection is that it obviates the need for deflecting

plates inside the tube and thus simplifies the tube construction. In magnetic deflection, the **deflecting force** is obtained by a magnetic field set up by coils placed around the neck of the tube (Fig. 35).

To understand how the electron beam can be deflected by a magnetic field, you must consider that the stream of **electrons** traveling in one direction constitutes an **electric (d-c) current.** It is immaterial whether the electrons flow in a wire **conductor** or in free space. One of the basic laws of electricity states that a **magnetic field surrounds every electric current.** When a current-carrying wire is placed in a uniform magnetic field, its own field interacts with the external magnetic field and the **wire experiences a force** that depends on the relative direction of the fields and the strength of the current. The same is true for a beam of electrons, which is an electric current, although not carried in a wire.

A simple way of determining the direction in which an electron current (beam or wire conductor) will be deflected is by the application of the so-called **right-hand rule** (Fig. 34): If the thumb, index finger and middle finger are held at right angles to each other, and the *index finger* is pointed in the direction of the **magnetic field** (from north pole to south pole), while the *middle finger* points in the direction the **electron beam** is traveling, then the *thumb* will indicate the direction of the **force** on the electron beam.

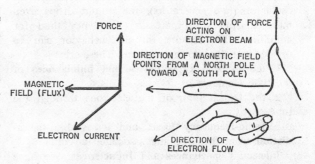

FORCE

DIRECTION OF FORCE ACTING ON ELECTRON BEAM

DIRECTION OF MAGNETIC FIELD (POINTS FROM A NORTH POLE TOWARD A SOUTH POLE)

MAGNETIC FIELD (FLUX)

ELECTRON CURRENT

DIRECTION OF ELECTRON FLOW

Fig. 34. *Determining Direction of Force on Electron Beam Moving through a Magnetic Field*

Fig. 35 illustrates the action of the vertical deflecting coils in a magnetic-deflection type cathode-ray tube. For horizontal beam deflection another pair of deflecting coils must be mounted alongside the neck of the tube at **right angles** to the vertical coils. The amount of deflection produced is roughly proportional to the product of the number of turns in each coil and the number of amperes flowing through the coil (i.e., the **ampere-turns).** By applying the right-hand rule, you can see that the beam illustrated in Fig. 35 will be deflected **upward** by the action of the magnetic field. In actual magnetic-deflection cathode-ray tubes, the deflecting coils are usually wrapped almost **flat** around the neck of the tube, but still produce parallel field lines at right angles to the beam, similar to those shown in Fig. 35.

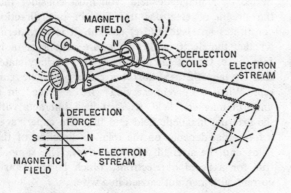

Fig. 35. *Action of Vertical Deflecting Coils on Electron Beam in Magnetic-Deflection Cathode-Ray Tube*

SPECIAL UHF TUBES

As the operating frequency is raised, conventional electron tubes become progressively less effective as amplifiers and oscillators. Above frequencies of about 100 megacycles (one-hundred million cycles per second) the output of conventional triode and pentode amplifiers begins to drop off rapidly until finally a frequency is reached where there is no amplification at all. When the tubes are operated as oscillators (a-c generators), the output drops even more sharply with increasing frequency than for amplifiers. The reasons for this behavior can be divided roughly into three groups. These are:

1. the **internal capacitances and inductances** of the tube elements

2. the **transit time** of the electrons through the tube

3. **radio-frequency losses and losses due to radiation** from the tube and associated circuit.

Internal Capacitances and Inductances. We have already become familiar with the interelectrode capacitances of conventional tubes. You will remember these capacitances exist between the grid, plate, and cathode electrodes of the tube and are labeled C_{gp}, C_{gk}, and C_{pk} respectively. As the frequency goes up, the **reactance** (opposition to the flow of a.c.) of these capacitances goes down and begins to short out part of the input and output signal voltages, as well as feed back increasingly more energy from the plate (output) to the grid (input). All this cuts down the amplification.

Furthermore, the **electrode leads** between the tube elements and the connecting base pins have a definite, though small, **inductance.** The reactance of this inductance **goes up** with increased frequency.

This means that the voltage applied to the base pins will not actually appear at the tube's electrodes, but will be reduced by the amount wasted in the series (inductive) reactance of the leads. Again, this cuts down the available gain from the tube.

Finally, you must consider that at radio-frequencies the tube elements are connected to external **tuned** or **resonant circuits,** consisting of inductances and capacitances. As the frequency is raised, the effect of the tube's **internal** capacitances and inductances begins to become progressively greater in relation to the external tuned circuit, and the external capacitances and inductances must be cut down to obtain resonance. (See Chapter 13.) Eventually a frequency is reached, where—to obtain resonance— the external capacitance must be made zero (that is, removed) and the external inductance becomes simply a shorting lead between the plate and grid electrode terminals. At this frequency, then, the **internal capacitances and inductances of the tube resonate with each other,** and the tube is said to be at its **resonant frequency.** Tubes cannot be operated above the resonant frequency.

The electrode lead inductances can be cut down in special tubes by making the leads of large diameter and as short as possible (bringing them straight out, instead of connecting them to a base) and, also, by reducing the physical size of the tube. Low interelectrode capacitances can be obtained by reducing the size of the electrodes and spreading them further apart. The latter, however, has the undesirable effect of increasing the electron transit time, which we must now consider.

Electron Transit Time. The time taken by an electron to travel from the cathode to the plate of an electron tube is known as the **transit time.** This time represents a negligible part of one cycle (two a-c alternations) at operating frequencies lower than 100 megacycles and, hence, the output plate current may be assumed to respond *instantaneously* to changes in the control-grid voltage. But as the operating frequency is raised into the UHF region (300–3000 megacycles), the transit time of the electrons from cathode to plate becomes an appreciable part of the a-c cycle. At a 1000 megacycles, for example, one cycle lasts only .001 microsecond, which is considerably less than the transit time in conventional electron tubes. Thus, a change in the control-grid (or other) voltage will not affect the plate current immediately, but there will be a definite time lag before the plate current can respond to the change in control-grid voltage. This time lag may be thought to be effectively the same as the lagging of the current behind the applied voltage in an **inductance.** (You will remember that this is expressed by a **lagging phase angle.**) Thus, the transconductance of the tube (i.e., the ratio of a change in plate current to a change in grid voltage) has a **lagging phase angle** at very high frequencies.

A more serious consequence of the finite transit time is that the grid of the tube **absorbs power** at high frequencies, **even when the grid bias is made negative** and no grid current flows. The mechanism by which this happens is somewhat too complex to be discussed here, but the fact is that power is actually absorbed by the grid circuit and this power consumption increases with the **square of the frequency.**

The transit time can be reduced appreciably by decreasing the physical size of the tube and, particularly, the spacing between cathode and control grid. Increasing the positive plate voltage increases the electron velocity and thus also assists in cutting down the electron transit time.

Radio-Frequency and Radiation Losses. Finally, there are certain power losses associated with the tube and its circuit which all tend to increase with frequency. At ultra-high frequencies all currents flow in thin surface layers on the conductor, a phenomenon that is called **skin effect.** Skin effect causes *increased* resistance. The skin effect and, hence, the conductor resistance and associated power loss, increase with frequency.

Electric fields produce molecular movements in glass and other insulating supports used in tubes, leading to heat and power losses. These **dielectric hysteresis** losses, as they are called, go up directly with the frequency.

As we shall learn later, even an exposed piece of wire (such as an electrode lead) will **radiate** radio-frequency power, if its dimensions are comparable to the wavelengths of the radio-frequency current through it. (Wavelength $= \dfrac{3 \times 10^{10}}{\text{frequency}}$ cm)

As a consequence, the tube and its associated circuit will have **appreciable radiation losses** at ultra-high frequencies. All the factors discussed above generally decrease the tube's efficiency as the frequency goes up.

The skin effect can be reduced and the consequent resistance losses may be made lower by increasing the surface area of the current-carrying conductors. Dielectric losses can be reduced by properly positioning the glass insulators with respect to the electric field. Radiation losses can be reduced by shielding the tube and its circuit in an enclosure or by using "concentric lines" in the tube construction to confine the electric fields of the tube and its associated circuit.

TYPICAL UHF TUBES

The factors we have discussed above, which make conventional tubes unsuitable for operation at ultra-high frequencies, have led to the design of special UHF tubes that minimize internal capacitances and inductances, cut down transit time, and reduce radiation and radio-frequency losses. This is done by reducing the physical size of the tubes, making the electrodes quite small and spacing them very close together, and by making special lead arrangements. Because of these constructional features, the power handling ability of these UHF tubes is smaller than that of conventional tubes used at lower frequencies. The tubes are, therefore, usually operated at high cathode emission currents and relatively heavy plate currents and plate voltages (large plate dissipation). Ultra-high-frequency tubes frequently have no tube base; connections to the electrodes are made through pins protruding through the tube envelope at the shortest possible distance from each electrode.

Some typical examples of commercial high-frequency tubes that incorporate these features, are illustrated in Fig. 36. The **acorn** tube, **doorknob** tube, **pencil** tube and **lighthouse** tube are so named because of their characteristic shapes and sizes. Acorn and doorknob tubes are available as diodes, triodes, and pentodes, while the pencil and lighthouse types come as triodes.

Acorn tubes are characterized by very small size (about 1½ inches in height and diameter), close spacing of the electrodes, and leads that come out through a ring seal instead of a base, thus minimizing their lengths and the capacity between them. The **resonant frequency** of acorn tubes is about 1500 megacycles and they operate at frequencies anywhere from 400 to 1500 megacycles.

The **doorknob tube** is an enlarged version of the acorn type and has an upper frequency limit of about 700 megacycles, when used as an oscillator. Note that the leads are brought out directly through the glass envelope and are widely spaced to reduce the capacitance between them. The tube elements are small and closely spaced. The tube shown in Fig. 36 is about two-thirds of actual size and has a plate dissipation of 30 watts.

The **pencil triode** is used as amplifier and oscillator at frequencies up to 1700 megacycles. It is characterized by a closely spaced cylindrical structure that is designed for operation with **coaxial and cavity resonators,** rather than ordinary tuned circuit elements. (See Chapter 11, Oscillators.)

The **lighthouse tube** represents a different approach to the problem of losses at ultra-high frequencies. The active parts of the cathode, grid and plate in the lighthouse tube are parallel planes and the leads are metal disks, as illustrated. Such a structure can be fitted into a system of concentric lines, used to form the tuned (resonant) circuit at ultra-high frequencies. By connecting the tube directly to its tuned circuit in this manner, the losses

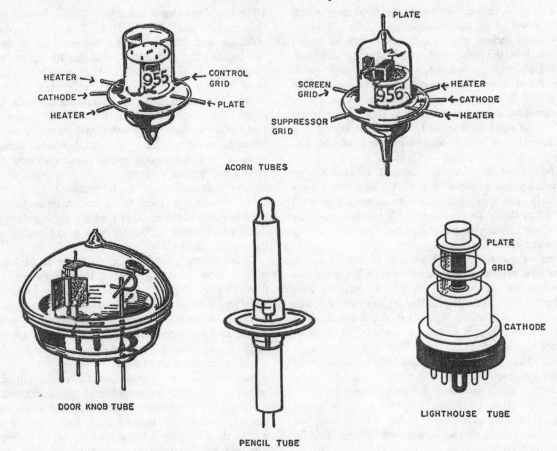

Fig. 36. *Typical Commercial Ultra-High-Frequency Tubes*

due to lead inductance of the connecting wires are practically eliminated. In addition, the parallel-plane construction permits extremely close spacing of the tubes electrodes. (The spacing between grid and cathode is about 1 mil.)

CERAMIC TUBES

Based upon the planar structure of the lighthouse tubes, the General Electric Company introduced in the late 1950s a line of tiny ceramic UHF tubes, measuring approximately ¼ × ½ inch over-all. Constructed of alternating layers of titanium electrodes and forsterite ceramic insulators, the ultra-small ceramic tubes have proven themselves capable of high-quality performance up to several thousand mega-cycles under severe environmental conditions, such as are encountered in space applications. The rugged performers have exceptional tolerance to shock and vibration, and function equally well in the icy regions of outer space and in high-temperature and high nuclear-radiation environments. Except for being much smaller in size, the appearance of ceramic tubes is similar to the lighthouse tube illustrated above. The Radio Corporation of America recently has introduced a line of Nuvistor UHF

tubes, which are similar in performance and characteristics to the G.E. ceramic tubes.

MICROWAVE TUBES

Frequencies higher than about 2000 megacycles are usually referred to as **microwave frequencies.** The UHF tubes just discussed will not operate well at these frequencies. Special microwave tubes operating on different principles, such as **magnetrons** and **Klystrons,** have been developed for this frequency region. These will be discussed in a later chapter (Chapter 11).

SUMMARY

In **gas-filled** tubes gases such as nitrogen, helium, neon, argon, or mercury vapor are purposely introduced into the tube envelope. This gives them the ability to handle heavy currents and makes them useful for rectifier, relay and switching applications.

The voltage at which ionization occurs in a gas tube is known as the **ionization potential, striking potential,** or **firing point.** Plate conduction commences with the firing point, and the tube voltage drop falls to a low, constant value.

Below the **de-ionizing** or **extinction potential** ionization cannot be maintained and conduction stops.

Grid control in gas-filled tubes is limited to *starting* conduction, but cannot be used to stop it. To stop conduction the plate voltage must be removed.

Cold-cathode, gas-filled tubes utilize **field emission** from an unheated cathode; **hot-cathode** or **thermionic** types have conventional heater-type cathodes; and **ignitrons** obtain electron emission from a pool of liquid mercury.

Cold-cathode gas diodes are used to indicate polarity of a.c., the presence of **r-f fields,** and serve as **voltage regulators.**

Mercury-vapor tubes are hot-cathode gas diodes that are chiefly used as **heavy-duty rectifiers.**

Gas-filled triodes and tetrodes are known as **thyratrons** and are used for relay applications and for sawtooth sweep generators.

Critical grid voltage in thyratrons is the value of the negative bias to **start conduction** at a certain plate voltage.

Ignitrons utilize pointed **ignitor** electrodes, dipped into a pool of mercury, to trigger off the mercury-vapor discharge and thus start rectification.

Photosensitive devices can be classified as **photoemissive** (or **photoelectric), photovoltaic,** and **photoconductive.**

Photoemissive or **photoelectric tubes** utilize the principle of photoelectric emission from a light-sensitive cathode surface.

Photovoltaic cells convert incident light energy *directly* into electric current.

Photoconductive cells contain a **semiconducting,** solid material whose resistance decreases in proportion to the light falling on it.

The **number of electrons emitted** from a photosensitive surface (regardless of type) is **proportional to the intensity** of the incident light.

The **maximum kinetic energy** of the released electrons is independent of the light intensity, but is **proportional to the frequency** of the incident light. The **initial electron velocity** is proportional to the **square root** of the incident light frequency.

Electrons are *not* emitted below the characteristic **threshold frequency** of the photosensitive material.

Gas-filled photoemissive tubes are **more sensitive** than high-vacuum types, but are more **erratic** in operation and their characteristic is **non-linear.** Their frequency response falls off at higher audio frequencies.

The **photocathode** determines **luminous sensitivity** (microamps per lumen) and **spectral** (color) **response.**

Photomultipliers (or **electron multipliers**) utilize the principle of secondary emission to multiply the electron current.

Cathode-ray tubes convert an electrical signal (current or voltage) into a visual one by shooting a beam of electrons at a **fluorescent screen** and **deflecting the beam** in accordance with the variations of the electrical signal.

Cathode-ray tubes consist of an **electron gun,** a **deflection system,** and a **fluorescent screen,** all housed in a glass envelope.

The **electron gun**—usually made up of a heater-type cathode, a control grid, an accelerating electrode or grid, a focusing anode, and a final accelerating anode—**produces, accelerates and focuses the emitted electrons into a narrow beam.** It acts as an **electron emitter** and **electronic lens system.**

The **fluorescent screen** is coated with a phosphor (Willemite, zinc oxide, zinc sulphide) that produces visible light when electrons impinge on it.

The **deflection system** deflects the electron beam either **electrostatically,** by means of deflecting plates, or **magnetically,** by coils placed around the neck of the tube, in accordance with the voltage or current waveform to be displayed.

Electrostatic deflection sensitivity, measured in volts per inch deflection, is **inversely proportional to the velocity** of the electrons (i.e., the final anode voltage) and to the **spacing** between the deflecting plates, and is **directly proportional** to the length of the deflecting **plates** and the length of the beam from the center of the deflecting plates to the screen.

Magnetic deflection is proportional to the **ampere-turns** of the deflecting coils.

Conventional tubes become inoperative at ultra-high frequencies because of the tube's **internal capacitances** and **inductances,** the **electron transit time,** and **radio-frequency and radiation losses.**

UHF tubes, such as **acorn, doorknob, pencil** and **lighthouse tubes,** overcome these effects by **reducing the physical size** of the tube and electrodes, **spacing the electrodes close together,** and by means of **special lead arrangements.**

Chapter Seven

TRANSISTORS AND SEMICONDUCTORS

In recent years the **transistor**—an entirely new type of electron device—has come into its own and bids to replace the bulky electron tubes in most, if not all applications. Transistors are far smaller than tubes, have no filament and hence need no heating power, and may be operated in any position. They are mechanically rugged, have practically unlimited life, and can do some jobs better than electron tubes, while catching up fast in other respects. In contrast to electron tubes, which utilize the flow of free electrons through a vacuum or gas, the transistor relies for its operation on the movement of charge carriers through a solid substance, a **semiconductor.** Transistors are only one of the family of semiconductors; many other semiconductor applications are becoming increasingly popular and new ones are constantly being discovered.

You will remember from Chapter 1 that materials are classed as semiconductors if their electrical conductivity is intermediate between metallic conductors, which have a large number of free electrons available as charge carriers, and non-metallic insulators, which have practically no free electrons available to conduct current. There are many varieties of semiconductors, but the two most frequently used in electronics and transistor manufacture, are **germanium** and **silicon.** Both elements have the same crystal structure and similar characteristics, so that the discussion that follows for germanium will also apply to silicon.

GERMANIUM CRYSTAL STRUCTURE

In Chapter 1 we described the structure of atoms and noted that only the **outermost** electron shell of an atom is of interest in electronics, since it contains the loosely held **valence** electrons, which are easily dislodged to become electric current carriers. Germanium has four valence electrons in its outer shell, and for our purposes, the atom may be pictured as containing only these electrons and four protons in the nucleus to keep it electrically neutral. When germanium is in crystalline form its atoms assume the typical diamond structure, illustrated in Fig. 37. In this structure adjacent germanium atoms share their valence electrons in a strong bond, so that effectively four orbital electron **pairs** are associated with each nucleus. These electron pairs are termed **covalent bonds** and they are bound so strongly to each other and to the nucleus that no

free electrons are available to conduct a current through the germanium. A pure germanium crystal, therefore, is practically a non-conductor of electricity. It is not completely non-conducting, since ordinary heat energy occasionally disrupts some of the covalent bonds, thus liberating free electrons as charge carriers.

If a small amount of an impurity is introduced into the germanium crystal, its current-conducting characteristics change radically. Thus, when atoms

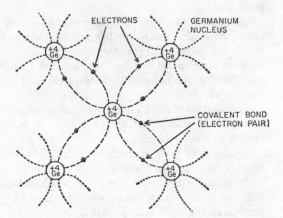

Fig. 37. *Germanium Crystal Structure Showing Covalent Bonds*

that have **five** electrons in their outer shell, such as antimony or arsenic, are introduced into the germanium atom (a procedure known as **doping),** the fifth electron of the impurity atom does not find a place in the symmetrical covalent-bond structure and, hence, is free to roam around through the crystal. These free electrons are then available as electric current carriers. By placing an electric field across the "doped" germanium crystal, as show in Fig. 38, the excess free electrons donated by the impurity atoms will travel toward the positive terminal of the voltage source. Relatively few impurity or "donor" atoms within the germanium structure permit fairly substantial electron currents through the crystal when an electric field is applied. Germanium that has been doped by **pentavalent donor atoms** (i.e., five electrons in the outer shell) is known as **N-type** germanium, because current conduction is carried on with **negative** charge carriers, or electrons.

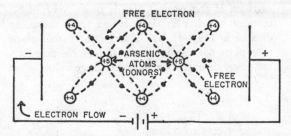

Fig. 38. *Electron Conduction through N-Type Germanium*

Consider now the situation when an impurity that has only **three** electrons in its outer shell, such as gallium or indium, is introduced into the pure germanium crystal. As shown in Fig. 39, the **trivalent** indium atoms take their place in the germanium structure, but one of the covalent bonds around each indium atom has an electron missing, or a **hole** in its place. Although the hole indicates the **absence** of an electron, it behaves like a real, **positively charged** particle when an electric field is applied across the crystal. Under the influence of the electric field, electrons within the crystal will tend to move toward the positive terminal of the voltage source and jump into the available holes of the indium atoms near the positive terminal. Since there are no free electrons available, the deficient indium atoms near the positive terminal "steal" electrons from their neighbors to the left (in Fig. 39) by disrupting their covalent bonds. This creates new holes in adjacent atoms to the left of those that have been filled. As electrons move to the right toward the positive terminal, the holes will move to the left toward the negative terminal, thus acting like mobile, positive particles. As the holes reach the negative terminal, electrons enter the crystal near the terminal and combine with the holes, thus canceling them. At the same time, the loosely held electrons that filled the holes near the positive terminal, are attracted away from their atoms into the positive terminal. This, of course, creates new holes near the positive terminal, which again drift toward the negative terminal. Current conduction may thus be considered to occur by means of holes

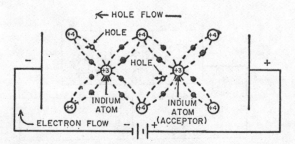

Fig. 39. *Hole Conduction through P-Type Germanium*

inside the crystal, and by means of electrons through the external connecting wires and battery.

An impurity that has three electrons in its outer shell (trivalent) is known as an **acceptor atom,** because it takes electrons away from surrounding germanium atoms. Germanium that has been doped with trivalent acceptor atoms is called **P-type** germanium, to specify that current conduction is carried on by holes, which are the equivalent of **positive** charges.

P-N JUNCTION DIODES

Conduction of electric current through P- or N-type germanium takes place equally well in either direction; hence, reversing the polarity of the battery in Figs. 38 and 39 will not affect the amount of current flow, although it reverses its direction.

Consider now what happens when P-type germanium is joined to N-type germanium and a voltage is applied across the junction, as illustrated in Fig. 40. In practice, such an abrupt P-N junction may be obtained in two ways. In the **grown junction** a single crystal is obtained from a melt which at first contains impurities of either the P- or N-type. In the middle of the growth process, impurities of the opposite kind are added to the melt, so that the remainder of the crystal abruptly grows into the opposite type.

In contrast, a **fused P-N junction** is obtained by pressing small "dots" of indium (P-type) on a wafer of N-type germanium. After a few minutes of heat treatment, the indium fuses to the surface of the germanium and produces P-type germanium for a thin layer below the surface. A P-N junction is thus formed between this P-region and the remainder of the N-type germanium wafer. Both the grown and the fused P-N junctions are extensively used in junction diodes and transistors, each having specific characteristics and limitations.

With the P-type germanium biased positively, as illustrated in Fig. 40, the (positive) holes are repelled by the battery voltage toward the junction between the P- and N-type material. Simultaneously, the electrons in the N-type germanium are repelled by the negative battery voltage toward the P-N junction. Although there is normally a potential **barrier** at the P-N junction that prevents electrons and holes from moving across and combining, under the influence of the electric field of the battery the holes move to the right across junction and the electrons move to the left. In the region of the P-N junction, therefore, electrons and holes meet and combine, thus ceasing to exist as mobile charge carriers. For each electron-hole combination that takes place near the junction, a covalent bond near the positive battery terminal breaks down, an electron is liberated and enters the positive terminal. This action creates

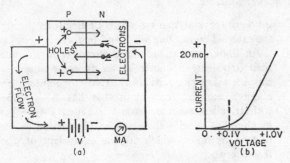

Fig. 40. *Current Flow Across P-N Junction with Forward Bias*

a new hole which moves to the right toward the P-N junction.

At the opposite end, in the N-region near the negative terminal, more electrons arrive from the negative battery terminal and enter the N-region to replace the electrons lost by combination with holes near the junction. These electrons move toward the junction at the left, where they again combine with new holes arriving there. As a consequence, a relatively large current flows through the junction. The current through the **external** connecting wires and battery is carried by **electrons,** in the direction shown in Fig. 40a.

The battery connection that permits current to flow across the P-N junction is known as **forward bias.** A minimum voltage of about 0.1 volt is needed to overcome the potential barrier at the junction and permit any current to flow. (See Fig. 40b.) The current then increases rapidly with increasing battery voltage and as little as one to two volts permit currents of 20 to 100 milliamps.

If the battery voltage is reversed in polarity, as illustrated in Fig. 41a, an entirely different situation prevails. The holes are now attracted to the negative battery terminal and move away from the P-N junction, while the electrons also move away from the junction because of the attraction of the positive terminal. Since there are effectively no hole and electron carriers in the vicinity of the junction, current flow stops almost completely. A small **reverse current** of a few microamperes still flows across the junction, however, as illustrated by the V-I characteristic (Fig. 41b). This reverse current is due to thermally generated electron-hole pairs within both the P- and N-type materials. As mentioned before, some covalent bonds always break down because of the normal heat energy of the crystal molecules. Electrons liberated by this process in the P-material move right across the junction under the influence of the electric field, while holes generated in the N-material move to the left into the P-material. Thus a small electron-hole combination current is maintained by these so-called **minority**

carriers. If the reverse bias is made very high, the covalent bonds near the junction **break down,** as indicated in Fig. 41b, and a large number of electron-hole pairs will be liberated; the reverse current then increases abruptly to a relatively large value.

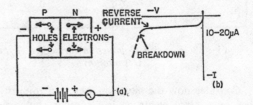

Fig. 41. *P-N Junction with Reverse Bias*

The unilateral current conduction characteristic of a P-N junction is seen to be similar to that of the conventional diode tube discussed in Chapter 3. It was pointed out there that this characteristic permits diode tubes to change alternating current into unidirectional current. Germanium and other semiconductor diodes (silicon, selenium, etc.) are, therefore, extensively used as rectifiers and detectors.

JUNCTION TRIODE TRANSISTORS

Just as the triode tube followed on the heels of the vacuum diode, you might expect that the logical extension of the semiconductor diode junction would be a triode junction, consisting of two P-N junctions. This is indeed the case, and the modern P-N-P or N-P-N **junction triode transistors** are in many respects analogous to triode electron tubes. A junction transistor can function as an amplifier or oscillator, as can a triode tube, but has the additional advantages of long life, small size, ruggedness and absence of cathode heating power.

Fig. 42 illustrates a P-N-P junction, made up of a sandwich of two P-N germanium junction diodes, placed back to back. Although exaggerated in the illustration, the center or N-type portion of the sandwich is extremely thin in comparison to the P-regions. The double junction may be either of the **grown** or **fused** crystal types, obtained in the manner discussed for junction diodes.

With the battery polarities as shown in Fig. 42,

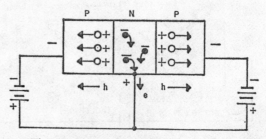

Fig. 42. *Non-Conducting P-N-P Junction*

the P-regions are negative with respect to the central N-region or conversely, the N-region (called the **base**) is positive with respect to the P-regions. The mobile electrons in the N-region, therefore, initially move away from both junctions in the direction of the positive connecting terminal. The holes in each of the P-regions also move away from the junctions and are attracted toward the negative terminals. After these initial displacements of holes and electrons the current flow stops.

Consider now the same P-N-P sandwich, but with the batteries connected as in Fig. 43. Note that the P-region at the left of Fig. 43 is biased positively, in the **forward** direction (see Fig. 40), while the P-region at the right is biased negatively, in the **reverse** direction (see Fig. 41). This is one of the basic

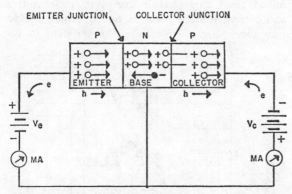

Fig. 43. *Basic Connection of P-N-P Junction Transistor*

operating connections of a P-N-P junction transistor.

The holes in the left P-region, known as **emitter**, are repelled by the positive battery terminal toward the left P-N or **emitter junction.** (The junction that is **forward** biased in a transistor is always termed **emitter junction.**) Under the influence of the electric field the holes overcome the barrier and cross the emitter junction into the N-type or base region. This region is very thin and only lightly "doped" with impurity atoms, so that the majority of the holes are able to drift across the base without meeting electrons to combine with. A small number of holes (about five percent), however, are lost in this area because of recombination with electrons. The remainder penetrate through the almost porous base region and flow across the right junction into the P-region or **collector.** (The junction with a reverse bias in a transistor is termed **collector junction.**) The negative collector voltage (V_c) aids in rapidly sweeping up the holes that pass into the collector region.

As each hole reaches the collector electrode, an electron is emitted from the negative battery terminal (V_c) and neutralizes the hole. For each hole that is lost by combination with an electron in the collector and base areas, a covalent bond near the

emitter electrode breaks down and a liberated electron leaves the emitter electrode and enters the positive battery terminal (V_e). The new hole that is formed then moves immediately toward the emitter junction, and the process is repeated. It is evident, therefore, that a continuous supply of holes are injected into the emitter junction, which flow across the base region and collector junction, where they are gathered up by the negative collector electrode. Current conduction **within** the P-N-P transistor thus takes place by hole conduction from emitter to collector, while conduction in the external circuit is carried on by electrons. Furthermore, the collector current must be *less* than the emitter current by an amount proportional to the number of electron-hole combinations occurring in the base area.

The ratio of collector current to emitter current is known as alpha (symbolized a) and it is a measure of the possible current amplification in a transistor. From the definition, alpha cannot be higher than 1, but practical values of 0.95 to 0.99 are attained in commercial transistors.

Because of the reverse bias no current can flow in the collector circuit, unless current is introduced into the emitter. Since a small emitter voltage of about 0.1 to 0.5 volt permits the flow of an appreciable emitter current, the input power to the emitter circuit is quite small. As we have seen, the collector current due to the diffusion of holes is almost as large as the emitter current. Moreover, the collector voltage (V_c) can be as high as 45 volts, thus permitting relatively large output powers. A large amount of power in the collector circuit may thus be controlled by a small amount of power in the emitter circuit. The **power gain** in a transistor (power out/power in) thus may be quite high, reaching values in the order of 1000.

A diagrammatic sketch, illustrating the structure

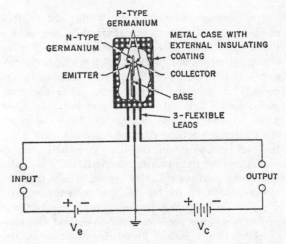

Fig. 44. *Sketch of P-N-P Junction Transistor and Associated Circuit*

of a P-N-P junction transistor and the associated input and output circuits, is shown in Fig. 44. The transistor shown is of the **fused** junction type, such as a RCA-2N109, or 2N175. To operate as an amplifier an a-c signal must be introduced into the input circuit and a load resistance must be connected across the output.

N-P-N JUNCTION TRANSISTOR

An N-P-N type junction transistor is sketched in Fig. 45. Note that the positions of the N- and P-type germanium have been interchanged and the battery polarities have been reversed with respect to the P-N-P transistor. The emitter junction is still **forward biased,** however, in accordance with our previous definition, since electrons are repelled from the negative emitter battery terminal (V_e) toward the junction. Likewise, the collector junction has **reverse bias** because electrons are flowing away from the collector junction toward the positive collector battery terminal (V_c). The main difference between the P-N-P and N-P-N transistor, therefore, is that current conduction in the latter is carried by **electrons,** while the charge carriers in the former are **holes.**

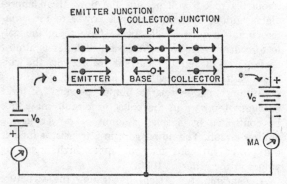

Fig. 45. *Basic Connection of N-P-N Junction Transistor*

The process of conduction within the N-P-N transistor is similar to that described before for the P-N-P type. With the emitter voltage, V_e, applied as shown, electrons are repelled from the negative terminal and injected into the emitter junction, overcoming the potential barrier there. Since the P-region (base) is only lightly doped and very thin, most of the electrons diffuse through the base and reach the collector junction. A few electrons (about 5%), however, combine with the holes present in the P-region and are lost as charge carriers. The remainder cross over into the collector region, where they are rapidly swept up by the positive collector voltage, V_c. For each electron flowing out of the collector and entering the positive battery terminal, an electron enters the emitter from the negative emitter battery terminal. Electron conduction thus takes

place continuously in the direction shown in Fig. 45.

The following conclusions about junction transistor operation may be drawn from the above analysis:

1. The major charge carriers in the P-N-P junction transistor are **holes.**

2. The major charge carriers in the N-P-N junction transistor are **electrons.**

3. The collector current in either type of junction transistor is always *less* than the emitter current because of the recombination of holes and electrons occurring in the base area. The ratio of the collector to emitter current is known as the **current gain** or **alpha.**

TRANSISTOR CHARACTERISTIC CURVES

The performance of transistors may be determined from characteristic curves of their voltage and current relations, just as for conventional electron tubes. Fig. 46 illustrates the variation of the

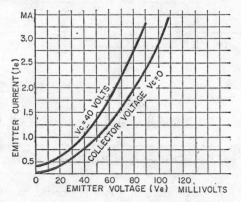

Fig. 46. *Emitter Current vs Emitter-to-Base Voltage for an N-P-N Junction Transistor*

emitter current as a function of the emitter-to-base voltage (V_e) for a typical N-P-N junction triode transistor.

Note that the shape of the emitter current-emitter voltage (I_e-V_e) curves is somewhat similar to the characteristic plate current-plate voltage curve of a vacuum diode. (See Fig. 6.) The emitter current increases rapidly with small increments in the emitter voltage and reaches a value of several milliamps with an emitter voltage no higher than about 0.1 volt (100 millivolts). Moreover, the emitter current is almost **independent** of the collector-to-base voltage (V_c), although there is a small interaction, as shown by the two curves for $V_c = 0$ and $V_c = 40$ volts. The main point to note is that a very small emitter voltage suffices to produce a large flow of emitter current. This also means that a small signal voltage variation at the input of the transistor produces a large emitter current variation, or equivalently, the **input resistance** to a small signal voltage impressed on the emitter is very **low.**

A family of collector current-collector voltage (I_c-V_c) characteristics for the same N-P-N junction transistor is shown in Fig. 47. The shape of the curves resemble very much the constant current I_b-E_b curves of a pentode tube, shown in Fig. 19 (Chapter 5). The main difference between the tube curves and transistor characteristics is that the **current is used as independent variable** for plotting transistor characteristics.

Each curve in Fig. 47 represents the variation in collector current with changing collector-to-base voltage (V_c) for a **constant** value of the emitter current (I_e). Note that almost the entire variation in the collector current takes place at very low values of the collector voltage. When the collector voltage is raised above a value of about one to two volts, it collects all the charge carriers that diffuse via the base to the collector junction, and the collector current becomes practically **independent** of the collector voltage. The collector current above this minimum potential is nearly, but not quite,

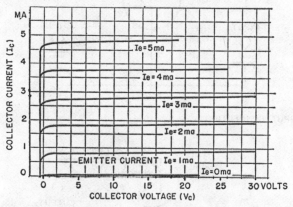

Fig. 47. *Collector Current vs Collector-to-Base Voltage for an N-P-N Junction Transistor*

equal to the emitter current for each curve. You will remember that the collector current cannot equal the emitter current because of the small percentage of charge carriers lost in the base due to electron-hole combinations. As seen from the curves the value of the current ratio, alpha, may be increased slightly by raising the collector voltage.

The I_c-V_c curves (Fig. 47) also show that a very large change in the collector voltage produces only a tiny change in the collector current, which means that the **output resistance** of the transistor (i.e., the ratio of the voltage-to-current change) is also **very high.** This sheds some light on the process of voltage and power amplification in a transistor, where the current gain (alpha) is necessarily less than one. We have already determined that a small signal voltage impressed in the low-resistance input (emitter) circuit of a transistor produces a relatively large emit-

ter current. Almost the same amount of current will flow in the high-resistance output (collector) circuit of the transistor, where the voltage may be very high. Evidently, then, both the output voltage and output power can be quite large, compared to the tiny input voltage and power present at the emitter.

TRANSISTOR SYMBOLS AND CONNECTIONS

You've met P-N-P and N-P-N transistors and found out something about their operation. In later chapters you will learn about their fascinating applications in various ingenious gadgets and compact electronic circuits. But in order to recognize a transistor when you see one (in a circuit, of course) you must become familiar with some of the forms, connections, and circuit symbols used by engineers to schematicize them.

Fig. 48 illustrates the common forms and circuit symbols used for N-P-N transistors (a) and P-N-P transistors (b). Note that in each case the emitter is distinguished from the collector by an arrow, which indicates the direction of "conventional" current flow with forward bias. You will remember from basic electricity that conventional or "Franklinian" current flow is *opposite* in direction to electron flow, because of an erroneous assumption made by Franklin concerning the direction of flow. Checking back (in Fig. 45) for the N-P-N connection, it is apparent that electrons flow *into* the emitter from the negative battery terminal; this means that conventional current flows *out* of the emitter, as indicated by the outgoing arrow in Fig. 48a. Similarly, you can determine from Fig. 43 that for the P-N-P transistor, electrons flow out of the emitter toward the positive battery terminal. Consequently, conventional current flow is *into* the emitter, as indicated by the inward arrow in Fig. 48b.

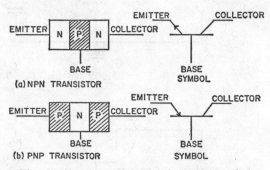

Fig. 48. *Transistor Forms and Circuit Symbols (a) NPN-type (b) PNP-type*

When transistors are operated as amplifier, three different basic circuit connections are possible, as illustrated in Fig. 49. These are (a) common-base, emitter-input; (b) common-emitter, base-input; and (c) common-collector, base-input. Each of these circuit configurations has specific advantages and limitations.

Note that regardless of the circuit connection the emitter is always biased in a *forward* direction, while the collector always has a *reverse* bias, in accordance with the basic connections shown in Figs. 43 and 45. This necessitates a *positive* emitter bias and a *negative* collector bias for the P-N-P transistor, while opposite polarities are required for the N-P-N transistor. Except for this polarity reversal, the connections for P-N-P and N-P-N transistors are identical.

Common-Base Emitter-Input Connection. The connections shown in Fig. 49 can be understood by considering the emitter of the transistor roughly analogous to the cathode of a triode tube, the collector analogous to the plate of a triode, and the base analogous to the grid of a triode. The common-base connection (Fig. 49a) is thus roughly analogous

to a **grounded-grid** triode tube circuit, whose specific characteristics shall be discussed in a later chapter. We have already become familiar with the common-base connection, since it is convenient for illustrating transistor physics. As we found out, this circuit provides a very low input resistance, a high output resistance and a current gain (alpha) of less than one. Despite this low current gain, the common-base circuit provides respectable voltage and power amplification, as was mentioned earlier.

Common-Emitter Base-Input Connection. The common-emitter circuit (Fig. 49b) is roughly analogous to the conventional **grounded-cathode** triode tube circuit and is the most flexible and efficient of the three basic connections. While its input resistance is somewhat higher and its output resistance is lower than that of the common-base connection, the common-emitter connection has the highest voltage and power gain of the three circuits. Like the equiv-

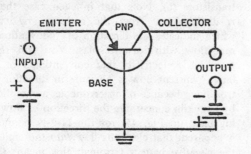
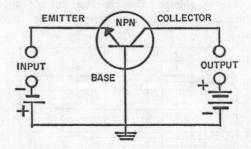

(a) COMMON-BASE, EMITTER-INPUT CONNECTION

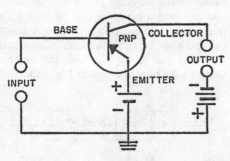
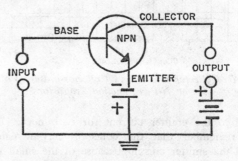

(b) COMMON-EMITTER, BASE-INPUT CONNECTION

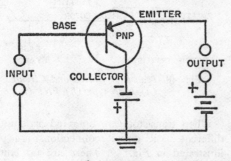
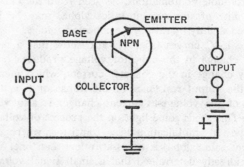

(c) COMMON-COLLECTOR, BASE-INPUT CONNECTION

Fig. 49. *Transistor Amplifier Circuit Connections*

alent vacuum-tube circuit, the common-emitter connection produces a **phase reversal** between the input and output signals. No such phase reversal occurs with the other two connections.

Common-Collector Base-Input Connection. The common-collector connection is the transistor equivalent of the grounded-plate triode tube (cathode follower circuit) that will be discussed in a later chapter. The circuit provides a relatively high input resistance, a low output resistance, and about the same current gain (alpha) as the common-emitter circuit. Its voltage gain, however, is always *less than one,* as is the case in the equivalent cathode-follower circuit. As with the latter, the common-collector circuit is primarily used for impedance matching and as a buffer stage.

JUNCTION TETRODE TRANSISTOR

The frequency response of conventional junction transistors is considerably limited because of the capacitance associated with the collector electrode and because of the finite time taken by the charge carriers to diffuse through the transistor base region. The collector capacitance acts as a shunt path for high frequencies, just like the interelectrode capacitances in electron tubes. The time taken by the charge carriers to arrive at the collector (**transit time**) places an upper frequency limit on the operation of the triode transistor. Obviously, if the polarity of the collector voltage reverses before the charge carriers have a chance to drift across to the collector, the current gain will fall off. The upper frequency limit for a transistor is termed the **alpha cutoff frequency** and it is defined as the frequency at which the current gain falls to 70.7% of its low-frequency value.

A **junction tetrode transistor** partially overcomes this unfavorable frequency-response characteristic of the conventional transistor triode. In the tetrode transistor, whose structural and circuit representations are illustrated in Fig. 50, a fourth electrode,

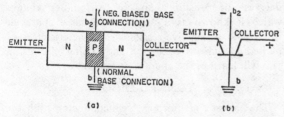

Fig. 50. *Junction Tetrode Transistor (a) Form, (b) Circuit Symbol*

b_2, is connected to the base region (P-layer) on the opposite side of the conventional base connection. When a negative d-c bias of about −6 volts is applied to this second base terminal, the emitter-to-collector current is forced by the field into a very

small region near the vicinity of the regular base connection, (b). This produces the equivalent of a transistor with a very thin P-region and low base resistance. Both these factors reduce the transit time of charge carriers from emitter to collector, resulting in vastly improved response at high frequencies.

POINT-CONTACT TRANSISTORS

The **point-contact transistor** is the forerunner of the modern junction transistor and is also related in some ways to the old-fashioned crystal detector, used in the early days of radio. Although still employed to advantage in some applications, the point-contact transistor has been largely replaced by the

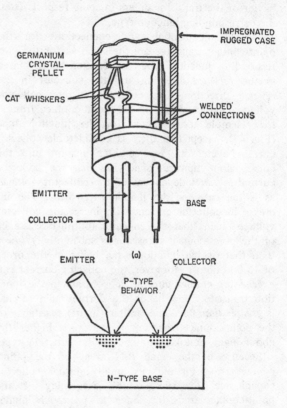

Fig. 51. *Point-Contact Transistor (a) Construction, (b) Behavior in Point Region*

junction transistor because of recent advances in junction design.

Fig. 51 illustrates the construction and approximate behavior of a typical N-type point-contact transistor. In contrast to the early crystal detector, **two** sharply pointed tungsten electrodes (often referred to as "cat whiskers") make contact close together on the surface of a single germanium crystal pellet. A third electrode, the base, is soldered to the N-type germanium pellet. The two cat whiskers are called **emitter** and **collector,** respectively, just as in the junction transistor. The entire assembly is enclosed in an impregnated plastic housing, sealed against the atmosphere.

Although the theory of the point-contact transistor is fairly complicated, you can gain a qualitative understanding of its operation by considering that the transistor must be "formed" initially by passing a current through its electrodes during manufacture. This effectively produces a small P-region underneath each of the points, as is illustrated in Fig. 51b. With the two P-regions imbedded within the N-type germanium base, a P-N-P type transistor results, which is similar to the P-N-P junction type. Because of the point contacts, however, the behavior of the transistor differs in some respects from the corresponding junction type.

The circuit symbols and connections for the point-contact transistor are the same as for the corresponding junction type, the emitter being biased in the *forward* direction and the collector in the *reverse* direction. The collector current collector voltage characteristics of the point-contact transistor resemble those shown for the junction type (Fig. 47), except that there is considerable collector current even with **zero** emitter current and the curves slope upward, signifying that the collector current is *not* independent of the collector voltage, as is the case for the junction transistor. The increase in collector current with increasing collector voltage means that the collector (output) resistance of the point-contact transistor is substantially *lower* than that of the junction type, being in the order of 15,000 ohms. Moreover, the collector current can *exceed* the emitter current, in contrast to the junction transistor, and hence the current gain (alpha) is *greater than one*. The emitter (input) resistance of the point-contact transistor, however, is higher (by about eight times) than that of the junction type.

Because of the small P-N areas of the point-contact transistor and its small internal collector capacitance, the **alpha cutoff frequency** of the point-contact transistor tends to be much higher than that of the corresponding junction transistor. (The alpha cutoff frequency is the point where the current gain drops to 70.7 percent of the low-frequency gain.) As a consequence, point-contact transistors can be easily made to amplify or oscillate up to 100 megacycles and above, which is far beyond the capabilities of *early* junction transistors. Recently developed junction transistors, such as the transistor tetrode and the "mesa" transistor, are rapidly making point-contact transistors obsolete, even at high frequencies.

UNIJUNCTION TRANSISTORS

As its name implies, the **unijunction transistor** has only one P-N junction, in contrast to the ordinary two-junction transistor. This single junction drastically modifies the electrical characteristics of the device and makes it suitable for a variety of oscillator applications.

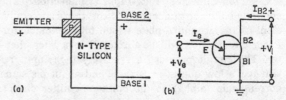

Fig. 52. *Unijunction Transistor* (*a*) *Construction* (*b*) *Symbol and Principal Voltages*

Fig. (52) shows the construction and symbol of the unijunction transistor. Two contacts, base 1 (B1) and base 2 (B2), are made at opposite ends of a small bar of N-type silicon. At the opposite end of the bar, close to base 2, is a single P-type rectifying junction, the emitter. A resistance of 5000–10,000 ohms exists between the two bases (the **interbase resistance,** R_{BB}). In normal operation, base 1 is grounded and a positive voltage, V_{BB}, is applied at base 2. In the absence of emitter current, I_e, the N-type silicon bar acts simply as a voltage divider. Thus a certain fraction of the positive base 2 voltage, V_{BB}, appears at the emitter junction. As long as the emitter voltage, V_e, is less positive than this fraction, the junction is reverse-biased and only a small emitter leakage current flows. When the positive emitter voltage becomes greater than the fraction of the base voltage at the emitter, however, the junction will be forward-biased and an emitter current, I_e, flows. This current consists of holes, injected into the silicon bar, which move from the emitter to base 1. As the holes reach base 1, an equal number of electrons are attracted into the region between base 1 and the emitter. The net increase in electrons is equivalent to a reduction in the resistance between emitter and base 1. The upshot is that an **increase** in emitter **current** results in a **decrease** of resistance and in the emitter **voltage,** which is known as a **negative-resistance** characteristic. Such a negative-resistance characteristic is exploited in various oscillator and timing circuits, as you will learn in the chapter on oscillators.

TUNNEL DIODES

The most glamorous semiconductor junction device that has come along in recent years is the **tunnel diode,** which promises to push semiconductor operating frequencies into the microwave range of several thousand megacycles (kilomegacycles or **gigacycles**). Invented by the Japanese scientist Leo Esaki, the tunnel diode—like the unijunction transistor—has of a negative-resistance characteristic (i.e., a **decrease** of current with an **increase** in voltage), which makes it useful both as high-frequency oscillator and amplifier. The tiny and reliable tunnel diode also enjoys great advantages in switching applications, such as are needed in electronic computers, since it can switch on and off in a few billionths (10^{-9}) of a second. Although the tunnel diode shares a negative-resistance characteristic with the unijunction transistor, its operating principles are completely different from the latter, as we shall see presently.

The increase in high-frequency and switching capability of the tunnel diode is achieved through a highly conductive, extremely narrow junction of P- and N-type germanium (or some other crystal, such as gallium-arsenide). Because of this extremely narrow junction, electrons are capable of "tunneling through" from one side of the junction to the other, though they have insufficient energy to surmount the "potential barrier," or wall, always present at such a junction. You can conceive of this tunneling effect in terms of a billiard ball that rolls over hump in the table, though it has hardly been pushed at all, and does not (or should not) have the energy to do so. Neither common sense nor classical physics can explain this surprising situation, but it is explained by quantum physics as "quantum-mechanical tunneling," a matter we cannot go into here.

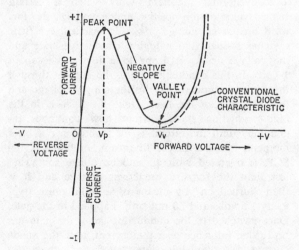

Fig. 53. *Tunnel Diode Characteristic Compared with Conventional Crystal Diode*

What occurs in a practical way is apparent from the characteristic of a typical tunnel diode, illustrated in Fig. (53), which also gives the characteristic of a conventional crystal diode for comparison. When a reverse (negative) bias voltage is applied to the anode of a conventional crystal diode, it does not conduct, while a tunnel diode, in contrast, **will conduct**. At low values of an applied forward (positive) anode voltage, the tunnel diode passes a

considerable current, which reaches a peak at a low voltage, V_p, when the conventional diode has not even begun to conduct as yet. As the applied forward voltage is increased, the tunnel diode current starts to decrease again and reaches a minimum at the valley point, for a voltage V_v. This decrease in tunnel diode current with increasing voltage causes a **negative slope,** or **negative-resistance** characteristic, which is typical for all tunnel diodes. It is this negative-resistance characteristic that permits the tunnel diode to be used either as an amplifier, an oscillator, or a switching device (gate or "flip-flop") in computers. When the forward anode voltage is further increased beyond the valley point, however, the tunneling effect ceases and the current increases in a manner similar to that of a conventional diode.

ZENER DIODES

Semiconductor P-N junction diodes, usually made of germanium or silicon, have been used for many years as power rectifiers. They are capable of rectifying (i.e., converting to pulsating d.c.) relatively large currents, but in comparison with tubes have the disadvantage that their **maximum safe inverse voltage** (the voltage opposite to the direction of current flow) is relatively low. What happens if this safe inverse voltage is exceeded? This question is best answered by looking at the current-voltage characteristic of a typical silicon P-N junction diode (see Fig. 54a).

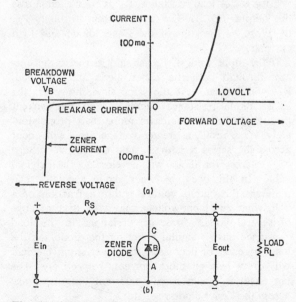

Fig. 54. *Current-Voltage Characteristic (a) and Circuit (b) of Typical Zener Diode Regulator*

As is apparent from the characteristic, a substantial current flows through the diode when a forward voltage of approximately 1 volt is applied.

When a reverse voltage is applied, making the anode negative, conduction stops and the diode blocks the reverse current, like any other rectifier. As the reverse voltage is increased, a small but substantially constant **leakage current** begins to flow, which is characteristic of all semiconductor diodes. The small leakage current does not prevent the diode from functioning as a rectifier. If the reverse potential continues to be increased, however, beyond the safe inverse voltage, a critical voltage, called **breakdown voltage**, is eventually reached, where the reverse current increases sharply to a high value. The breakdown region is the "knee" of the characteristic curve in Fig. 54a, and the sudden increase in current is known as **avalanche** or **Zener current** (after an explanatory theory by the scientist C. Zener). Conventional diodes and rectifiers never operate in the breakdown region, but the so-called **Zener diode** makes a virtue of it and operates at this very point.

The advantage of the strangely behaving Zener diode is that it makes a superb regulator at the output of a power supply. Consider the Zener regulator circuit shown in Fig. 54b. The input to the circuit is a d-c voltage, E_{in}, whose voltage variations are to be regulated. (E_{in} may represent the output voltage of a power supply, such as described in Chapter 14.) The Zener diode is reverse-connected across the input voltage, so that its cathode is positive and its anode is negative. Since E_{in} is greater than the diode's breakdown voltage, it conducts and draws a relatively large Zener current through the series resistor R_S. A load resistance, R_L, is connected across the output (E_{out}). The total current through R_S, therefore, is the sum of the Zener diode and load currents.

If the input voltage, E_{in}, should increase, the current through both the Zener diode and the load, R_L, will increase. At the same time, however, the Zener diode resistance decreases and the current through the diode increases more than proportionately. As a result, a greater voltage drop will occur across the series resistor R_S and the output voltage (E_{out}) across the diode will be close to the original value. In this way, a Zener diode regulator can maintain the output voltage within a fraction of a volt, when the input voltage may vary over a range of several volts. Moreover, variations in the load resistance have a similar effect on the diode. As the load increases or decreases, the Zener shunt element will draw less or more current, respectively. The net result is a substantially constant output voltage across the Zener diode regulator.

SILICON-CONTROLLED RECTIFIER

The solid-state equivalent of the gas-filled switching tube, or thyratron, we have become acquainted with in Chapter 6 is the **Silicon-Controlled Rectifier**, or SCR. This extremely useful PNPN silicon device consists essentially of an alloyed P-N rectifier junction and a diffused P-N-P silicon pellet with a separate contact, known as **gate.** The schematic symbol and external construction of a 16-ampere SCR is shown in Fig. 55.

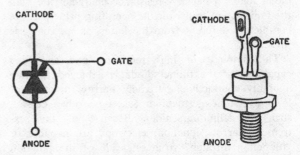

Fig. 55. *Symbol and Construction of Silicon-Controlled Rectifier*

The addition of the gate drastically changes the characteristics of the rectifier. When the SCR is operated with reverse voltage (anode negative and cathode positive), it blocks the flow of current until the breakdown or avalanche voltage is reached, as we have seen in the Zener diode (Fig. 54). When a positive voltage is applied to the anode, the SCR also blocks the flow of forward current, in contrast to conventional rectifiers. Only when a certain critical value of the positive anode voltage, the **forward breakover voltage** (V_{BO}), is reached, the SCR switches suddenly to a highly conductive state and the voltage across it drops to a low value (about 1 volt). In this conducting state, the current through the device is limited only by the supply voltage and the resistance of the load. Once the SCR is in the high-conduction state, current flow continues indefinitely until the circuit is interrupted for a brief moment, just like in the thyratron. In practice, the SCR is operated with an anode voltage somewhat less than the forward breakover voltage and it is then "turned on" by means of a small pulse (typically 1.5 volts and 30 milliamps) applied to the gate. Once switched to the conducting state, the gate has no further influence over the current, until the anode current is completely interrupted for about 20 microseconds.

Since the SCR permits control over powerful currents (30–100 amps) by means of a small gate pulse, you can easily see that the device is very useful in many relay, switching, and control applications. Typical applications of silicon-controlled rectifiers are in regulated power supplies, d.c.-to-a.c. inverters, radar modulators, servo systems, latching relays, electronic ignition systems, etc. The important advantage of the SCR is, of course, that switching is extremely rapid and requires no moving parts.

SUMMARY

Atoms within a pure germanium (or silicon) crystal are strongly bound together by means of electron-sharing or **covalent bonds** in a diamond-like structure. Each germanium atom completes its outer valence shell by combining its four electrons in electron pairs with those of adjacent germanium atoms.

A pure germanium crystal is practically an insulator.

When pure germanium is "doped" with atoms that have **five** electrons in their outer shell (antimony or arsenic), the impurity or **donor** atoms make their excess electrons available as negative charge carriers.

When pure germanium is "doped" with **trivalent** impurity atoms, such as indium or gallium, the impurity atoms **(acceptors)** borrow electrons from surrounding germanium atoms, leaving a deficiency of electrons, or **holes,** in their place. The holes act like mobile, **positive** charge carriers.

Germanium doped with pentavalent donor atoms is called **N-type** germanium; current conduction takes place through **negative** charge carriers, or electrons.

Germanium doped with trivalent acceptor atoms is called **P-type** germanium; current conduction takes place through **positive** charge carriers, or **holes.**

A **P-N junction,** consisting of wafers of P-type and N-type germanium, may be either **grown** or **fused,** depending on manufacturing technique.

When a P-N junction is biased in the **forward** direction, by applying a positive potential to the P-region and negative potential to the N-region, the holes and electrons are repelled toward the junction area and overcome the potential barrier there. Current conduction then takes place by means of electron-hole combinations in the vicinity of the junction.

With **reverse bias** applied to a P-N junction (P-region negative and N-region positive), holes and electrons are attracted away from the junction area, and current conduction stops except for a small **reverse current.** If the reverse bias is made very high, the junction breaks down, and a relatively large reverse current flows.

A **junction triode transistor** is a sandwich made up of two P-N junctions, either in P-N-P form or N-P-N form. The central region is called the **base** and the two outer layers are called the **emitter** and **collector,** respectively.

The emitter junction of a transistor is always biased in a **forward** direction, while the collector junction is biased in **reverse** direction.

Current conduction in a P-N-P transistor takes place by **hole conduction** from emitter to collector, while current conduction in a N-P-N transistor is carried on by **electrons as majority** charge carriers.

The collector current in a junction transistor is **less** than the emitter current by an amount proportional to the number of electron-hole combinations occurring in the base area.

The **ratio** of collector-to-emitter current is known as **current gain** or **alpha** (α) and is always less than 1 (practical values to 0.99).

Transistors may be connected into either of three basic (amplifier) circuits. These are: 1. **common-base, emitter-input;** 2. **common-emitter, base-input;** and 3. **common-collector, base-input.**

The **common-base** connection is analogous to the **grounded-grid** triode tube and it provides a **very low input** resistance, a **high output** resistance, and a current gain of **less than 1.** There is no phase reversal.

The **common-emitter** connection, the most flexible and efficient of the three basic connections, is analogous to the **grounded-cathode** triode tube, and like the latter it reverses the phase of the output signal with respect to the input. Its input resistance is higher and its output resistance is lower than those of the common-base connection, but it provides the **highest voltage and power gain.**

The **common-collector** connection is the transistor equivalent of the grounded-plate triode (cathode-follower) and is used primarily as a buffer and for impedance matching. The connection provides a **high** input resistance, a **low** output resistance and a **voltage gain of less than 1.**

A **junction tetrode** transistor has an extra, **negatively** biased connection to the base. This effectively extends the upper frequency limit **(alpha cutoff frequency)** of transistor operation.

A **point-contact** transistor uses two "cat whiskers" electrodes to form a type of P-N-P (or N-P-N) transistor. Its current gain (alpha) is greater than 1.

A **unijunction transistor** is a P-N junction with two bases, which has a **negative-resistance characteristic** that is controlled by the positive emitter voltage. This characteristic makes it useful in oscillator and timing circuits.

Tunnel diodes consist of extremely narrow P-N junctions that permit operation at microwave frequencies of several thousand megacycles or switching speeds of a few billionths of a second. Because of their negative-resistance characteristic, tunnel diodes are useful as oscillators and bistable switching devices, as well as high-frequency amplifiers.

The **Zener diode** consists of a reverse-biased silicon P-N junction, which is operated in the breakdown region. The large resulting reverse **avalanche** or **Zener current** is useful in voltage regulator applications.

The solid-state equivalent of the gas-filled thyratron is the **Silicon-Controlled Rectifier** (SCR). It consists of a P-N rectifier junction and a P-N-P silicon pellet to which a contact (the gate) is made. When reverse-biased, the SCR blocks the flow of current like conventional rectifiers. It also blocks current flow when forward-biased until the **forward break-over voltage** (V_{BO}) is reached, at which point it switches to a high-conduction state. The SCR may be turned on by means of a small positive voltage pulse applied to the gate. Once switched on, a large current (30–100 amperes) flows and the gate looses control. SCR's are useful whenever large currents must be controlled by means of a small trigger, such as in relay, switching, and other control applications.

Chapter Eight

AUDIO AMPLIFIERS

We are finally ready to apply the knowledge we gained in previous chapters about electrons, vacuum tubes, transistors, and so on, to some practical matters. In the following chapters we shall consider a variety of circuits employing electron tubes, or equivalently, transistors. **Circuits** are combinations of tubes (or transistors) with other components, such as resistors, capacitors and inductors, and form the basic building blocks of **electronic systems:** radio, radar, television, and so on. To understand the systems, you must be familiar with the circuits that make them up.

In this chapter we shall discuss amplifier circuits, or more specifically, **audio amplifiers. An amplifier is an electron tube or transistor circuit, which builds up an a-c signal applied to its input.** It is called a **voltage amplifier** if the magnitude of the output voltage from the amplifier is considerably greater than that of the input voltage. As a matter of fact the **ratio** of the output voltage to the input voltage is called the **amplification** or **gain** of the amplifier. There are also so-called **power amplifiers.** These are similar to voltage amplifiers, except that their main purpose is to supply a considerable amount of power (i.e., voltage *times* current) to the **output** or load circuit, although the a-c input signal may not draw any grid current and, hence, the **input power** may be zero. A power amplifier may also build up the voltage to some extent, but this is of secondary interest. When a number of amplifiers are hooked up in series (called **casade),** so that the output of one serves as the input to the next amplifier stage, the function of the early stages is usually to build up the **voltage** to a high level, while the last stage builds up the **power** to a level sufficient to operate a headset, loudspeaker, or similar output device. In the present chapter we shall talk about these early stages in an amplifier chain—the voltage amplifiers—and in a later chapter we shall describe the final stage, or power amplifier.

What about the "audio" in audio amplifier? This term refers to sound or human hearing. As every hi-fi enthusiast knows, the range of human hearing extends from about 20 to 15,000 cycles per second, but varies considerably with age and individual. This means that we can *hear* air pressure changes or vibrations that vary in **pitch** from about 20 times per second to about 15,000 times per second. **Pitch** describes the subjective sensation of hearing an air pressure vibration, or sound. The pitch of a bass fiddle is low, that of a flute or coloratura soprano is high. The more air pressure vibrations per second, the higher the pitch.

But neither tubes nor transistors will amplify air pressure vibrations, or sound, directly. It is necessary, therefore, to first convert the air pressure vibrations into equivalent **electrical vibrations** by means of a microphone or similar acoustic-electrical **transducer.** This means that if you sing a pure tone into a microphone that has a pitch of, say, 1000 vibrations per second, the microphone output voltage will be an electrical a-c signal, or sinewave, that has a frequency of 1000 cycles per second (abbreviated cps, or simply, cycles).

Audio amplifiers, therefore, amplify electrical a-c signals that have a frequency range corresponding to the range of human hearing, or from about 20 to 15,000 cycles per second. In recent years, however, hi-fi "purists" have greatly extended the frequency range of audio amplifiers, so that some commercial models are available with a range from about 5 cps to over 100,000 cps. You may wonder at this, since nobody can hear these extreme frequencies; but as we shall see in a later chapter it really *does* make a difference.

BASIC VOLTAGE AMPLIFIER

In Chapter 4 we saw that a triode acts as an amplifier because the plate current is affected to a much greater degree by a change in the grid voltage than by a change in the plate voltage. We then defined the **amplification factor** (μ) as the ratio of a small change in the plate voltage to the small change in grid voltage required to produce the *same* change in the plate current (but in opposite direction, of course). We also placed variable d-c voltages on the grid and plate of a triode (Fig. 9) and obtained a series of curves known as the static characteristics of a triode. From the latter, in turn, we were able to calculate the amplification factor (μ), the plate resistance (r_p) and the transconductance (g_m) of the tube.

GRID CIRCUIT

Now let us see what happens under **dynamic** conditions when an actual a-c signal voltage (e_g) is applied to the grid of the tube and a load resistor (R_L) is introduced into the plate circuit to extract useful power from the tube. Fig. 56 shows this basic triode amplifier circuit together with the notation employed to designate the various voltages and currents.

In Fig. 56 the capital letters indicate steady or d-c **supply** voltages or currents, while the lower-case

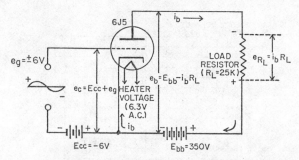

Fig. 56. *Basic Triode Amplifier Circuit and Notation*

letters indicate varying or a-c **signal** voltages and currents. Some typical circuit values have been assumed to make the example more illustrative. Thus, an a-c signal voltage (e_g, indicated by the sinewave) equal to 6 volts peak amplitude has been introduced in the grid circuit of the 6J5 triode, in series with a **fixed** bias voltage (E_{cc}) from a 6-volt battery. You may well wonder why the bias battery is necessary. Assume for a moment, that it isn't there, and the 6-volt signal voltage is applied to the grid alone. The voltage on the grid of the tube will then vary between zero, +6 volts, and −6 volts, as the a-c signal goes through one cycle or two alternations. Whenever the input signal goes **positive** and hits its +6-volt peak value, a **grid current** will flow out from the grid of the tube, as you will remember from our earlier discussion. This grid current not only distorts the linearity of the tube's characteristic curves, thus leading to distorted amplification, but it also requires a certain amount of **power** (i.e., grid voltage times grid current), which it extracts from the input signal. If the input signal is too weak to supply the power (such as a tiny radio signal), further distortion takes place, with undesirable results.

For this reason, voltage amplifiers are always operated so as to consume *no* power in the grid circuit, particularly since the input signal is generally too weak to supply the required power. Moreover, it is the great advantage of electron tubes that they are capable of being purely **voltage-operated in the grid circuit, although power may be available from the plate circuit.** To avoid consuming power in the grid circuit it is essential, therefore, that **no grid current should flow** during any portion of the a-c input signal. This, then, is the function of the bias voltage. The bias voltage in a voltage amplifier is always of sufficient magnitude and of such polarity as to maintain the grid **negative** (or at zero volts) during the **positive** peak of the a-c input signal. In our example (Fig. 56) the bias battery (another source could be used) has been connected to keep the grid at a potential of −6 volts, in the absence of an a-c input signal.

When an a-c signal voltage is now applied in series with the bias voltage, the total **instantaneous** voltage between grid and cathode of the tube (symbolized "e_c") is equal to the algebraic sum of the a-c signal voltage and the d-c grid bias. Thus,

$$e_c = E_{cc} + e_g$$

In our example, since $E_{cc} = -6$ volts and the signal voltage (e_g) varies between +6 and −6 volts, the total instantaneous grid voltage, e_c, will vary between 0 and −12 volts. Whenever the signal voltage reaches its positive peak of +6 volts, the total grid voltage will go to zero (+6 − 6 = 0) and whenever the signal reaches its negative peak of −6 volts, the total grid voltage falls to −12 volts (−6 − 6 = −12).

PLATE CIRCUIT

We have been concentrating on the grid circuit thus far. Let's see now what happens in the plate circuit of the 6J5, while the grid-to-cathode voltage goes through its cycle from 0 to −12 volts. Assuming that a voltage is applied to the heater of the tube (6.3 volts a.c., in this case), a plate current (i_b) will flow. This plate current will continue to flow as long as the a-c signal voltage is not sufficiently large to drive the tube to plate-current cutoff. The bias, signal voltage and plate-supply voltage (E_{bb}) in our example (Fig. 56) have been so chosen that the tube does *not* reach plate-current cutoff, even for the most negative value of the grid voltage, or −12 volts. (You can verify this for yourself by consulting the 6J5 characteristics, Fig. 10.) Plate current will, therefore, flow at all times during the a-c input signal cycle. As a matter of fact, this is a necessary condition for all voltage amplifiers to obtain linear, **distortionless** amplification. You will remember that the tube characteristics become very distorted or non-linear near plate-current cutoff and, hence, the bias must be chosen to maintain the tube well above plate-current cutoff for the most **negative** value of the a-c input signal.

With plate current (i_b) flowing at all times, a voltage drop ($i_b \times R_L$) is produced across load resistor R_L in the plate circuit (25,000 ohms, in this case). The voltage drop across this fixed load resistor depends on the value of the plate current and the plate current, in turn, is controlled by the grid voltage. Because of the amplifying action of the tube, a small change in the signal or grid voltage produces a large change in the plate current and, hence, in the voltage across the load resistor. (The voltage across the load resistor, $e_{RL} = i_b R_L$.) By making the load resistance sufficiently high a large voltage drop is produced across it, resulting in high voltage amplification. Amplification, or gain, is the **ratio** of the output voltage across the load to the input signal voltage, you will remember. A large load voltage and, hence, high amplification, may be obtained even with quite small plate currents, since there is no theoretical limit for the size of the load resistor. The output **power,** therefore, may be quite small.

Since the total plate current, i_b, never becomes negative (or even reaches zero), it is *not* alternating current. Actually, it consists of **two** components, one d.c. and one varying, like the grid voltage (e_c). The d-c component of the plate current, identified as I_b, is the normal plate current that flows with the plate voltage and bias applied, but with the a-c input signal voltage (e_g) **absent.** The a-c component of the plate current (identified as i_p), on the other hand, is the **variation** in the total plate current caused by the a-c input signal at the grid. The total instantaneous plate current i_b, thus, is the sum of the d-c (I_b) and a-c (i_p) components.

Note further (in Fig. 56) that the plate current (i_b) flows *out* of the plate of the tube, through the load resistor, and then toward the positive terminal of the plate-supply battery (E_{cc}). Since electrons flow from minus to plus, the top of the load resistor is more negative than the bottom or battery-connected end. The load voltage, thus, is in **opposition** to the plate supply voltage, and since it is connected in series with it, **subtracts** from it. The greater the plate current, the greater is the load voltage ($i_b R_L$) and, hence, the **less** plate-supply voltage, is left over to reach the plate of the tube. The total instantaneous voltage on the plate (identified by e_b), therefore, is the **difference** between the plate-supply voltage, E_{bb}, and the voltage across the load, e_{RL}. Expressed mathematically,

total instantaneous plate voltage $e_b = E_{bb} - i_b R_L$

From the above expression it is clear, that the total plate voltage (at any instant) becomes **smaller** as the plate current becomes **larger.** Furthermore, since the plate current increases directly with the input signal voltage, the total plate voltage **decreases** with increasing grid or signal voltage. Thus the plate voltage (e_b) is a **maximum** when the grid voltage (e_c) is a **minimum,** and vice versa. The grid and plate voltages, therefore, are in **phase opposition** or equivalently, the tube is said to produce a **180-degree phase reversal** of the plate voltage with respect to the grid voltage. This is a very important fact to keep in mind. The amplification is, of course, not affected by this phase reversal.

GRAPHICAL ANALYSIS

The word picture we have just given about the voltage and current relations in a basic triode amplifier circuit can be considerably simplified by plotting the input and output waveforms against the characteristics of the amplifier tube, as shown in Fig. 57.

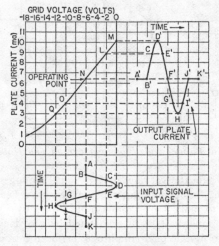

Fig. 57. *Typical Input and Output Waveforms Obtained from Dynamic Transfer Characteristic of 6J5 for 25,000-Ohm Load. (Courtesy "Basic Vacuum Tubes and Their Uses" by John F. Rider and Henry Jacobowitz, John F. Rider Publisher, Inc.)*

The 6J5 plate current-grid voltage characteristics are shown in the upper left-hand portion of Fig. 57. These are *not* the static characteristics with which you have become familiar, but the **dynamic transfer** characteristic, so called because it has been obtained under actual operating conditions for a specific plate-supply voltage (350 volts) and a (25,000-ohm) plate-load resistor. A new characteristic must be obtained each time the load resistor or the plate-supply voltage is changed. We have assumed an **operating point** (*N*) on the dynamic **transfer** characteristic of —6 volts grid voltage, equal to the value of the bias battery (E_{cc}) in Fig. 56. The operating point determines the **quiescent** values of the grid voltage (bias) and plate current in the absence of a signal voltage.

Now let us apply a 6-volt a-c input signal to the grid of the tube and see what happens at the output. The input signal sinewave A-B-C-D-E-F-G-H-I-J-K is plotted against time at the lower left of Fig. 57. The waveform of the output plate current may be obtained by drawing lines from the points of the input waveform to their intersections with the dynamic transfer characteristic at points N, L, M, O, Q and then extending the lines from the intersections to the right to plot the corresponding points A'-B'-C'-D'-E'-F'-G'-H'-I'-J'-K' of the output current waveform. By choosing the same time scale for both input and output waveforms, the two waveforms may be directly compared. It is seen from Fig. 57 that the output plate current is a reasonably pure sinewave and not distorted in comparison with the input signal. This excellent amplifier characteristic has been achieved by choosing the operating point (*N*) so that the largest "swing" of the input signal (from 0 to −12 volts on the grid) is accommodated within the **linear portion** of the dynamic transfer characteristic. If the operating point had been chosen near the curved portion of the characteristic, say at point *Q*, the output plate current waveform would have been considerably distorted with respect to a pure sinewave. Note, however, that the bottom portion of the dynamic transfer characteristic (Fig. 57) is much less curved than the corresponding portion of the **static** plate current-grid voltage characteristic (Fig. 10) and, hence, operation is more linear (less distorted) even for large signals extending into the lower portion of the characteristic. This "straightening out" of the characteristic curve is achieved solely by the insertion of the load resistor in the plate circuit.

AMPLIFICATION

Besides the graphic portrayal of the plate current waveform, Fig. 57 also permits us to compute the amount of amplification or **voltage gain** of the 6J5 triode, when operated in the circuit and with the voltages shown in Fig. 56. You will remember that the voltage gain is the **ratio** of the output voltage across the load to the input signal voltage on the grid. Since we are dealing with a ratio, a simple way of obtaining the gain is to compare the **peak** value of the a-c output voltage with the peak value of the signal voltage. The peak output voltage is the product of the **peak** a-c plate current (i_p) and the load resistor (R_L). To find the peak value of the a-c plate current we must determine the total **change** in the plate current from its **quiescent** or no-signal value to its full value at maximum input signal. From Fig. 57 we obtain a quiescent value (at point *N*) of

about 6.4 ma and a full-signal value at point D' of the plate current (or point *M* of the characteristic) of about 10.1 ma. Hence, the change in plate current is $10.1 - 6.4 = 3.7$ ma, which, thus, is the peak amplitude of the alternating current.

Knowing the peak value of the current, the output voltage

$$e_{out} = i_p \times R_L = 0.0037 \text{ (amps)} \times$$
$$25,000 \text{ (ohms)} = 92.5 \text{ volts}$$

The peak value of the input signal we have arbitrarily fixed at 6 volts and, hence, the

$$\text{voltage gain} = \frac{e_{out}}{e_g} = \frac{92.5 \text{ volts}}{6 \text{ volts}} = 15.4$$

If we had chosen any other value of the input signal and computed the output voltage for this signal, the voltage gain would have come out about the same. Note also that the voltage amplification is somewhat less than the amplification factor (μ) of the 6J5, which we have previously determined (in Chapter 4) to be about *20*. This is always true for any practical amplifier.

PHASE RELATIONS

We have mentioned the 180-degree phase reversal that occurs in an electron tube between the input signal and the plate voltage. It is of interest to compare the phase relations between all major input and output waveforms in the basic triode amplifier circuit, of Fig. 56. These phase relations are illustrated in Fig. 58 for our previous example of −6 volts bias, 350 volts plate-supply voltage and 25,000 ohms load resistance. Although this specific example has been chosen, the phase relations shown in Fig. 58 hold true regardless of the particular values of the voltages and currents in the amplifier. The dashed vertical lines passing through the four waveforms permit comparing corresponding waveforms at the *same* instant in time.

Curve (A) represents the instantaneous signal voltage (e_g), which is seen to vary between +6 and −6 volts. Curve (B) shows the variation in the total grid voltage (e_c) with the signal. When the signal is absent, or zero, the grid voltage is equal to the bias, or −6 volts. When the signal is present, the total grid voltage rises to zero volts at maximum positive (+6-volt) signal swing, and it drops to − 12 volts for the maximum negative (−6-volt) signal swing. Curve (C) shows the variation in the instantaneous value of the total plate current (i_b) with signal. As noted before, the plate current has a quiescent value (I_{bo}) of 6.4 ma for zero signal voltage and varies between 3 ma and 10.1 ma for the full (±6 volt) signal swing. Since the plate current *increases* with *increasing* grid voltage, waveforms (A), (B), and (C) are

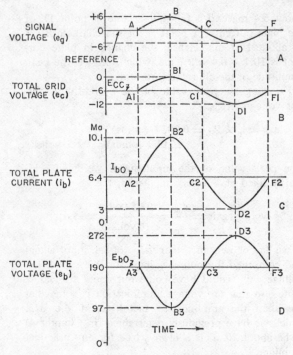

Fig. 58. *Phase Relations in Triode Amplifier Circuit of Fig. 56*

all **in phase** (that is, they reach their positive and negative peaks at the same time).

Finally, curve (D) illustrates the variation in the instantaneous value of the total plate voltage (e_b) with the signal voltage. Note that the quiescent value of the plate voltage (E_{bo}) with the signal zero or absent, is 190 volts. You can verify that this must be the correct value by subtracting the voltage drop across the load resistor (with the signal absent) from the plate-supply voltage. The drop across the load resistor (E_{RL}) is the product of the quiescent (no-signal) value of the plate current (I_{bo}) and the load resistance (R_L), or $I_{bo} \times R_L = 0.0064 \times 25{,}000 = 160$ volts. The quiescent value of the plate voltage, therefore, is

$$E_{bo} = E_{bb} - E_{RL} = 350 - 160 = 190 \text{ volts}$$

Now when the plate current rises to its maximum value of 10.1 ma at full signal (and develops an output voltage across the load of about 92.5 volts), the plate voltage is seen to fall in direct proportion to the rise in plate current. It reaches a **minimum** value of 97 volts, when the input signal, grid voltage, and plate current reach their **maximum** values. The drop in total plate voltage from 190 to 97 volts (i.e., 93 volts) is seen to be just about equal to the peak value of the load voltage (92.5 volts). (Theoretically, it is exactly equal, of course.) Since the plate voltage reaches its negative peak when the input signal reaches its positive peak, and vice versa,

the two waveforms are one-half cycle, or 180 degrees, out of phase with respect to each other. This is the reason why an electron tube is said to introduce a 180-degree phase reversal between input and output voltage. Summing up the relations shown in Fig. 58, we see that the plate current is **in phase** with the grid voltage, but the **plate** voltage is **180 degrees out of phase** with the grid voltage.

SUPPLY VOLTAGE SOURCES

Fig. 56 shows for convenience batteries as sources of all tube operating potentials. In practice, other supplies are usually employed for the d-c operating voltages. The plate-supply voltage (E_{bb}) and fixed grid bias (E_{cc}) are most frequently secured from a rectifier power supply, with which we shall become acquainted in a later chapter. A low-voltage winding on the power transformer of this supply provides the heater voltage for the tubes of the amplifier, since a.c. is satisfactory to heat the filaments. A fixed voltage is not always used for the grid bias, but the tube sometimes supplies its own bias. This method, known as **self-bias,** makes use of the insertion of appropriate resistors either in the cathode or grid circuit of the tube. Plate current flowing through the cathode resistor, or sometimes grid current flowing through a grid resistor, will then develop the necessary voltage drop to bias the grid of the tube.

AMPLIFIER COUPLING METHODS

The output voltage from a single amplifier stage may not be sufficient to be applied directly to an output device, such as a loudspeaker. Additional amplification over two or three stages is usually necessary. To accomplish this the output voltage of each amplifier stage must be coupled in some way to the grid of the succeeding amplifier tube. Four coupling methods are in general use. These are **(1) resistance-coupling, (2) impedance-coupling, (3) transformer-coupling, and (4) direct (d-c) coupling.** Since only the varying or a-c component of the tube's plate voltage is needed to couple the output signal to the next stage, the d-c plate voltage is generally prevented from reaching the grid of the following tube. This is true for all but method (4) listed above.

RESISTANCE COUPLING

Resistance coupling is the most popular of all coupling methods because it is cheap and provides excellent audio fidelity over a wide frequency band. Typical resistance-coupled triode and pentode amplifier circuits are illustrated in (a) and (b), respectively, of Fig. 59. The voltage supplies for the heaters of the tubes are not shown, but it is understood that they are there.

Note that the first stage of the triode circuit (Fig. 59a) is the same as the basic amplifier circuit of Fig.

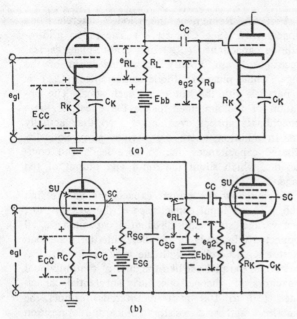

Fig. 59. *Resistance-Coupled Amplifier with Self-Bias*
(a) Triode Circuit, (b) Pentode Circuit

56, except that the bias battery (E_{cc}) has been replaced by a self-bias resistor, R_k, in the cathode circuit of the tube. The cathode-resistor is shunted by a so-called bypass capacitor, C_k. You can easily see that this device enables the tube to develop its own bias. When plate current flows from the negative terminal of the plate-supply battery (E_{bb}), through resistor R_k, and from cathode to plate of the tube, the bottom end of bias resistor R_k becomes **negative** and the top (or cathode) end becomes positive, since electrons flow from minus to plus. Since the bottom end of R_k is connected to the control grid in series with signal voltage e_g, the grid becomes **negative** with respect to the cathode, thus furnishing the necessary bias voltage. This method of biasing is to some extent self-regulating. If the d-c plate current should increase for some reason (such as a fluctuation in the plate-supply voltage), the negative bias developed across R_k will also increase, which, in turn, tends to reduce the plate current again. If the d-c plate current should decrease, on the other hand, the bias voltage will also decrease, which will tend to increase the plate current again. Self-bias, thus, assures stable amplifier operation.

Although we have shown, for convenience, a battery for the plate-supply voltage E_{bb}, the d-c plate voltage is usually obtained from the positive output of a rectifier power supply, as was previously mentioned. The a-c component of the first tube's output voltage, e_{RL}, is coupled to the grid of the fol-

lowing stage through a coupling capacitor, C_c. The coupling capacitor prevents the d-c plate voltage of the first tube from reaching the grid of the second tube and, thus, overloading it. (Remember that a positive grid voltage will result in grid current and distortion.) The input signal to the second stage, e_{g2}, is developed across a grid resistor, R_g, which also acts as a grid return for the bias voltage developed across R_k in the cathode circuit.

You may wonder about the function of bypass capacitor C_k across bias resistor R_k. Think for a moment what would happen if it was not there. The plate current flowing through R_k would then develop an output signal across it, just as across load resistor R_L. This output signal is in **phase opposition** to the grid input signal, as we have seen before, and since the voltage across R_k is in series with the input signal, e_g, the two signals would tend to partially cancel themselves out, a process known as **degeneration.** The total input signal to the grid of the first stage would then be considerably reduced, and so would be the available voltage amplification. By placing capacitor C_k across the bias resistor, no a-c signal voltage is developed across it, since the capacitor is large enough to present practically a short circuit to a-c. The a-c voltage across R_k, thus, is **bypassed** to ground and cannot cancel the input signal at the grid. Care must be taken, however, that capacitor C_k is sufficiently large (i.e., has a low **reactance** in relation to the bias resistance) to offer a short circuit to the **lowest** frequency present in the a-c output signal. A value of about 25 microfarads is usually needed. Sometimes the capacitor is purposely left off to obtain degeneration of the input signal, which also has some advantages, as we shall see later on. (For one thing, degeneration is the same for all frequencies.)

Pentode Circuit. The pentode circuit shown in Fig. 59b is essentially the same as the triode amplifier, except for the method of supplying the voltages to the additional electrodes. The suppressor grid (Su) of the pentode is generally operated at cathode potential and, hence, is simply connected to the cathode. The screen grid (Sc), however, requires a positive operating potential, which is generally less than that of the plate. For convenience, a separate screen-supply battery, E_{SG}, is shown in the figure, which supplies the screen with the proper positive operating voltage through screen-dropping resistor R_{SG}. In practice, the screen and plate are usually supplied from the same source, either battery or power supply, and the correct voltages are obtained by choosing suitable values for the screen and plate dropping resistors, R_{SG} and R_L. The screen bypass capacitor, C_{SG}, serves to bypass the a-c signal of the voltage developed across R_{SG}, similarly as does the cathode bypass capacitor C_k.

DESIGN CONSIDERATIONS
AND FREQUENCY RESPONSE

One good way of designing a resistance-coupled audio amplifier is to choose a tube type likely to give the desired voltage amplification and look up the values of all circuit components and voltages in the manufacturer's tube manual. Design and performance data are generally listed for a variety of typical operating conditions and they are backed by the manufacturer's long years of experience. The theoretical design and prediction of amplifier performance is a somewhat complicated mathematical procedure into which we cannot here go, but it may be of interest to point out some of the factors that affect the design and performance of a resistance-coupled amplifier stage. These considerations are equally true for the triode and pentode circuits, although there are some differences between the two in performance.

The easiest part is determining the bias resistor R_k. From the tube's characteristics (in the tube manual) we know the quiescent plate current that flows for the available plate voltage and the desired grid bias voltage. By Ohm's law, the bias resistance R_k is then simply the quotient of the grid bias divided by the d-c plate current—(that is, $R_k = E_{cc}/I_b$, at the operating point). The bypass capacitor C_k should have a value of 20 to 25 microfarads to assure effective bypassing at the lowest audible frequencies, as we have mentioned before.

The choice of the plate-load resistor R_L is somewhat more complex. Three main factors determine it—namely, the available plate-supply voltage (E_{bb}), the desired voltage gain and the frequency response. These factors conflict with each other to some extent and a reasonable compromise must be made in each case. As we have seen before, the tube's output voltage and, hence, the amplification (gain) increase in direct proportion with the value of the load resistor R_L. We have also seen that the d-c drop across the load resistor *subtracts* from the available plate-supply voltage. If the plate-load resistor is too large, the voltage available at the plate of the tube (E_b) becomes too low for proper operation of the circuit. It then becomes necessary either to reduce the value of the load resistor or to increase the plate-supply voltage. Since the latter is generally fixed, the **maximum** value of the plate-load resistor is also determined.

It is not immediately apparent why the value of the load resistor should have any effect on the frequency response of the amplifier (i.e., the relative amplification throughout the audio frequency range). We know that inductors and capacitors are frequency-sensitive but resistors are not supposed to be. Well, they are not. But there is more to the circuit of the audio amplifier than we have shown in Fig. 59. Remember the triode's interelectrode capacitances? (See Chapter 4.) The tube's grid-to-cathode capacitance (C_{gk}) and grid-to-plate capacitance (C_{gp}) are directly across the input of the stage, while plate-to-cathode capacitance (C_{pk}) is in parallel with the output of each stage. The operation of the amplifier makes the effect of these capacitances appear even larger than they actually are. In addition, there are various **stray** and **distributed** capacitances due to wire leads and components, which shunt the input and output of the stage.

Imagine now the output capacitance of the first tube, *plus* the input capacitance of the second tube, plus various stray and distributed capacitances, all connected effectively in **parallel** with the plate-load resistor R_L, as well as with the grid resistor R_g. Capacitors in parallel act like the sum of the individual capacitors and, hence, one fairly substantial capacitance (50 to 100 micromicrofarads) shunts the plate-load and grid resistors. Since the opposition to the flow of alternating current (reactance) of a capacitor *decreases* as the frequency *goes up*, the shunting effect of these various capacitances will make itself felt at the **higher** audio frequencies and will effectively cut off all frequencies above a certain limiting frequency. The frequency at which this cutoff happens depends on the plate-load resistor R_L and to some extent also on the grid resistor R_g. If the plate-load resistor has a relatively low value, most of the a-c output signal will be developed across it and relatively little will flow through the parallel capacitance. The frequency at which capacitive shunting becomes excessive will then be high and possibly beyond the audible range. If we attempt to make the load resistor too large, however, we will force a substantial part of the a-c output signal to flow through the relatively low shunt reactance of the capacitor and this shunting effect will *increase* with frequency (since the capacitor's reactance decreases with frequency). The cutoff frequency beyond which the amplifier gain becomes insufficient will then be relatively low.

The desired frequency response of the amplifier is perhaps the most important consideration in limiting the values of the load resistor and grid resistor. Here again a compromise must be made. If we desire large voltage amplification, large values of plate-load and grid resistors are necessary (about 500,000 ohms in practice), but then the frequency response at the higher audio frequencies will be defective. On the other hand, if we desire a "flat" frequency response curve over the audible range (see Fig. 60), the plate-load and grid resistor must be made *small* at the expense of the available voltage amplification. The grid resistor (R_g), however, must always be sufficiently large not to shunt the plate-load resistor appreciably. Fig. 60 shows some typical frequency response curves of a resistance-

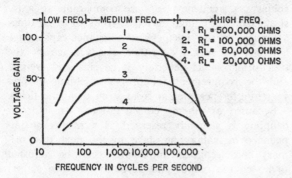

Fig. 60. *Frequency Response of Resistance-Coupled Amplifier for Various Values of the Load Resistence R_L*

coupled audio amplifier for various values of the load resistance. Note in Fig. 60 that the voltage gain of the resistance-coupled amplifier also drops off at low frequencies, below approximately 100 cycles. Again some capacitors are to blame. This time it is primarily the effect of the coupling capacitor C_c, but also to some extent the combined effect of the cathode and screen bypassing capacitors, C_k and C_{SG} (if present). We have already seen that the bypass capacitors must have a value sufficiently large so as not to cause any degeneration and resultant loss of gain at the lowest audio frequency we wish to hear. The effect of the coupling capacitor (C_c) is the converse of the tube's input and output capacitances, since the coupling capacitor is in **series** with the input voltage to the second stage. This being so, a part of the signal voltage of the second stage is wasted across the coupling capacitor, while the remainder is developed across grid resistor R_g. Since the reactance of the capacitor *decreases* with frequency, the voltage developed across it (which is proportional to the reactance) is almost always negligible in comparison to that across R_g at medium frequencies, above about 200 cycles. At low frequencies, however, the reactance of C_c may become appreciable and cause a dropping off in voltage gain. The value of the coupling capacitor must, therefore, be made large enough so that it has a negligible reactance at the lowest audio frequency to be reproduced. This generally requires values of about 0.01 to 0.1 microfarad.

This leaves only the design of the screen resistor R_{SG} in a pentode circuit. It is simply made sufficiently large so that the voltage drop developed across it, when subtracted from the screen-supply voltage (E_{SG}), will leave the correct voltage on the screen. For example, if a d-c supply voltage of 300 volts is available for both plate and screen, and the screen voltage should be 100 volts for proper operation at 0.5 ma screen current, a voltage drop of

200 volts (i.e., $300 - 100$) is required to be developed across screen resistor R_{SG}. The value of the screen resistor, thus, is 200 volts/0.0005 amps, or **400,000 ohms.** Since it shunts a large resistance, the value of the screen bypass capacitor C_{SG} can be relatively small, and is generally in the order of 0.1 to 0.5 microfarad.

IMPEDANCE COUPLING

Fig. 61 illustrates an impedance-coupled amplifier, a coupling method which is rarely used. Since only the coupling method is of interest, batteries have been shown as sources for all tube operating potentials, although a self-bias resistor for the grid bias and rectifier power supply for the plate voltage are actually used in practice.

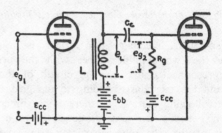

Fig. 61. *Impedance-Coupled Triode Amplifier*

As is evident from Fig. 61, impedance coupling is obtained by substituting an inductor coil, *L*, for the plate-load resistor in the resistance-coupled amplifier of Fig. 59. The circuit derives its name from the **impedance** of the coil, which you will remember is made up of the coil resistance and the **inductive reactance** ($2\pi fL$), the latter **increasing directly with frequency.** Since the d-c resistance of the inductor coil (L) is low and the inductive reactance (opposition to a.c.) is relatively high, a large value of the a-c load impedance can be obtained at the output of the first stage without an excessive d-c voltage drop, as occurs with resistance-coupling. Because of the low d-c voltage drop, the tube can be operated at higher plate voltages than a resistance-coupled stage and, hence, a greater voltage amplification can be obtained. The price paid for this increase in gain is a non-uniform frequency response. Since the inductive reactance of the coil and hence the load impedance goes up with frequency, both the output voltage (e_L) and amplification will rise with increasing frequencies. The frequency response of an impedance-coupled stage, therefore, shows a slowly rising amplification with increasing frequencies; or equivalently, a marked falling off in response at lower frequencies. At very high frequencies the amplification begins to drop off because of the tube's and coil distributed capacitances, as in the case of the resistance-coupled circuit. In other respects the impedance-coupled circuit is the same as the resistance-coupled type.

TRANSFORMER COUPLING

The transformer-coupling method, which used to be more popular than it now is, is shown in Fig. 62.

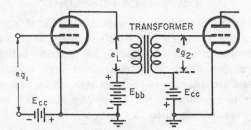

Fig. 62. *Transformer-Coupled Triode Amplifier*

A so-called **interstage transformer** is used to couple the two amplifier tubes with the primary winding connected in the plate circuit of the first tube and the secondary winding to the grid circuit of the following tube. Since there is no direct connection between the two windings, the d-c plate voltage of the first tube is isolated from the grid circuit of the second tube. As in impedance coupling, the primary reactance of the transformer has a high inductive reactance and a low d-c resistance, and thus wastes little of the d-c supply voltage. This permits operation at high plate voltages, resulting in large amplification. In addition, transformer coupling has the advantage that the output voltage (e_L) of the first stage may be stepped up by the **ratio** of the secondary turns to the primary turns. This ratio (symbolized N) may in practice be about 3 : 1. If the primary reactance of the transformer is large, to obtain a large output voltage (e_L), the amplification of a transformer-coupled stage may approach the *product* of the tube's amplification factor and the secondary-to-primary turns ratio (i.e., $A \cong \mu \times N$).

The frequency response of a transformer-coupled circuit is generally rather poor. At low frequencies the primary reactance of the transformer (which is proportional to frequency) begins to fall off, resulting in decreased gain below a certain limiting frequency. At high frequencies the distributed shunting capacitances of the windings and the tube capacitances go into resonance with the transformer secondary reactance, which results in a pronounced peak or rise in the frequency response at the resonant frequency. Above and below this resonant frequency, the response falls off rapidly. Nevertheless, in a properly designed, expensive transformer, the various factors can be so balanced as to achieve a fairly uniform response over the

audio-frequency range. But you had better underline the word *expensive,* since a transformer that achieves a frequency response comparable to a resistance-coupled stage may cost from 10 to 20 times as much as the inexpensive resistance-capacitance coupling network. A further disadvantage of the transformer is its generally large size and weight, and the requirement for proper shielding to prevent interference pickup from stray magnetic fields.

Because of these considerations, transformer coupling is generally reserved as a means of **impedance matching** one stage to the next (the primary and secondary impedances can be adjusted to match the output and input of any two stages) and for **phase inversion** to drive a push-pull power amplifier, which we shall become acquainted with in Chapter 10. Transformer coupling is also employed in tuned radio-frequency and i-f amplifiers, which will be discussed later.

DIRECT COUPLING

In a direct-coupled amplifier the plate of one tube is connected directly to the grid of the next tube without any intervening capacitor, transformer, or other coupling device. (See Fig. 63.) As a result both the d-c plate voltage and the alternating component of the first tube's output signal pass on to the grid of the next stage. As you can imagine this causes some complications. Since the plate of the first tube must have a positive voltage with respect to its cathode and the grid of the next tube must have a negative voltage with respect to its cathode, a special voltage divider must be used to obtain the voltages for proper circuit operation. Fig. 63 shows a simple direct-coupled amplifier, called the **Loftin-White circuit,** in which this neat trick is successfully accomplished.

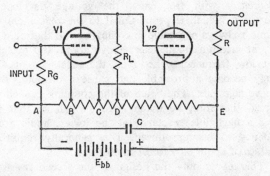

Fig. 63. *Direct-Coupled Loftin-White Amplifier*

The proper voltage distribution is obtained through voltage divider A-B-C-D-E, which is connected across the d-c supply (E_{bb}). The most positive point of the supply (E) is connected to the plate of the second tube through plate-load resistor R. The plate of the first tube obtains its d-c operating voltage from a somewhat less positive point, *D,* on the divider through plate-load resistor R_L. The output voltage of the first tube is directly coupled to the grid of the second tube so that the latter is at plate-voltage potential. However, by connecting the cathode of tube 2 (V2) to a potential **more positive** than the plate of V1 and the grid of V2 (at point *C* of the divider), the grid of V2 is actually **negative** with respect to its cathode. Although it appears that the cathode of V2 is connected to a point (*C*) less positive than the plate of V1, this is not actually the case, since the plate voltage of the first tube is diminished by the large voltage drop across R_L. Finally, the cathode of V1 is connected to a point (*B*) on the voltage divider that is more positive than the grid of V1, which is connected to the negative terminal of the supply at point *A*. This assures the necessary grid bias for stage 1.

Although the voltage divider network looks deceptively simple, it is quite difficult to design and adjust, since the various tube currents flowing through it must be taken into account. When the tube voltages are adjusted properly, the circuit serves as a distortionless amplifier with a uniform frequency response over a wide range of frequencies, from d.c. to considerably above the audio range. The direct-coupled amplifier is especially useful for amplifying very slow variations in the input voltage, since the impedance of the coupling elements does not vary with frequency. Furthermore, since its response is practically instantaneous (because of the absence of time constants), the d-c amplifier is also useful for amplifying **pulse** signals.

TRANSISTOR AMPLIFIERS

For variety let us look at a couple of transistorized audio amplifiers, which undoubtedly will soon replace electron tube amplifiers, as transistors become further improved. Fig. 64 shows a two-stage resistance-coupled amplifier (a) and a transformer-coupled circuit (b), which correspond to the circuits shown in Figs. 59 and 62, respectively. The outstanding advantages of these circuits are, of course, that no heater voltage supply is required and that a small (6 to 12 volt), low-current d-c source takes care of all transistor current needs, thus making battery operation quite economical.

The two-stage resistance-coupled amplifier (Fig. 64a) makes use of two 2N190 (PNP-type) transistors in a common-emitter connection, which is quite similar to the conventional triode-tube amplifier circuit. The main difference between this circuit and the resistance-coupled tube amplifier shown in

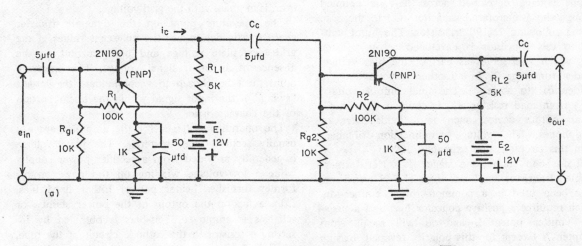

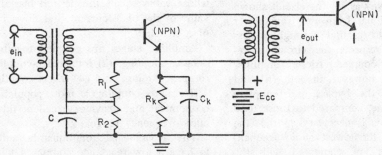

Fig. 64. *Transistor Amplifiers (a) Two-Stage Resistance-Coupled Circuit, (b) Transformer-Coupled Circuit*

Fig. 59a is the method used to obtain correct bias-ing for the emitter and collector electrodes. You will recall from Chapter 7 (Figs. 43 and 49) that in a P-N-P transistor the emitter junction must be for-ward (positively) biased, while the collector junc-tion is reverse (negatively) biased. In the circuit of Fig. 64a the necessary bias voltage and polarity is obtained from a single voltage source in a rather interesting manner. Typical values are shown to illustrate the operation. A voltage divider, consist-ing of resistors R_1 and R_{g1} (in stage 1), is connected across the battery and the junction of the two re-sistors is connected to the base of the transistor. Because of the ratio of the two resistors (10K to 100K, or 1 : 10), the base is only about 1.1 volts **negative** with respect to ground and the emitter, or conversely, the emitter is about 1.1 volts **positive** with respect to the base and, thus, is furnished the required forward bias. The current in the emitter circuit is then essentially the bias voltage divided by the 1000-ohm emitter resistor, or about $1.1/1000 = 1.1$ ma. To prevent degeneration of the a-c input signal, the emitter resistor is bypassed with a large (50-microfarad) capacitor.

The required **negative** collector voltage is ob-tained through load resistor R_{L1}, which is connected directly to the negative terminal of the battery. The output voltage developed across R_{L1} is coupled through the 5-microfarad capacitor (C_c) to the base of the following 2N190 transistor. The input volt-age to this transistor is developed across resistor R_{g2}, which also serves as part of the bias voltage divider for stage 2. In all other respects stage 2 is identical to stage 1. The final output voltage (e_{out}) from load resistor R_{L2} is coupled through C_c to an output device, such as a loudspeaker or headphones. The voltage amplification obtained from this circuit is approximately 5000.

Figure 64b shows one stage of a transformer-coupled transistor amplifier. Here N-P-N transistors have been used in a common-emitter connection, which requires a **positive** collector bias and a **nega-tive** emitter potential, as you will recall from Chapter 7. Except for this polarity reversal, biasing is obtained in the same way as for the circuit shown in (a). The a-c input signal from the transformer secondary is returned to ground through bypass capacitor C. In other respects the circuit corre-sponds to transformer-coupled triode amplifier shown in Fig. 62. Note, however, in both (a) and (b) that the values of the bypass and coupling capacitors are rather large, compared to those used with tube amplifiers. This is necessary so that the capacitive reactances at the lowest audio frequen-cies of interest will be low compared with the relatively small associated resistors and transistor impedances.

SUMMARY

In a voltage amplifier, the amplification or **gain** is the **ratio** of the output voltage across the load resistor to the input voltage on the grid.

Audio amplifiers amplify electrical a-c signals that have a frequency range corresponding to the range of human hearing, from about 20 to 15,000 cycles per second.

Voltage amplifiers are always operated so as not to consume any power in the grid circuit. To avoid grid current and the resulting power consumption and distortion, a negative bias voltage equal to or greater than the **positive** peak of the a-c input signal is applied to the grid. The total grid voltage, thus, is equal to the difference between the input signal and the bias. For distortionless amplification, the bias is chosen to maintain the flow of plate current during the entire a-c input cycle.

The amount of voltage amplification depends primarily on the **amplification factor** and the value of the **plate-load resistor.**

The total instantaneous plate voltage is the dif-ference between the d-c plate-supply voltage and the a-c signal voltage across the load.

The input signal at the grid and the output signal or plate voltage are **180 degrees out of phase** (i.e., in phase opposition). The plate current, how-ever, is **in phase** with the grid voltage.

The **operating point on the dynamic transfer characteristic** determines the **quiescent values** of the grid bias, plate voltage and plate current, in the absence of an input signal voltage. The operating point must be chosen to accommodate the largest "swing" of the input signal within the linear portion of the characteristic.

The operating voltages for the amplifier are *not* usually secured from batteries. The plate voltage is generally provided from a rectifier power supply, while a low-voltage winding on the power trans-former furnishes heater power. Either fixed bias, from a tap on the output of the power supply, or self-bias is employed. Self-bias is obtained by in-serting a resistor in the cathode circuit of the tube, whose value equals the desired bias divided by the value of the plate current that flows in the absence of a signal.

Amplifier stages are coupled together by (1) re-sistance-coupling, (2) impedance-coupling, (3) trans-former-coupling, and (4) direct coupling.

Resistance-coupling is most popular because it is economical and provides excellent fidelity over the audio-frequency range.

The **bypass capacitors** must be sufficiently large to have a low reactance compared with the associ-ated resistances at the **lowest** audio frequency of interest.

The **high-frequency response** of an R-C amplifier is limited by the tube's input and output (interelectrode) capacitances and by the stray wiring and distributed capacitances.

The **low-frequency response** of an R-C amplifier is limited by the coupling capacitance and also by the value of the bypass capacitors.

The choice of the plate-load and grid resistors in an R-C amplifier is a compromise between **large voltage gain** (for large resistors) and **flat high-frequency response** (for small resistors).

Impedance and transformer coupling permit an **increase in gain** at the price of uniform frequency response. Transformer coupling produces a **resonant peak** in the medium-frequency range. The frequency response falls off above and below this peak. Transformer coupling is used primarily for **impedance matching** the output of one stage to the input of the next and to provide **phase inversion** for a push-pull power amplifier. It is also used in tuned r-f and i-f amplifiers.

In a **direct-coupled** amplifier the plate of one tube is connected directly to the grid of the next without intervening capacitors, transformers, or other coupling devices. Consequently, both the d-c plate voltage and the a-c component of the output voltage pass on to the grid of the next stage.

Direct-coupled amplifiers are useful for amplifying **slow variations** in the input voltage and for amplifying **pulse signals.**

Chapter Nine

WIDEBAND (VIDEO) AMPLIFIERS

We have already mentioned that some audio amplifiers reproduce frequencies far beyond the audible range. This extended response becomes of great importance in the so-called **wideband** or **video** amplifiers. A wideband amplifier must have a frequency response from a few cycles per second to several megacycles (1 megacycle = 1,000,000 cycles), in order to be called wideband. (See Fig. 65.) The term **video** amplifier originated from the fact that wideband amplifiers were first developed to handle the picture (video) signals of television systems, which require an almost "flat" frequency response from about 60 cycles per second to 4 megacycles per second. At the present time, however, video amplifiers are used in a variety of applications besides television, such as radar systems and other devices employing **pulse-type** signals.

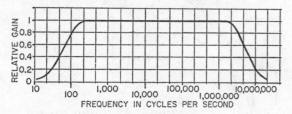

Fig. 65. *Frequency Response of Typical Video Amplifier*

VIDEO AMPLIFIER PULSE RESPONSE

We have already referred to the utility of direct-coupled and wideband amplifiers for **pulse-type** signals. Let us for the moment explore the connection (if any) between pulse signals and amplifier frequency response.

Thus far we have been dealing exclusively with **sinewaves,** which are **gradually changing** alternating currents or voltages, consisting of a positive and negative alternation, each. Two such alternations make up one cycle and the **frequency** of the voltage or current refers to how many such cycles are completed each second. In contrast, a **pulse is an abruptly changing voltage or current,** which may or may not repeat itself. The simplest non-repetitive pulse is the **step voltage** (or current) shown in Fig. 66a. Such a step voltage may be obtained, for example, by connecting a voltmeter across a battery through a switch and then suddenly closing the switch. Thus, the voltmeter would read zero voltage up to time t_1, when the switch is closed, whereupon the voltage would suddenly rise to its maximum value and stay there. But even the most sensitive voltmeter would take a certain amount of time to indicate the maximum value of the battery voltage, while the *ideal* voltage step takes *no time at all* to reach its maximum. This sudden rise to maximum amplitude is characteristic of all square or rectangular-shaped pulses. In practice it is somewhat easier to obtain a square or rectangular pulse waveform, as illustrated in Fig. 66b (at left), than a step voltage. By applying such a square (or rectangular) waveform to the input of an amplifier and repeating it a number of times per second, the response of the amplifier to the pulse waveform may be conveniently studied.

The pulses shown in Fig. 66 are, clearly, not

sinewaves but rather **non-sinusoidal** waveforms. A great French mathematician by the name of Fourier demonstrated long before the advent of video amplifiers, that an abruptly changing non-sinusoidal waveform may be made up mathematically of a series of sinewaves of varying amplitudes and harmonically related frequencies. (A **harmonic** relation may consist of frequencies in the ratio of 1, 2, 3, 4, . . . etc., 2, 4, 6, 8, 10, . . . etc., or 1, 3, 5, 7, 9, . . . etc.) As a matter of fact, an **infinite number** of sinewave frequencies are needed to make up or reproduce the abruptly changing pulses, shown in Fig. 66. If only a finite number of sinewave frequencies are available to compose the pulses, either mathematically or in an amplifier, the pulse waveforms will be distorted in various ways and the distortion will become more serious the fewer frequencies are available. This, then, is the startling relation between amplifier frequency response to sinewaves and the reproduction of pulse waveforms: **The greater the frequency range of the amplifier (called bandwidth), the smaller is the distortion of pulse waveforms.**

Now it is a very laborious procedure to determine the frequency response of an amplifier, by plotting it step by step. In contrast, it is very simple to observe the response of an amplifier (on the cathode-ray tube oscilloscope) to square or rectangular pulses applied to its input. Since the relative distortion of the pulses, as reproduced by the amplifier, tells a great deal about its frequency response, the second method is much preferred. Moreover, since most video amplifiers are used for the amplification of complex pulse waveforms (such as in radar and television) rather than sinewaves, the square-wave method of amplifier testing and design directly gives the amplifier response to abruptly changing waveforms (pulses).

Fig. 66b shows some of the things that can happen, when an *ideal* square wave (good approximations of which are easily obtained) is applied to the input of an amplifier. The distorted waveform present in the output of the average amplifier is shown at right of Fig. 66b. The principal changes introduced by the amplifier are: (1) the amplitude of the pulse rises and falls at a **finite** rate, rather than instantaneously; (2) the amplitude of the reproduced pulse initially **overshoots** the correct value, and when falling, it will **undershoot** the minimum value; (3) the top of the reproduced square wave falls off or sags with time, instead of being flat. Any or all of these defects may be present simultaneously in the amplifier output.

The rate at which the output voltage rises when

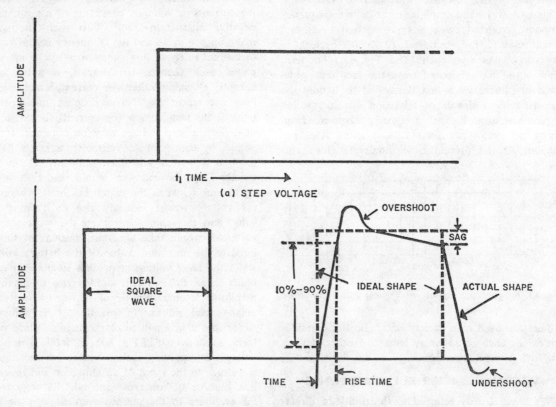

(b) IDEAL SQUARE WAVE & TYPICAL AMPLIFIER RESPONSE

Fig. 66. (a) *Step Voltage,* (b) *Ideal Square Wave and Typical Amplifier Response*

an input pulse is suddenly applied, called the **rise time,** represents the amplifier's capacity to reproduce **sudden** changes in amplitude of the applied signal. The initial response of an amplifier to the sudden rise of a pulse is termed the **transient response** and, as we have seen, it is related to the sinewave frequency or **bandwidth** of the amplifier. The rise time of a pulse reproduced by the amplifier is usually measured by the time it takes the output voltage to rise from **10 percent** to **90 percent** of its maximum value (not considering the overshoot). As thus defined, the rise time is inversely proportional to the bandwidth of the amplifier and it is approximately given, for **zero** or **small overshoots,** by the relation

$$\text{Rise time in seconds} = \frac{0.35}{\text{Bandwidth}}$$

The bandwidth in the relation just stated is defined as the highest frequency, in **cycles per second,** for which the overall response of the amplifier does not fall below **70.7 percent** of its maximum response at medium frequencies. Of course, the above relation can be worked the other way; the bandwidth in cps can be computed by **measuring** the 10–90% rise time of a square-wave or rectangular pulse applied to the amplifier and dividing the result into 0.35 (i.e., bandwidth = 0.35/rise time).

The amount of **overshoot** present in the amplifier output depends on the sharpness with which the high-frequency response of the amplifier falls off and also on the amount of phase shift introduced by the various capacitive and inductive elements. A response curve that falls off only moderately at the high-frequency end, together with small phase shift, produces only a small or zero overshoot. When overshoot is present, however, undershoot is also there.

Finally, the amount of falling off of the top of the reproduced pulse, or **sag,** depends on the phase shift and frequency response of the amplifier at the low-frequency end. If the low-frequency response is good, sag will be small, resulting in the desired **flat top.**

HIGH-FREQUENCY COMPENSATION

The resistance-coupled amplifier described in the last chapter represents a fair attempt at a wideband amplifier, provided the plate-load resistor (R_L) of each stage is made small enough to achieve the desired wide frequency response. You will recall that by sacrificing voltage gain and using low plate-load and grid-coupling resistors, a frequency response can be achieved that is uniform to frequencies far beyond the audio response range. The amplification at medium frequencies of such an uncompensated resistance-coupled amplifier is approximately equal to the product of the tube's transconductance (in mhos)

and the plate-load resistance. Using, for example, a type 6AK5 miniature tube, which has a transconductance of 5100 micromhos, and making the plate-load resistor (R_L) of the amplifier equal to 10,000 ohms, we obtain for the maximum mid-frequency amplification = $g_m R_L$ = .0051 × 10,000 = 51.

Such an arrangement has a **rise time** in microseconds equal to approximately 2.2 times the product of the plate-load resistance and the combined input and output capacitance (in μf) of the tube. The manufacturer lists an input capacitance for the 6AK5 of 4 $\mu\mu$f and an output capacitance of 2.8 $\mu\mu$f, making a total of 6.8 $\mu\mu$f.

Thus, for our present example,

the rise time = $2.2 R_L C_{Tot}$ = 2.2 × 10,000 × (6.8 × 10^{-6} μf) = **0.15 microsecond**

Consequently, from our previous relation, the

$$\text{Bandwidth} = \frac{0.35}{\text{rise time (sec)}} = \frac{0.35}{0.15 \times 10^{-6}}$$
$$= 2.33 \times 10^6 \text{ cps,}$$

or **2.33 megacycles,** which is rather good for an R-C amplifier.

While an uncompensated resistance-coupled amplifier provides a fair bandwidth and rise time without overshoot, its characteristics fall short of the requirements of television and pulse systems. As we have seen, the resistance-coupled amplifier is beset with deficiencies at both ends of the frequency band. At high frequencies, the tube's combined input and output capacitances together with stray capacitances shunt the plate-load resistor, thus limiting the amplification at these frequencies. At low frequencies, the reactance of the coupling capacitor becomes large compared to the grid input resistor, which again limits the gain and leads to a falling off in response at low frequencies. Equivalently, the R-C amplifier reproduces pulses with a relatively long rise time and with a pronounced sag at the top of the waveform.

SHUNT COMPENSATION

It is possible to improve the high-frequency response of a resistance-coupled amplifier markedly by adding a small inductance L_1 in series with the plate-load resistor R_L. (See Fig. 67a.) This arrangement is known as shunt compensation or **shunt peaking,** because the increased reactance of the shunt coil L_1 at high frequencies serves to peak the response at these frequencies to compensate for the falling off in response due to the total shunt capacitance (C_{Tot}). Shunt peaking improves the high-frequency response and the rise time, provided the bandwidth demands are not too great and there are only a few stages of amplification. The amount of peaking introduced by the coil depends on a quality or "Q" factor, which is defined as

$$\text{shunt-peaking "Q"} = 2\pi fL_1/R_L = 6.28fL_1/R_L$$

where L_1 is the shunt-peaking inductance
R_L is the plate-load resistance

and f is the frequency at which the response of the **uncompensated** amplifier (i.e., **without the**

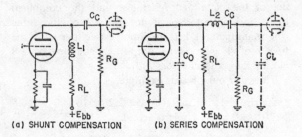

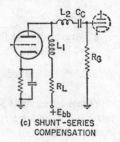

Fig. 67. *High-Frequency Compensation*

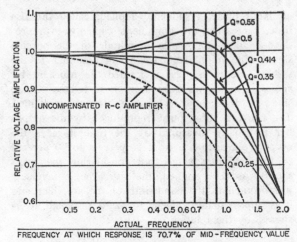

Fig. 68. *Relative Frequency Response of Shunt-Peaking Amplifier for Various Value of "Q"*

coil) has dropped to 70.7 percent of the mid-frequency gain.

Figure 68 shows a universal response curve for a shunt-peaked amplifier, which graphically illustrates the dependence of the frequency response on the "Q"-factor, as defined above. The vertical axis shows the relative voltage amplification, while the horizontal axis gives the ratio of the actual frequency to the frequency at which the amplification is down to 70.7% of the mid-frequency value, when no compensation is used.

For comparison, the case of the **uncompensated** R-C amplifier has been dashed in. By definition, its response is down to 70.7% of the mid-frequency value, when the frequency ratio is 1. You can see how the response at that frequency is strikingly lifted as the "Q" of the shunt-peaking coil in a compensated amplifier is increased, so that for a Q of 0.55, the response is even greater than at the mid-frequency. This large amount of peaking is, however, not desirable, since it, too, represents a distortion of the frequency response and, also, because it leads to large overshoots in the reproduction of pulse waveforms. The curve for Q = 0.5 gives nearly constant amplification up to the compensated fre-

quency (f), while the curve for Q = 0.414 is of **maximum flatness,** having no peak at all. In the curves, for "Q" equal to 0.35 and 0.25, the response drops off moderately at high frequencies, but these are characterized by small or zero overshoot in the pulse rise and excellent phase-shift characteristics. Even for Q = 0.25 the speed of rise is 1.4 times as great as that of the uncompensated amplifier, while for Q = 0.414, it is 1.7 times as great.

SERIES COMPENSATION

Another method of boosting the high-frequency response of a resistance-coupled amplifier, shown in Fig. 67b, is known as **series compensation** or **series peaking.** In this case a small inductance coil, L_2, is connected in series with the coupling capacitor (C_c). At high frequencies the coil L_2 resonates with the input capacitance (C_i) of the next stage, thus causing an increased voltage across C_i. Since both C_i and grid resistor R_g are across the input of the next tube, the input voltage and, hence, the gain is boosted.

Fig. 67c shows a shunt-series compensation circuit that combines the high-frequency peaking of shunt coil L_1 with the resonant effect of series-peaking coil L_2. The high-frequency response of an amplifier can be further improved by so-called **four-terminal compensating networks,** which use additional elements besides series and shunt-peaking coils.

LOW-FREQUENCY COMPENSATION

We have seen that the low-frequency response of an uncompensated resistance-coupled amplifier falls off because of the increasing reactance of the coupling, grid and screen bypass capacitors at low frequencies. These low-frequency deficiencies make themselves felt in the inability of the amplifier to reproduce the flat top of a square or rectangular wave without a pronounced sag. (Fig. 66b.) The amount of sag that is permissible depends on the application of the amplifier. For example, in television video amplifiers the sag is not allowed to exceed *5 percent* (of maximum amplitude) at 60 cycles, the time of one picture field being 1/60 second.

Fortunately, the loss of gain at low frequencies and the resulting sag can be compensated for fairly easily by the addition of a low-frequency compensating filter in series with the plate-load resistor, as shown in Fig. 69. The filter, consisting of R_F and C_F, has a double effect. First, it introduces a phase shift in the plate circuit that compensates for the phase shift in the coupling circuit R_gC_c. Second, by increasing the plate-load impedance at low frequencies because of the increased reactance of C_F, the low-frequency voltage gain of the amplifier is maintained practically constant. This, of course, eliminates sag. The actual proportioning of the filter

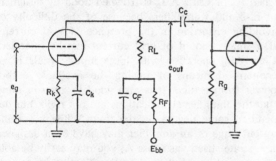

Fig. 69. *Low-Frequency Compensation in Resistance-Coupled Amplifier*

elements, R_F and C_F, is usually a compromise between the allowable voltage drop across R_F and the filter values needed for perfect compensation.

SUMMARY

An amplifier having a relatively **uniform** frequency response from a few cycles to several megacycles is called a **wideband** or **video** amplifier: video amplifiers are used in television and pulse-type systems.

A **pulse** is a non-sinusoidal, **abruptly changing** voltage or current waveform, which may be distorted by an amplifier in various ways. The response of an amplifier to a pulse or **square wave** is indicative of the amplifier's **transient response** and its total frequency range (**bandwidth**).

The principal modifications of a square wave, applied to the input of an amplifier, are (1) the **finite rise time** (measured from 10 to 90 percent of maximum amplitude), (2) an initial **overshoot** and final **undershoot** of the correct square-wave amplitude, and (3) a **sag** or falling off of the top of the reproduced square wave.

The **rise time** in seconds for a reproduced pulse is **inversely proportional to the bandwidth** of the amplifier, and for small overshoots is approximately given by 0.35/bandwidth, where the bandwidth is measured by the highest frequency for which the response does not fall below 70.7 percent of the mid-frequency value.

The amount of **overshoot and undershoot** present depends on the sharpness of the **high-frequency** drop-off in amplifier response.

The amount of **sag** present depends on the phase shift and low-frequency response of the amplifier.

The **high-frequency** response of a resistance-coupled amplifier may be compensated for by adding a small inductance coil either in series with the plate-load resistor (**shunt compensation**), or in series with the coupling capacitor (**series compensation**), or in series with both.

The **low-frequency response** of an amplifier may be compensated for by adding a suitable R-C filter in series with the plate-load resistor.

Chapter Ten

POWER AMPLIFIERS, DISTORTION AND FEEDBACK

We have learned quite a bit about **voltage amplifiers,** both of the audio and video kind, but thus far we have not talked much about the process of obtaining useful power from an amplifier to operate a loudspeaker or other output device. Of course, you understand that the distinction between voltage and power amplification is somewhat artificial, since useful power (i.e., the product of voltage and current) is always developed in a load resistance through which current is flowing and across which a voltage drop exists. The distinction between the two types is really one of degree: It's a question of *how much* voltage and *how much* power. If you have a tiny radio signal, which may have a magnitude of a ten-millionths of a volt, you could never hope to obtain sufficient amplification from one tube to operate a loudspeaker *directly;* and there are other problems you would run into. However, if you magnified the amplitude (or voltage) of the signal first by a number of voltage amplifier tubes, you could apply the strengthened signal to the input of a final power amplifier stage, especially designed to provide large power output to a loudspeaker. Not only are there amplifiers designed to provide ample voltage amplification in the early stages and adequate output power in the final stage, but the tubes themselves are specially designed for each job, voltage amplifiers having large amplification factors and a relatively high plate resistance, while power amplifier tubes provide large plate currents through relatively low plate resistances and with small amplification.

In the present chapter we shall explore some of the means and circuits used to secure large power amplification. We shall also have a closer look at the various distortions a signal may undergo when passing through an amplifier and a powerful method of combating this distortion—**negative feedback.**

AMPLIFIER CLASSES

Besides being classified as to frequency (audio, video, r-f, etc.) and type (voltage or power), amplifiers are also frequently classified according to their mode of operation as either Class A, Class AB, Class B, or Class C. This classification depends on the portion of the input signal cycle during which plate current is expected to flow, and it is especially applicable to power amplifiers.

In a **Class A amplifier** the signal voltage and grid bias are so adjusted that plate current flows at all times, throughout the full cycle of the applied signal voltage. This type of amplifier is characterized by excellent fidelity and low distortion, relatively low power output for a given tube type and low (20–35%) efficiency (i.e., the a-c power output for a given d-c input power).

A **Class B amplifier** is adjusted so that the grid bias is approximately equal to the **cutoff** value of the plate current, when no input signal is applied. Thus, when a signal voltage is applied, the plate current in *one* tube of a Class B amplifier (usually two are used) flows only for about one-half of each cycle. Class B amplifiers have somewhat higher distortion than Class A, and provide intermediate power output and efficiency (50 to 60%).

A **Class AB amplifier** is designed to operate with an input signal voltage and grid bias that permits the plate current to flow for considerably more than half, but less than the entire cycle of the input voltage. Since these amplifiers automatically operate as Class A at low signal levels, they have the advantage of low distortion for small signals and medium power output and efficiency at high signal levels.

In a **Class C amplifier** the grid bias is adjusted to be **greater** than the value required for plate-current cutoff. No plate current flows, therefore, in the absence of a signal, and the plate current in each Class B tube flows for **less than half of each cycle** of the input voltage, when a signal is applied. Class C amplifiers are characterized by relatively high distortion, high power output and excellent efficiency (70 to 75%). Because of their inherent distortion Class C amplifiers are never used as audio amplifiers, but they are primarily employed in the final stages of radio transmitters. Audio amplifiers use almost exclusively Class A or Class AB amplification, if high-fidelity reproduction is desired, while Class B amplifiers are occasionally used in conventional units or automobile radios.

You may sometimes find the suffix "1" or "2" added to the amplifier class; this denotes the presence or absence of grid current flow during the input cycle. Thus, a Class AB_1 amplifier does not draw any grid current during any part of the cycle, while the grid of a class AB_2 is driven positive during a part of the input signal cycle, so that grid current flows during this portion of the cycle. The use of Class A_2 or Class AB_2 is frowned upon by designers of hi-fi audio amplifiers because of the difficulty of avoiding distortion in the presence of grid current. If a small amount of grid current can be tolerated, Class AB_2 operation with large input signals is an economical means of raising the available output power from a tube. It is also good to keep in mind that the amplifier classifications are not rigid, but depend to some degree on the user. While a specific output stage of an amplifier may have been designed to operate strictly as Class A, you may easily be able to overload it and change its classification by operating it from a large input signal source. When turning up the volume for such a large input signal, the power stage may at first operate as Class AB_1 with increased power output and hardly noticeable distortion, but if you turn the gain up high enough, it will eventually operate as Class AB_2, with high output and high distortion.

CLASS A POWER AMPLIFIERS

Let us now study the operation of Class A power amplifiers in more detail, since this is the type most frequently used in well-designed audio amplifiers. The discussion that follows also applies to some extent to Class AB and Class B amplifiers, since it is usually only necessary to modify the grid bias and plate voltage slightly to change from Class A to Class AB or B.

The output of a power amplifier stage in an audio amplifier generally supplies power to the voice coil of a loudspeaker, designed to convert the electrical audio signal variations into air pressure vibrations, or sound. In this application, sufficient power to move the coil of the loudspeaker through relatively large displacements is more important than high voltage amplification. Consequently, the power tubes used for this final stage are designed to sacrifice voltage amplification for high power handling capacity. As we have seen, this generally means high plate currents and high voltages, a high transconductance, a low amplification factor and a relatively low plate resistance. Either triodes, pentodes, or beam-power tubes can be used for Class A power amplifier service. The triodes have the least amount of distortion, but also the smallest power output for a given signal and the lowest efficiency. Pentodes have higher power output with a small input signal and also somewhat more distortion. Finally, beam-power tubes have still higher power output for a given signal and increased efficiency than either type, with about the same distortion as pentodes. Although hi-fi purists favor triodes, pentodes and especially beam-power tubes are quite popular in high-quality audio power amplifiers, because they require less voltage amplification and give more output than triodes. By using beam-power tubes in a **push-pull circuit** and applying **negative feedback,** the distortion can be practically eliminated, as we shall see later on.

Fig. 70 shows a typical triode power output stage, which is transformer-coupled to the voice coil of a loudspeaker, represented by the load impedance R_L. A pentode or beam-power tube circuit would be similar, except that provisions would have to be made to supply the screen with a d-c operating potential through a screen resistor.

Transformer coupling is *always* used in the output stage, rather than resistance coupling, to **match the plate resistance of the power amplifier tube to the voice coil impedance of the loudspeaker.** The voice coil of a loudspeaker consists of a few turns of wire, which have a d-c resistance of a few ohms and an a-c **impedance** (opposition to the flow of a.c.) between four and 16 ohms, depending on the speaker. Since the plate resistance of the power tube may be anywhere from about 1000 ohms to over 100,000 ohms, you can imagine how little power could be developed across the tiny impedance of the voice coil, if it were directly inserted into the output circuit of the tube. Theoretically, maximum power would be transmitted to the voice coil of the speaker, if the voice-coil impedance could somehow be made equal to the tube's plate resistance. For various reasons, primarily excessive distortion, this is not practical for pentodes. However, the voice-coil impedance must be made comparable in value to the plate resistance, if useful power is to be extracted from the tube. This impedance matching is easily achieved by the output transformer, which makes the voice-coil impedance "look" as if it were

multiplied by the **square of the turns ratio** (N^2), as you may remember from basic theory. The transformer turns ratio is thus chosen to achieve the desired impedance matching between the tube's plate resistance and the voice coil impedance.

The performance of a power amplifier stage, such as that shown in Fig. 70, is usually determined graphically from the tube's characteristic curves. As an example, let us assume that a loudspeaker with an eight-ohm (impedance) voice coil is to be operated Class A from a type 2A3 power amplifier in the circuit of Fig. 70. The available plate-supply voltage (E_{bb}) is 250 volts, and it is desired to determine the required signal input voltage to the grid and the available power output, when the tube is

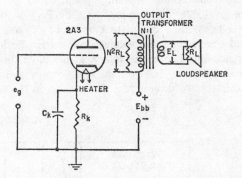

Fig. 70. *Triode Power Output Stage*

operated under the conditions recommended by the manufacturer for Class A amplification. From the manufacturer's tube manual, we find that the 2A3 power triode has a plate resistance of 800 ohms, an amplification factor of 4.2 and a transconductance of 5250 micromhos. This combination of low amplification factor and low plate resistance with high transconductance is typical for power triodes, which must produce high output currents. For a plate voltage of 250 volts and Class A operation, the manufacturer recommends a negative grid bias of —43.5 volts (for a d-c operated filament) and a load resistance of 2500 ohms to obtain low distortion at reasonable power output. The choice of the load resistance is an important factor in determining the power output as well as the distortion, and we shall learn later how to compute both for a given load resistance.

Accepting for the moment the manufacturer's recommendations, we can determine immediately the turns ratio of the output transformer to make the 8-ohm loudspeaker impedance appear as if it were 2500 ohms. The load across the transformer secondary (R_L) "looks into" the primary as if it were multiplied by the square of the turns ratio ($N^2 R_L$). Thus, the **square** of the turns ratio is equal to the ratio of the primary to the secondary impedance, or

$$N^2 = \frac{2{,}500 \text{ ohms}}{8 \text{ ohms}} = 313. \text{ Hence, the turns ratio,}$$

$N = \sqrt{313} = 17.7$ (stepdown primary-to-secondary)

LOADLINE

Fig. 71 is a reproduction of the 2A3 characteristics, upon which we have placed a 2500-ohm **loadline** to determine the dynamic performance of the tube for the chosen operating conditions. To draw this loadline, we start out with the chosen operating point "O," for a plate voltage of 250 volts and a grid bias of −43.5 volts. The quiescent or d-c plate current (I_o) at this point is seen to be 60 ma. We have talked rather loosely about a plate voltage (E_b) of 250 volts at the operating point, although the entire available plate-supply voltage (E_{bb}) is only 250 volts. With a d-c plate current of 60 ma we should expect some d-c voltage drop across the load resistance and, hence, the voltage at the plate should be the difference between the plate-supply voltage and this voltage drop across the load, as in the case

of resistance coupling. As a matter of fact, there is a small voltage drop across the **d-c winding resistance** of the transformer primary winding, but this is only a few volts and may be neglected. The 2500-ohm load impedance that is "reflected" into the primary from the 8-ohm speaker voice coil is only present for the **alternating component** of the plate current (i.e., the output signal), while the resistance to *d.c.* is only that of the primary transformer winding. Thus, we see that practically the entire plate-supply voltage appears at the plate of the tube and only a negligible amount is lost in the primary winding of the transformer. This is the reason that we may use the terms "plate-supply voltage" and "plate voltage" synonymously for transformer coupling.

Now let us get back to the loadline. As we shall see, the loadline is a very handy device for **graphically portraying the distribution of the plate-supply voltage across the load and across the internal (plate) resistance of the tube.** By tracing the path of the input signal along the loadline, we obtain the tube's performance in a nutshell. To construct the

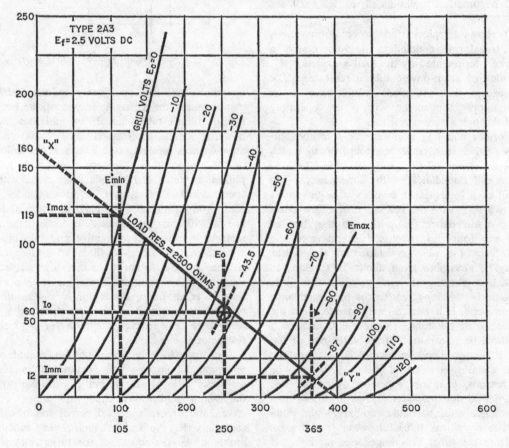

AVERAGE PLATE CHARACTERISTICS

Fig. 71. *2A3 Characteristics for 2500-ohm Load*

loadline, you only have to keep in mind that the load resistance represents the **ratio** of the a-c load voltage (E_L) to the changing or a-c component of the plate current. But we have seen before (in the discussion of resistance coupling) that the load voltage is exactly equal to the **change** in plate voltage, since the two together make up the plate-supply voltage. Thus, we must construct a line whose **inverse slope or ratio of plate-voltage change to plate-current change is everywhere equal to the load resistance** (2500 ohms, in this case). Since the operating point must be on the loadline, we need only one other point to fix the line. The simplest procedure is to determine the intersection of the loadline with either the plate-current or plate-voltage axis. Let us find the intersection with the plate-current (I_b) axis first. At the axis, the plate voltage is zero and the **change** in the plate current equals the change in plate voltage (with respect to the operating point) divided by the load resistance. Hence,

$$\text{change in plate current} = \frac{\Delta E_b}{R_L} = \frac{250}{2,500} = 0.1 \text{ amps}$$

or *100 ma.*

As the plate voltage goes to zero, therefore, the plate current increases by 100 ma—from 60 ma at the operating point to *160 ma* at the intersection with the plate-current axis. We can thus mark off point "X" at the plate-current axis next to 160 ma. Point "X" together with the operating point "O" determines, of course, the loadline completely, but for exercise let us now determine the intersection of the loadline with the plate-voltage axis. At that point the plate **current** must be zero and, hence, the **change** in plate current (with respect to the operating point) is 60 ma, or 0.06 amp. The **change** in plate voltage to produce this current must, therefore, equal the product of the load resistance times the change in plate current ($\Delta E_b = \Delta I_b R_L$).

Hence, the change in plate voltage
$$= \Delta I_b \times R_L = .06 \times 2,500$$
$$= 150 \text{ volts.}$$

Since the plate voltage at the operating point is 250 volts, the plate voltage for zero plate current must be 250 + 150, or *400 volts.* We therefore mark off point "Y" at the plate-voltage axis, next to 400 volts. We can now draw a 2500-ohm loadline from point "X" to point "Y," which will pass, of course, through the operating point "O."

The actual **path of operation** does not extend throughout the entire length of the loadline, however, since we do not want to draw grid current on the positive swing of the input signal; nor do we wish to drive the tube to plate-current cutoff on the negative swing of the signal. We therefore limit the positive swing of the input signal arbitrarily to +*43.5 volts,* which is just sufficient to overcome the −43.5 volts grid bias and, thus, drive the grid to **zero grid voltage** ($E_c = 0$). An a-c input signal that has a positive peak of +43.5 volts will, of course, have a **negative** peak of −43.5 volts and thus drive the grid of the 2A3 to −*87 volts* grid voltage (−43.5 − 43.5 = −87) on the negative swing. The **path of operation** on the loadline, consequently, extends to zero grid voltage on the positive swing of the input signal and to −87 volts on the grid during the peak of the negative swing of the a-c input signal, as is shown by the solid line in Fig. 71. To obtain this grid swing, an a-c input signal of *43.5 volts peak amplitude* is required.

Having laid out the path of operation on the loadline, we can now determine some useful information from it. Of special interest are the plate voltages and currents at the extremes of the input signal swing, since these extreme values permit us to compute the power output and distortion of the amplifier. Thus, at the extreme **positive** swing of the a-c input signal, when the total grid voltage is zero ($E_c = 0$), we find from Fig. 71 that the maximum plate current (I_{max}) is about 119 ma, while the plate voltage has dropped to its minimum value (E_{min}) of about 105 volts. This means, of course, that the voltage across the 2500-ohm load (i.e., the primary of the output transformer) has reached its negative peak equal to the difference between the plate-supply voltage and the plate voltage, or 250 − 105 = *145 volts.* The load voltage is **negative** because of the 180-degree phase reversal occurring within the tube.

Similarly, at the extreme **negative** swing of the a-c input signal, when the total grid voltage is −87 volts, we obtained from Fig. 71 a minimum plate current (I_{min}) of about 12 ma and a maximum instantaneous plate voltage (E_{max}) of about 365 volts. Again we see that the plate voltage is 180 degrees out of phase with the input signal, the latter reaching its negative peak when the plate voltage is at its positive peak, and vice versa. The voltage across the transformer primary (load) during the maximum negative swing of the input signal now reaches its **positive** peak, equal to 365 − 250 or *115 volts.* Since the negative peak of the load voltage is 145 volts, while the positive peak is only 115 volts, the output voltage waveform is clearly not quite as symmetrical as a sinewave. This leads us to suspect strongly the presence of some **distortion** in the signal output, a suspicion which we shall confirm a little later on.

From the operating point "O" in Fig. 71 we can determine the d-c input power to the tube, called the **plate dissipation,** which is an important factor determining the tube life. Multiplying the (quiescent) values of the d-c plate voltage and the d-c plate current at "O" we obtain a plate dissipation of $250 \times .06 = 15\ watts$. This "happens" to coincide with the maximum rating listed in the tube manual. (Of course, we chose it this way.)

A-C POWER OUTPUT

Although the a-c power developed across a resistor is simply the product of the rms (root-mean-square) values of the current and the voltage, the output power in the loadline construction of Fig. 71 is more easily determined by using the maximum and minimum current and voltage values at the extremes of the input signal swing. You can easily derive yourself the following approximate relation for the power output of a Class A amplifier:

$$\text{Power Output} = \frac{(I_{max} - I_{min}) \times (E_{max} - E_{min})}{8}$$

Substituting the values obtained from Fig. 71, after converting milliamps to amperes, we obtain for the 2A3 triode:

$$\text{Power Output} = \frac{(0.119 - 0.012) \times (365 - 105)}{8}$$
$$= 3.48\ watts$$

Although this power, when applied to a loudspeaker, is sufficient to pierce the eardrums of any sensitive listener we shall see a little later on that the power output may easily be more than doubled by employing two 2A3s in a **push-pull circuit.**

DISTORTION

We all seem to have a pretty good idea of what distortion is, since we all have listened to the tinny, squeaky and unnatural sounds of miniature radios; the noisy, crackling and tortured ones emanating from an overloaded table radio; and the boomy, and screechy ones being projected from some so-called high-fidelity phonographs. Being aware of this intuitive knowledge of distortion of the average radio listener (or should we say, sufferer), we have glibly talked about distortion being produced in amplifiers by "non-linear" operation. But in order really to know what we are talking about we have to analyze the phenomenon of distortion somewhat more closely and classify it according to the various forms it may take.

Ideally, if some waveshape that is an exact electrical representation of a sound passes through an audio amplifier completely unmodified, except as to amplitude, there is no distortion and the amplified output from a speaker will sound exactly like the original (provided, of course, that the speaker does not introduce any distortion). In practice, a reproduced sound never sounds like the original, even with an ideal amplifier and speaker. For one thing, a speaker is not a musical instrument and can never reproduce the spatial sound distribution of a symphony orchestra, or even a grand piano. Another reason is that sound is usually reproduced **monaurally**, that is, by a single channel from one microphone through one amplifier and one speaker, while we listen **binaurally** with two ears. The time difference with which sound strikes each ear of the listener produces a spatial realism in depth, which permits us to tell approximately where each sound is coming from. This effect cannot be reproduced by monaural, or single-channel reproduction, but has been simulated quite successfully by the so-called **stereophonic** or binaural reproduction devices.

Leaving aside these refinements, let us define the main modifications, or distortions, of an input waveform caused by a deficient voltage or power amplifier. These are **frequency distortion, phase or time-delay distortion, and amplitude or non-linear distortion.**

Frequency Distortion. We have already met frequency distortion: It is the limitation in the frequency range or bandwidth reproduced by an amplifier, caused by various coupling elements and unavoidable shunt capacitances. Frequency distortion exists whenever the full audio range from about 20 to 15,000 cycles per second is not completely reproduced. If the high-frequency response is deficient, you will not be able to hear the "overtones" of some high-pitched instruments, such as the violin, flute, or horn, and they will all sound very much alike. The sound from the speaker will then appear very boomy or bassy. On the other hand, if the low-frequency response of the amplifier is deficient, you may not hear at all the gorgeous sound of the large (30–50 cycles) organ pipes and the bass drum may sound no different from the kettle drum. In general, the sound will then be tinny and squeaky.

Phase or Time-Delay Distortion. You have become acquainted with phase or time-delay distortion during the discussion of the pulse response of wideband amplifiers. The various capacitive and inductive coupling elements of an amplifier cause **phaseshift** because the current in an inductance **lags behind** the applied voltage (in time or **phase),** while the current in a capacitor **leads** (in time or phase) the applied voltage. This in itself would not cause distortion if the resultant time delay of the output with respect to the input would be the *same for all frequencies.* Unfortunately, this is usually not the case, and different frequencies are delayed by different amounts, resulting in output waveforms that differ in appearance from the input waveform. The primary effect of phase or time-delay distortion is the production of **overshoots** and **undershoots** in a pulse waveform, such as was illustrated in Fig. 66b. Although this is important in the exact reproduction of pulse waveforms in a video amplifier, it does not appear to have much **audible** effect in an audio amplifier. However, the phase-shift characteristics and frequency response of an amplifier *together* determine the **transient response,** that is, the response to sharply rising and suddenly changing waveforms. Combined frequency and time-delay distortion may thus lead to **poor transient response** in an audio amplifier, which, in turn, may wash out the sharp "attack" of a piano or percussion instrument.

Amplitude or Non-Linear Distortion. Amplitude or non-linear distortion is perhaps the most serious form of distortion occurring in an amplifier and it is usually this type we have in mind when we say that a sound is "terribly distorted." As the term implies, amplitude distortion causes a modification in the relative amplitudes of a signal. It does *not* mean that the amplitudes of the two half-cycles of a sinewave cannot be enlarged, as indeed they must be in an amplifier, but it means that they are *unequally* amplified with respect to each other and the input waveform. Thus, amplitude distortion exists if the top or bottom of a sinewave is "clipped off" or reduced in amplitude, as shown in Fig. 72. The pure sine-waveform of the applied input voltage is shown in Fig. 72a. If the amplifier is operated Class A, in the linear portion of the characteristic, the waveform of the plate current and the output voltage should look exactly as in (a), except that it will be larger, of course. However, the characteristics of amplifier tubes are never completely linear and, hence, some "non-linear" distortion is always present. Fig. 72b shows the slightly flattened plate-current waveform of an amplifier that is driven into the curved bottom portion of the characteristic during the negative swing of the input signal. Finally, Fig. 72c shows the severe distortion of the plate-current waveform, when the grid of the amplifier draws grid current during the positive swing of the applied signal and the plate current is cut

off during the negative swing of the input signal. The result is a waveform that is clipped at the peak of each half-cycle and looks more like a square wave than a sinewave. The distortion of the plate-voltage waveform (output signal) is, of course, the same in each case, except that it occurs reversed in phase because of the 180-degree phrase reversal between input and output.

Fig. 72 illustrates graphically that non-linear operation of an amplifier results in first slightly flattening the sinewave amplitude and finally "clipping" or "squaring" it with severe distortion. The square-appearing waveform of Fig. 72c is similiar to the pulse waveforms illustrated in Chapter 9, and like those it can be thought to be made up of a **fundamental** sinewave frequency (equal to the frequency of the input sinewave, shown in Fig. 72a) and a number of multiples or **harmonics** of that frequency. Thus, the waveform of Fig. 72c, for example, may consist of a fundamental 1000-cycle sinewave, equal to the input voltage (Fig. 72a), plus a **second harmonic** of 2000 cycles that is perhaps 20% in amplitude compared to the fundamental, plus a **third harmonic** of 3000 cycles, which is perhaps 10% of the fundamental amplitude, plus a **fourth harmonic** of 4000 cycles, which may be 5% of the fundamental amplitude, and so on. The more *severe* the distortion, the *greater* is the **number** and the **amplitudes** of the harmonic frequencies present in the output signal. Because of the production of harmonic frequencies **not present in the input signal,** this form of distortion is sometimes called **harmonic distortion.** The presence of harmonic frequencies might by itself not be so unpleasant, if it were not for the fact that these new frequencies interact with each other, the high-frequency components being **modulated** by the low frequencies, and the resulting **intermodulation distortion** gives rise to sounds that are guaranteed to make you leave the room quickly.

As an example, let us determine the amplitude distortion of the Class A 2A3 amplifier, whose characteristics are shown in Fig. 71. Although it is

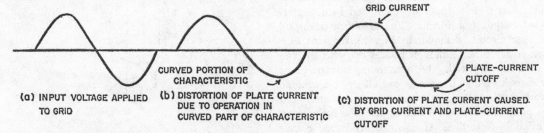

Fig. 72. Amplitude Distortion of Plate-Current Waveform in a Triode Amplifier—(a) Input Voltage Applied to Grid, (b) Distortion of Plate Current Due to Operation in Curved Part of Characteristic, and (c) Distortion of Plate Current Caused by Grid Current and Plate-Current Cutoff

possible to calculate the amplitude of all the harmonics present in the amplifier output from the characteristics of the tube in conjunction with the loadline, the **second harmonic** distortion is usually most severe, and hence its amplitude is of greatest importance. The amplitude of the second harmonic (in percent of the fundamental) can be determined by the following simple formula:

Percent 2nd harmonic distortion =

$$\frac{\dfrac{I_{max} + I_{min}}{2} - I_o}{I_{max} - I_{min}} \times 100$$

Substituting the values previously obtained from the loadline,

Percent 2nd harmonic distortion =

$$\frac{\dfrac{0.119 + 0.012}{2} - 0.06}{0.119 - 0.012} \times 100 = 5.1\%$$

A second-harmonic distortion of 5.1% (of the fundamental amplitude) in an audio amplifier is generally considered quite tolerable for ordinary music and speech reproduction, although it is far too noticeable to be accepted as **high-fidelity reproduction.** The latter usually requires the total harmonic distortion (of all harmonics together) to be less than *one percent.* Further reduction of distortion may be accomplished by operating strictly Class A, limiting the amplitude of the input signal and, consequently, the power output, by using a **push-pull circuit** and by the application of **negative feedback.** Let us consider push-pull operation next.

PUSH-PULL POWER AMPLIFIER

One way of obtaining more power output from an amplifier is to connect two identical tubes in parallel, plate-to-plate, grid-to-grid, and cathode-to-cathode. This will assure roughly *twice* the power output of a single tube, but the total distortion will also go up. A far better way is to connect two identical tubes in such a manner that their grids are excited by equal but 180-degree out-of-phase input signals and their outputs are combined in a **center-tapped** output transformer. The resulting push-pull amplifier circuit is shown in Fig. 73 for two 6L6 beam-power tubes, although triodes or pentodes can be used equally well. The outstanding advantage of this circuit is that it **cancels out all even harmonic distortion** (i.e., 2nd, 4th, 6th harmonic, etc.) and thus permits more power output *per tube* for a given permissible distortion. As a consequence the circuit will give *more than twice* the power output of a single tube with considerably less distortion. For example, a single 6L6 beam-power tube, operated

at 250 volts plate voltage under optimum conditions, will give a power output of about 6.5 watts with 10 percent total harmonic distortion, while the 6L6 push-pull circuit shown in Fig. 73, operated Class A under similar conditions, will give a power output of about 17.5 watts with only 2 percent total harmonic distortion.

Although for simplicity batteries are shown in Fig. 73 as sources of bias and plate potential, in practice the plate and screen potentials are derived from a rectifier power supply, while grid bias is obtained from a cathode-biasing resistor that is inserted into the common cathode circuit of the two tubes. Cathode biasing has the advantage that **no bypass capacitor** is required, since in a push-pull circuit the a-c signal currents do *not* flow through the common cathode return or the plate supply.

The symbol notations in Fig. 73 are the same as previously used, except that the numerical subscripts refer to tube 1 (VI) or tube 2 (V2), respectively. As mentioned, the circuit requires two input

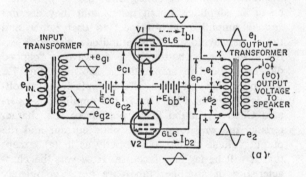

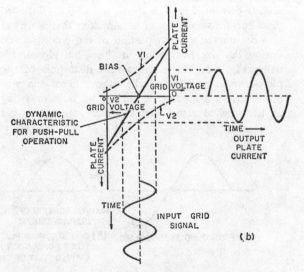

Fig. 73. *Push-Pull Amplifier Circuit (a) and Composite Characteristic (b). (Courtesy "Basic Vacuum Tubes and Their Uses" by John F. Rider and Henry Jacobowitz, John F. Rider Publisher, Inc.)*

signal voltages, e_{g1} and e_{g2}, which are equal in amplitude, but **180 degrees out of phase** with respect to each other. Two out-of-phase voltages are easily secured by means of a center-tapped input transformer, as shown, since whenever the top of the transformer secondary is positive with respect to the center tap ($+e_{g1}$), the bottom is negative with respect to center tap ($-e_{g2}$), and vice versa. When resistance coupling is preferred because of the improved frequency response, a **phase inverter** stage must be used to obtain the out-of-phase input voltages. Such phase inverter arrangements are described in the next section. Because of the necessity for two equal input voltages, the total a-c input voltage, $e_{g1} \times e_{g2}$, is twice that required for operating a single tube. The plate-supply voltage, E_{bb}, provides plate and screen potentials to both tubes. For Class A operation, the plates and screens may be operated at the same positive d-c voltage.

In the absence of input signals to the grids of the two tubes, each tube draws the same quiescent (d-c) plate current and both the output voltage and output current (i_o) in the output transformer secondary are zero, since a d-c current induces no voltage in a transformer. Furthermore, since the two d-c plate currents flow out of the tubes and through the transformer primary in **opposite directions,** there is no resulting **magnetizing current** in the transformer. When a d-c magnetizing current is present in a transformer, it tends to **saturate the core** and thus **lower its inductance.** To obtain the required inductance under these conditions a large transformer with a large core is needed. The absence of the magnetizing current in a push-pull circuit permits the use of a smaller, more economical output transformer.

When two a-c signal voltages, e_{g1} and e_{g2}, are now applied to the respective grids of the two tubes, sinusoidal plate currents, I_{b1} and I_{b2}, will flow in the plate circuits of V1 and V2, respectively. Plate current I_{b1} is 180 degrees out of phase with I_{b2} (i.e., when one is positive, the other is negative) because the two input signals to the grids are 180 degrees out of phase with respect to each other. Assume that e_{g1} is initially positive and hence the plate current of I_{b1} of V1 increases during the positive grid swing. This results in an increased voltage drop across the top half of the transformer primary and a decreased plate voltage of V1. (You will remember that the plate voltage drops, when the plate current increases.) Consequently, point "X" at the top of the transformer primary, which is at plate potential, becomes **negative** with respect to the primary center tap (point "Y").

At the same time, the grid of V2 is being driven negative by its negative input signal, e_{g2}, and its plate current I_{b2} decreases with respect to its qui-

escent value. This in turn results in a lowered voltage drop across the bottom half of the output transformer primary and in increased plate voltage at V2. As a consequence, point "Z" at the bottom of the transformer primary becomes **positive** by the same amount with respect to center tap (point "Y"), as point "X" is made negative.

The foregoing analysis shows that the top half of the transformer primary becomes negative with respect to center tap by a voltage e_1, while bottom half becomes positive with respect to center tap by an equal voltage, e_2. These two voltages are added in series across the transformer primary, resulting in the total primary voltage, e_p, which is exactly twice the amplitude of e_1 or e_2. The total a-c primary voltage induces an output voltage (e_o) in the transformer secondary and an output current (i_o) flows through the speaker. A half-cycle later all the input and output voltage polarities reverse, of course, and the primary voltage e_p is of opposite phase, though still equal to the sum of e_1 plus e_2.

The power output and distortion analysis of the push-pull circuit is somewhat complex, but the basic effect of push-pull operation is clearly shown by the combined dynamic transfer characteristic (Fig. 73b). This **composite** push-pull characteristic is obtained by adding together the individual transfer characteristics of V1 and V2, and it is seen to be completely straight. Because of the resulting linear Class A operation, the plate-current waveform in the output is practically undistorted.

PHASE INVERTERS

If resistance coupling is desired to the input of a push-pull amplifier, a **phase inverter** must be used to supply the two 180-degree out-of-phase grid input voltages. Fig. 74 illustrates two popular circuits out of a great variety of existing phase inverters. The circuit shown in (a) requires only one tube to provide the two out-of-phase voltages and thus is quite economical.

Phase inversion in the circuit of Fig. 74a is obtained by splitting the total plate-load resistance and, hence, the output voltage equally between the plate and cathode circuits of tube V1. Since R_L and R_k are equal in value and the plate current flows through both of them, voltages of equal amplitude are developed across the two resistors. But because of the phase reversal taking place within the tube, the output signal taken off the plate-load resistor R_L is **180 degrees out of phase** with the input signal, while the output signal taken off the cathode-load resistor R_k is **in phase** with the input signal. The two output signals, consequently, are 180 degrees out of phase with respect to each other. You can verify for yourself that for a positive input signal, the plate voltage of V1 will drop, thus making the

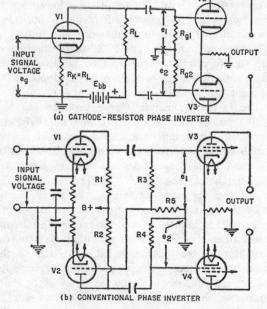

Fig. 74. *Cathode-Resistor Phase Inverter (a) and Conventional Phase Inverter (b)*

plate-connected end of R_L **negative,** while the cathode-connected end of R_k simultaneously becomes more **positive** (since electrons flow from minus to plus).

The two out-of-phase output voltages of V1 are coupled through d-c blocking capacitors and grid resistors R_{g1} and R_{g2} to push-pull amplifier tubes V2 and V3, where they develop out-of-phase grid input voltages e_1 and e_2. The push-pull amplifier is biased by a common cathode resistor. The output circuit of the push-pull stage is not shown.

As in the case of an unbypassed cathode-biasing resistor, a degenerative feedback action takes place across R_k, which is common to both the plate and grid circuits of V1. In effect, the output voltage developed across R_k "bucks" the signal input voltage to the grid of V1 and so prevents the tube from realizing its amplification. Each of the output voltages, e_1 and e_2, are therefore slightly *smaller* than the input signal voltage, although the total output voltage ($e_1 + e_2$) is slightly less than *twice* the value of the input voltage. Since phase inversion is the desired aim, the lack of amplification does not matter. Besides, the cost of tube V1 and the associated coupling components is still less than the expense of a high-quality push-pull input transformer.

Fig. 74b shows the conventional phase-inverter circuit that is usually employed to drive a push-pull power amplifier. In this circuit triode V1 drives the upper tube, V3, of the push-pull amplifier as a conventional resistance-coupled stage. Phase inversion is provided by triode V2. A portion of the out-

put voltage of V1, developed across resistor R5, is applied to the grid of phase inverter V2. Since the tube introduces a 180-degree phase reversal, V2 drives the lower push-pull amplifier V4 with a voltage e_2 that is 180 degrees out of phase with respect to voltage e_1 applied to V3. If the ratio of R5 to R3 is properly chosen, the input voltage applied to V2, when multiplied by the gain of the stage will just equal the output voltage e_1 of stage V1. To attain equal output voltages, the ratio (R3 + R5)/R5 must equal the voltage gain of V2. Since the gain of the phase inverter tube V2 is wasted by applying only a small input voltage, the amplification of V1 and V2 together is just equal to that of a single tube. Again, the price of phase inversion is the sacrifice of one tube.

You can visualize the process of phase inversion in the circuit of Fig. 74b somewhat better from the following consideration. Assume the input signal voltage is such as to drive the output voltage of V1 in a **positive** direction. A positive input signal, developed across R5, is then applied to the grid of phase inverter V2 and its plate current *increases*. This action *increases* the voltage drop across plate-load resistor R2 and, hence, swings the plate of V2 in a **negative** direction. Thus, when the output voltage of V1 swings positive, the output voltage of V2 swings negative, or vice versa. The two output voltages are therefore 180 degrees out of phase with each other.

FEEDBACK AMPLIFIERS

The man who first thought of feedback should have earned the undying gratitude of humanity, but he is no longer here to receive it. The principle of feedback is probably as old as the invention of the first machines, but it has only dawned upon us rather recently that feedback is not only marvelously useful, but applicable to almost everything. Technically, **feedback simply means transferring a portion of the energy from the output of some device back to its input.** By making the input thus dependent on the output of the device, very fine control of any process becomes possible. In effect, this amounts to spying on the end result (output) of some process and controlling the input to the process in accordance with the effectiveness of the output. Such an arrangement, in which the input depends on feedback from the output, is known as a **closed-loop control system** and it is the most effective control system possible. (See Fig. 75.) The **governor** of a steam engine, which regulates the flow of steam in accordance with the speed of the output drive wheel and thus controls the speed, is an early example of such a closed-loop feedback system. The **thermostat,** which controls the fuel flow to an oilburner or the operation of an air-conditioner in accordance with the desired room temperature, is another excellent example. You could easily devise such a closed-loop feedback system yourself, by regulating for example the heat output of your toaster in accordance with the brownness of the toast.

Although the applications of feedback are myriad, it has only been about 20 years that feedback has come into use in connection with electronic amplifiers. Here it has been found eminently useful in reducing **distortion** caused within amplifiers and making amplifier operation more **stable** in respect to variations in gain due to line voltage changes, tube differences, aging, etc. To understand how this is possible, we must first of all distinguish between **two basic types** of feedback: **regenerative** or **positive** feedback and **degenerative** or **negative** feedback. When the feedback energy (voltage or current) is **in phase** with the applied signal and thus **aids it,** **regenerative** or **positive** feedback takes place. **Degenerative** or **negative** feedback is the term used, when the feedback energy is **out of phase** with the applied signal and thus **opposes** it. Regenerative feedback **increases** the gain (and also the distortion) of an amplifier, and if sufficiently large, leads to **oscillations,** as we shall see in the chapter on oscillators. In contrast, degenerative feedback decreases the gain, as well as the **distortion** of an amplifier, and for the latter reason, we shall be primarily interested in degenerative (negative) feedback.

The basic principle underlying the operation of a feedback amplifier can be understood from the schematic diagram shown in Fig. 75. Here the amplifier gain in the **absence** of feedback is **A.** Feedback is then applied by feeding a portion β of the output voltage e_o back to the amplifier input. The actual input to the amplifier, thus, consists of the sum of the signal voltage, e_g, plus the feedback

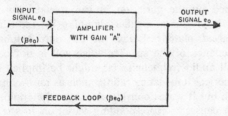

Fig. 75. *Schematic Diagram of Feedback Amplifier*

voltage, βe_o, or a total of $(e_g + \beta e_o)$. This total input voltage, amplified by the gain A of the amplifier, must be equal to the output voltage. Thus we may write

$$(e_g + \beta e_o) \times A = e_o$$

But since the voltage gain of an amplifier, by definition, is the ratio of the output voltage to the input voltage (e_o/e_g), we can solve the above expression

for this ratio and, thus, obtain the gain in the **presence** of feedback:

$$\frac{e_o}{e_g} = \text{Voltage gain with feedback} = \frac{A}{1 - \beta A}$$

where A is the gain of the amplifier **without feedback.**

We have obtained the above expression by assuming that the portion fed back, β, is positive and, hence, positive feedback or regeneration takes place. The quantity βA represents the amplitude of the feedback voltage (sometimes called **feedback factor**) superimposed upon e_g. The larger this feedback factor, the smaller is the denominator of the above expression and, hence, the larger is the gain *with* feedback. Unfortunately, increasing the gain by positive feedback also increases the distortion and noise in the same proportion and, hence, positive feedback is rarely used except for oscillators (Chapter 11).

For **negative feedback,** the fraction β and hence βA becomes negative (i.e., $-\beta A$) and the

$$\text{Voltage gain with negative feedback} = \frac{A}{1 + \beta A}$$

The greater the feedback factor βA, the smaller is the feedback gain of the amplifier, but also the smaller is the distortion introduced by the amplifier. When the feedback factor βA is made large compared to 1, as is the case for large amounts of feedback, the denominator of the above expression reduces simply to βA, and the previous expression becomes

$$\text{Voltage gain with large negative feedback} = \frac{1}{\beta}$$

The gain of the amplifier in this case is quite small, but—as is apparent—depends only on the feedback fraction β and is thus substantially **independent** of the actual gain A of the amplifier. This remarkable fact can be explained by considering that the voltage *actually* applied to the amplifier is the **difference** between a relatively large input signal (e_g) and a large feedback voltage (βe_o). If the amplification should increase for some reason (such as line voltage fluctuations or tube changes), the feedback voltage (βe_o) will be larger and hence the **difference** between it and the input signal will be much *smaller.* This tends to oppose the increase in amplification and maintain it stable. The same is true, if the amplification should tend to decrease.

As a result, the gain of an amplifier with large **negative** feedback is extremely **stable** and is practically independent of fluctuations in the supply voltage, aging of tubes, or differences between tubes caused by replacements. Moreover, since the gain depends only on the feedback fraction β, the **variation in gain with frequency** (i.e., the frequency response) is entirely controlled by the nature of β. If the feedback fraction is obtained through a resistive network, as is usually the case, the feedback, and hence the gain, **does not vary with frequency** as long as the feedback is **negative.** The frequency response of an amplifier may thus be considerably improved by using large amounts of feedback. This improvement is not absolute, because phaseshifts caused within the amplifier by various coupling elements, tend to **change the sign** of the feedback from negative to positive at the extremes of the audio band, thus nullifying the degenerative feedback action.

By designing the feedback loop with capacitive and inductive elements, rather than with resistors, the feedback may be made to vary in any desired manner with frequency and, thus, any required frequency response may be achieved. Feedback circuits are, therefore, very popular to boost the bass and treble response of an audio amplifier. The recording curves of various phonograph disks may be compensated for easily by a properly designed feedback loop.

REDUCTION OF DISTORTION AND NOISE

Hum, noise and distortion introduced within an amplifier are reduced with negative feedback by the same factor $(1 + \beta A)$ as the gain. Since the loss in gain can always be compensated for by increasing the amplitude of the input signal, the net effect of negative feedback is a reduction in distortion, noise and hum. To understand how this reduction takes place, refer to Fig. 76.

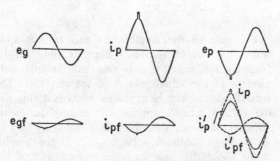

Fig. 76. *Reduction in Distortion Due to Negative Feedback*

Let us see first what happens without the use of feedback. Assume that when an input signal e_g is applied to the grid of an amplifier tube, the plate current i_p is afflicted with a pronounced irregularity during its positive half-cycle due to highly nonlinear operation. (For the purposes of illustration we have exaggerated the irregularity beyond anything likely to happen in practice.) The plate voltage, e_p, and the output signal will be, of course, reversed in phase and thus have the irregularity in its negative half-cycle.

Now suppose negative feedback is applied to the amplifier. The voltage fed back to the grid of the amplifier, e_{gf}, has the same waveform and phase as the plate or output voltage, but is smaller in amplitude. Since the original input signal, e_g, and the feedback signal, e_{gf}, are opposite in phase, it is evident that the two input signals oppose each other and produce degeneration. The feedback voltage (e_{gf}) produces a component of the plate current, i_{pf}, that is opposed in phase to the original plate current, i_p. Since the irregularity appearing in the feedback component of the plate current (i_{pf}) is out of phase with the irregularity in the original plate current, i_p, the two oppositely phased irregularities will tend to cancel each other. The greater the feedback component, the more cancellation will take place. The final result is the algebraic addition of the two plate-current components i_p and i_{pf} (shown dotted in Fig. 76) to produce the resultant plate current, i_p'. Although smaller in amplitude than the original plate current, the irregularity in the resultant plate current has also been substantially reduced.

The same reduction that is attained with negative feedback for non-linear distortion also takes place for extraneous noises and power supply hum introduced within the amplifier. As a consequence, less hum filtering of the power supply is required with a feedback amplifier than would otherwise be necessary.

OUTPUT DISTORTION AND SPEAKER DAMPING

When applied to the power output stage of an audio amplifier, negative feedback has two important beneficial effects: it reduces the distortion caused by the **variation in the speaker impedance** and it produces speaker *damping* by reducing the **apparent plate resistance** of the output tube. The impedance of the speaker voice coil is not constant at all audio frequencies (as might be implied by the numerical impedance rating, such as an 8-ohm voice coil), but it varies considerably over the audio range. This causes a corresponding variation in the plate-load impedance "seen" by the tube, which leads to distortion in the output. This distortion is especially serious for pentodes, which have an inherently high plate resistance. Negative feedback *lowers* the plate resistance of the tube and, thus, the distortion produced by the varying voice-coil impedance.

The high plate resistance of pentodes has another deleterious effect on speaker reproduction. The source impedance "looking back" from the voice coil into the pentode plate circuit is *very high*. This causes **oscillations** of the speaker voice coil (with resultant distortion) when sharp transient waveforms are applied to the speaker, an effect known as **hangover.** The cure for this is to **damp** the speaker by **lowering the source impedance** of the output tube. Again negative feedback does the trick by making the high plate resistance characteristics of a pentode look like those of a low plate resistance triode. Triode output tubes do not require negative feedback, since they have already a sufficiently low source impedance. Negative feedback applied to a power output tube is shown in Fig. 77a.

PRACTICAL DEGENERATIVE FEED-BACK CIRCUITS

Fig. 77 illustrates a few typical amplifier circuits employing negative or degenerative feedback.

The application of voltage feedback to a 6L6 power amplifier stage is shown in Fig. 77a. This type of feedback is termed **voltage feedback** because it is derived from the output voltage. An a-c voltage divider, consisting of R_1, R_2 and C_1, is connected across the output (plate circuit) of the 6L6. A fraction of the a-c output voltage ($e_t = \beta e_o$), developed across R_2, is applied to the grid of the tube through the secondary of the input transformer. Capacitor C_1 blocks the d-c plate voltage from the grid. Since the feedback voltage is developed across the output of the tube, it is automatically 180 degrees out of phase with the input voltage and, hence, is degenerative. The feedback voltage (e_t) in this case is approximately equal to the output voltage multiplied by the fraction $R_2/(R_1 + R_2)$. (This fraction is the "β" we have previously referred to.) We have already mentioned the beneficial effects of negative feedback in a power output stage, which consist of a reduction in the distortion caused by the varying speaker voice coil impedance and the prevention of "hangover" effects by increased speaker **damping.** Both these improvements are attained by a reduction in the source impedance (i.e., *effective* plate resistance) of the power output tube. Feedback is especially popular in beam-power tubes because full power output can be attained with a comparatively small input driving signal. As was explained before, the reduction in gain caused by feedback must be compensated for by increasing the grid input voltage to the tube. In triodes and some pentodes the increase in driving voltage required to produce full power output may be inconveniently large, requiring extra stages of amplification.

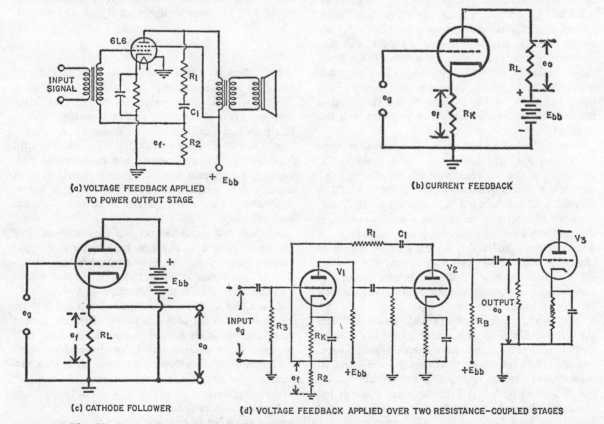

(a) VOLTAGE FEEDBACK APPLIED TO POWER OUTPUT STAGE

(b) CURRENT FEEDBACK

(c) CATHODE FOLLOWER

(d) VOLTAGE FEEDBACK APPLIED OVER TWO RESISTANCE-COUPLED STAGES

Fig. 77. *Typical Degenerative Feedback Amplifier Circuits—(a) Voltage Feedback Applied to Power Output Stage, (b) Current Feedback, (c) Cathode Follower, and (d) Voltage Feedback Applied Over Two Resistance-Coupled Stages*

Current Feedback. The circuit of Fig. 77b is the familiar cathode-biased resistance-coupled amplifier, with the bypass capacitor left off. We have already discussed the degenerative action caused by the feedback voltage e_f (across R_k) being out of phase with the input signal e_g. This type of feedback decreases the gain and distortion of the amplifier, but it *increases* the plate resistance of the tube. It is known as **current feedback** because the feedback voltage e_f is proportional to the plate **current** of the tube, rather than the output voltage. Current feedback improves the high-frequency response of an amplifier, but makes it *worse* at low frequencies. Also, by increasing the source impedance of the stage, loudspeaker hangover effects in a power amplifier are accentuated.

Cathode Follower. The circuit of Fig. 77c is known as a **cathode follower** because the output voltage, taken off cathode resistor R_L, is **in phase** with and **follows** the input signal. The input voltage is applied between grid and ground and the output is taken off between cathode and ground. You can easily verify that the output is in phase with the input by considering that an increasing positive input signal (e_g) will increase the plate current and, hence, produce an increased voltage drop, e_f, across cathode-load resistor R_L. Since electrons flow from minus to plus, the cathode-end of resistor R_L becomes more positive than before and the rise in output voltage is in phase with the positive-going input signal. Similarly, when the input signal goes negative, the cathode-end of R_L becomes less positive (or more negative) by the amount of the drop across R_L and the output signal is again in phase with the input.

Since cathode-load resistor R_L is common to both the output and grid circuits, the output voltage e_f across R_L is also in series with the input signal (e_g) and thus causes **degenerative current feedback.** It is **current** feedback because the amount of the feedback voltage depends on the plate current and it is **degenerative** because a positive-going cathode for a positive input signal will make the grid **negative with respect to the cathode** and thus oppose the positive input signal. Because of the large amount of degeneration, the voltage amplification of the cathode follower is *always less than unity*.

Although the gain of the cathode follower is less than 1, the circuit is capable of producing a large output voltage at large plate currents and, hence, considerable **power gain.** Furthermore, the effect of degenerative current feedback is to *increase* the input impedance to the grid of the tube, while at the same time lowering the plate resistance and output impedance by a factor $\mu + 1$. This permits the use of the cathode follower as an **impedance matching** device (similar to a transformer) for coupling a high-

impedance source to a low-impedance output, such as a transmission line. When used as a **power amplifier,** the cathode follower produces extremely **low distortion** and **excellent loudspeaker damping** because of the very low source impedance of the circuit. Because of the high input impedance, cathode followers are also frequently used as isolation devices to produce low **loading** of a circuit, as might be required in a vacuum tube voltmeter or an oscilloscope.

Multiple Stage Feedback. Feedback may be applied over several stages and is sometimes used from the secondary of an output transformer back to an early voltage amplifier stage. When this is done the benefits of feedback accrue to **all the stages included in the feedback loop,** but the correct design is somewhat difficult due to excessive phase-shifts. Fig. 77d shows voltage feedback being applied over two resistance-coupled stages. The feedback voltage is derived from an output voltage divider, C_1-R_1-R_2, in the same manner as shown in Fig. 77a. However, because of the **double phase reversal** occurring over two stages, the feedback voltage (e_f) is **in phase** with the input signal (e_g) and, hence, **cannot be injected into the grid circuit.** By injecting the feedback voltage into the cathode circuit, in this case, it is opposed in phase to the input signal, as required.

SUMMARY

In a **Class A** amplifier the signal voltage and grid bias are adjusted so that the plate current flows throughout the entire cycle of the applied input voltage. Class A amplifiers have excellent fidelity, relatively low distortion and low power output.

A **Class B** amplifier is adjusted so that the grid bias is **equal** to the **plate-current cutoff** value and, hence, plate current flows only for one-half of each input voltage cycle. A Class B amplifier has higher distortion than Class A, intermediate power output and efficiency.

A **Class AB** amplifier permits plate-current flow for **more than half but less than the entire input voltage cycle.** Class AB amplifiers have low distortion at low signal levels and medium power output and efficiency at high signal levels.

In a **Class C** amplifier the grid bias is adjusted to be **greater** than the **plate-current cutoff** value, so that plate current in each tube flows for **less than half** of each input cycle. Class C amplifiers have high power output and efficiency together with **high distortion,** and for the latter reason, are not used as audio amplifiers.

Transformer coupling is always used in a power output amplifier to **match** the plate resistance of the tube to the voice coil of a loudspeaker. The impedance ratio equals the **square** of the turns ratio.

The loadline is a graphic presentation of the distribution of the plate-supply voltage across the load and the internal resistance of the tube. It permits determining the power output and distortion of the tube.

The three main types of distortion are **frequency distortion, phase** or **time-delay distortion** and **amplitude** or **non-linear distortion.**

Frequency distortion is the limitation in the audio bandwidth caused by the amplifier coupling elements.

Time-delay distortion is the **unequal delay** of different audio frequencies caused by phaseshifts in the amplifier coupling elements. Frequency distortion in conjunction with time-delay distortion produces **poor transient response** in an amplifier (i.e., distorts pulses).

Amplitude distortion of the output waveform of an amplifier is caused by **non-linear operation,** such as plate-current cutoff and grid current flow. **Curved tube characteristics** also cause amplitude distortion. Amplitude distortion results in the production of harmonic frequencies **not present** in the input signal. These cause distorted sound.

In a **push-pull amplifier** the grids are excited by oppositely phased input signals and the output is combined in a **center-tapped output transformer.** The out-of-phase exciting voltages are supplied either by a center-tapped **input transformer** or by a **phase inverter.** Push-pull operation **cancels out even harmonic distortion** and produces **more than twice** the power output of a single tube.

In a **feedback amplifier** a portion of the output voltage (or current) is fed back to the input. If the feedback voltage is **in phase** with the applied signal and aids it, **regenerative** or **positive** feedback takes place. If the feedback voltage is opposed to the input signal (i.e., out of phase), **degenerative, negative** or **inverse** feedback is said to occur. Regenerative feedback **increases the voltage gain** and may lead to **oscillations.** Degenerative (negative) feedback **decreases the amplifier gain,** but also **decreases distortion, noise,** and **power supply hum** within the feedback loop. Large negative feedback makes an amplifier almost **independent** of voltage fluctuations and aging of tubes. It also improves the frequency response.

Voltage feedback is proportional to the output voltage, while **current feedback** is proportional to the plate current of the tube.

Chapter Eleven

OSCILLATORS

Rapid to-and-fro motion of electrons in a conductor results in radiation of electromagnetic (radio) waves —the basis of all radio communication. It is the purpose of electric oscillators to generate these rapidly alternating electron currents from a direct-current supply. When Heinrich Hertz devised the first radio transmitter in 1887, he utilized an electric spark in a tuned circuit to produce very rapid electron oscillations and, consequently, radio waves. Higher powers than possible with spark transmitters were obtained in turn by the Poulsen electric arc oscillators and the Fessenden-Alexanderson high-frequency electric generators. Although these devices eventually attained good efficiency, they were noisy, often had poor frequency stability, and usually produced several radio frequencies, in addition to the desired one.

The invention of the De Forest audion triode in 1907 finally made possible the development of powerful transmitting oscillators in their present form, with excellent frequency stability, silent operation, and ease in changing frequencies. Because of this combination of merits electron tube oscillators are the most widely used generators of radio frequencies,

and they will be the only ones discussed in this chapter. It is, however, important to understand that oscillations are not confined to electron tubes, but that they occur in a wide variety of electrical and mechanical devices. A pendulum, an ordinary watch, a "walking doll," windshield wipers, etc., are all excellent examples of mechanical oscillators. You can visualize the basic principles underlying all oscillating systems most clearly by considering such a mechanical oscillator.

MECHANICAL OSCILLATOR

Fig. 78(a) shows a simple spring pendulum, consisting of a weight suspended from one end of a coiled spring. If the weight is pulled downward from its resting position and then released, as in (b), it will swing back beyond its original position to that shown in (c). It will then continue to oscillate up and down until it gradually comes to rest. It stops when all the energy initially imparted to it is dissipated in heat because of the friction in the spring and bearings. One oscillation or **cycle** is completed, when the weight has moved from its initial position (a), down to position (b), then up to position (c), and finally back to its original position (a). The **frequency** (f) is the number of oscillations or cycles completed in one second. The **period** is the time required to complete one oscillation, and is equal to 1/f, the reciprocal of the frequency. It is easily confirmed that it takes exactly the same time for each cycle to be completed, regardless of whether the weight swings through large or through small distances in respect to its original position.

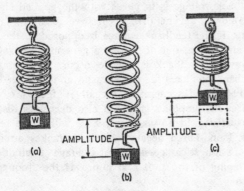

Fig. 78. *Simple Mechanical Oscillator*

The maximum distance the weight moves from its original point, either up or down, is called the **amplitude.** The amplitude of the spring pendulum becomes progressively smaller as the oscillations die down, and reaches zero when the pendulum comes to a stop. It is possible to obtain an accurate record of the action by attaching a fine ink-fed brush to the weight and drawing a paper tape horizontally past the swinging pendulum, in a slow and even motion. The result, shown in Fig. 79, is a graph of the displacements of the weight from its original position (the axis) against time. Since the oscillations eventually die down, the waveform of Fig. 79 is called a **damped oscillation.**

An analysis of the action described above discloses that two elements are required for oscillations

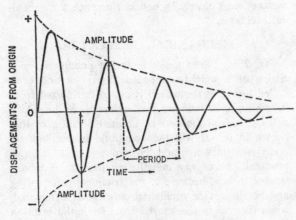

Fig. 79. *Waveform of Damped Oscillations*

to occur, **the spring and the weight.** Neither a weightless spring nor the weight alone can produce an oscillation. When the spring is extended by pulling the weight down, **potential energy** is stored in it in the form of tension. When the weight is let go, this tension or energy is released. The spring pulls back the weight and assumes its original slack position. During the motion, however, the potential energy of the spring has been transformed into the **kinetic energy** of motion of the weight. Because of its **inertia** or **flywheel effect,** the weight resists any sudden change in its motion. It does not stop, therefore, when the spring is slack again, but on the contrary continues to **compress** the spring until all its kinetic energy is again stored as potential energy in the compressed spring. Now the spring releases its tension again in the form of kinetic energy of motion of the weight, and the process is repeated all over again. The action would continue indefinitely if it were not for the fact that some energy is lost in the form of heat because of friction in the spring and bearings. As a result, a little less energy is stored in the spring during each cycle, until the oscillations finally die out.

If the weight of the spring pendulum is *increased,* the oscillations will take place more *slowly,* and if the weight is *decreased,* the oscillations occur at a *more rapid rate.* In other words, the frequency of oscillations is **inversely proportional** to the weight. If, on the other hand, the spring is shortened, the oscillations take place at a more rapid rate, and if it is elongated, the oscillations will be slower. Consequently, the frequency is also inversely related to the **length** of the spring. Instead of lengthening the spring, a more elastic or compliant spring could be substituted, and instead of shortening it, a stiffer spring could be used. Thus, the frequency can also be said to be inversely proportional to the **elasticity** of the spring.

The observations above can be stated as a general rule for all oscillators, whether mechanical or electrical. **Every oscillating system must have two elements, inertia and elasticity, which can store and release energy from one to the other at a natural frequency determined by the dimensions of the elements.** In one form or another inertia and elasticity are present in every oscillating system, and given an initial impulse they are sufficient to produce damped oscillations of the type described.

Many oscillators, such as the balance wheel of an ordinary watch or a radio-frequency generator, must be capable of producing continuous, undamped **oscillations** of the type shown in Fig. 80. In the spring pendulum this can be accomplished by pulling the weight each time it reaches bottom by just a sufficient amount to overcome losses due to friction. If the same result is to be achieved automatically, a mechanical source of energy must be supplied and some sort of synchronous trigger mechanism, which will release the energy to pull the weight at the moment it reaches the bottom. This is generally true for all oscillators producing continuous oscillations. An excellent example of a continuous mechanical oscillator is an ordinary clock or watch. Here, the balance wheel and hair spring represent the inertia and elasticity, respectively, of the oscillating system. The main spring is the source of energy, and the escapement is designed to release energy from the main spring at the proper instant during each oscillation of the balance wheel. An electrical oscillator, producing a continuous, undamped output, contains elements exactly analogous to those of a mechanical oscillator.

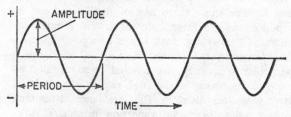

Fig. 80. *Waveform of Undamped Oscillations*

ELECTRON TUBE OSCILLATORS

Oscillators with a continuous output take energy from a unidirectional (d-c) source and transform it into undamped oscillations or alternations. An electrical oscillator, therefore, acts as an energy converter which changes direct-current energy into alternating-current energy. Because of their ability to amplify, electron tubes are very efficient energy converters, and are for this reason universally used as electrical oscillators. If the damped oscillations occurring in a resonant circuit containing inductance and capacitance are applied to the grid of a triode, the plate current of the tube varies in accordance with the grid signal, resulting in an **amplified** reproduction of the oscillations. Because of this amplification, more energy is available in the plate circuit than in the grid circuit. If part of this plate-circuit energy could be **fed back** by some means to the grid circuit in the proper **phase** to aid the oscillations of the resonant L-C circuit, its losses would be overcome and sustained, undamped oscillations would take place. This is in fact accomplished by a regenerative **feedback circuit,** which permits the combination of triode and resonant L-C circuit to function as a continuous self sustaining oscillator.

ESSENTIAL PARTS OF TRIODE OSCILLATOR

To produce electrical oscillations with an electron tube the following elements must be present:

(1) An oscillatory (resonant) circuit, containing inductance (L) and capacitance (C) to determine the frequency of oscillation. Such a circuit is called a **tank circuit** (See Fig. 81).

(2) A **source of (d-c) energy** to replenish losses in the tank circuit.

(3) A **feed-back circuit** for supplying energy from the source in the right phase (timing) to aid the oscillations; i.e., **regenerative** feedback. An electron tube can function as oscillator, if it has sufficient amplification (μ) *and* if a sufficient amount of energy is fed back to the tank to overcome all circuit losses. If the losses in the plate and the tank circuit are completely overcome, the **effective circuit resistance is zero,** and oscillations take place. However, if *either* the amplification of the tube *or* the amount of energy fed back are insufficient to overcome the circuit losses and make the effective resistance zero, the tube will not oscillate. Both conditions must be fulfilled for sustained oscillations to occur.

OSCILLATIONS IN A TANK CIRCUIT

A **tank circuit,** consisting of a capacitor and inductance coil in parallel (Fig. 81), is the simplest type of electrical oscillating system. Its action is analogous to that of the simple spring pendulum, just described, and like the latter it can generate damped oscillations.

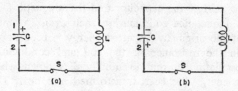

Fig. 81. *Operation of Tank Circuit*

To understand how this comes about, assume that the capacitor (C) of Fig. 81a has been charged from a d-c voltage source, and is now discharged through the inductance coil L, by closing the switch S. Assume further that plate 2 of the capacitor is initially charged negative (that is, it has an excess of electrons), while plate 1 is charged positively (that is, it has a deficiency of electrons). Consequently, an **electric field** exists between plates 1 and 2.

When the switch is closed, electrons move rapidly from plate 2 through the inductance *L* to plate 1. As soon as the electron current flows through *L*, a **magnetic field** begins to be established around the coil. In building up the field, the magnetic flux lines cut across the turns of the coil, and induce a **counter Emf which opposes the increasing current flow.** This slows down the rate of discharge of electrons from the capacitor. As plate 2 loses its surplus of electrons by discharge, the current tends to die down, but is prevented from doing so by the inductance which now opposes the *decrease* in current flow, and keeps the electrons moving. Consequently, plate 2 not only loses its original *excess* of electrons, but gives up more electrons resulting in a *deficiency,* or positive charge. Simultaneously, an *excess* of electrons is pushed onto plate 1, so that it acquires a negative charge, as shown in Fig. 81b. The process stops momentarily when all the energy that was stored in the magnetic field is used up in pushing an excess of electrons to plate 1. At that moment the magnetic field collapses, but the energy has been stored again in the electric field of the capacitor, which is now charged in the opposite direction. The action then continues with electrons moving out of plate 1 through the coil to plate 2. Again, a magnetic field is built up, which stores the energy, and prevents the electron current from dying down. This time, however, the current is flowing in the opposite direction, from plate 1 to 2, and the flywheel effect of the inductance now pushes an excess of electrons onto plate 2, recharging it negatively. A cycle is completed when the capacitor is again fully charged to the initial polarity.

The sequence of charge and discharge results in an alternating motion of electrons, or an oscillating current. The energy is alternately stored in the electric field of the capacitor and the magnetic field of the inductance coil. During each cycle, a small part of the originally imparted energy is used up as heat in the resistance of the coil and conductors. The oscillating current eventually dies down when all the energy has been transformed into heat. The waveform of these damped oscillations is exactly the same as that shown in Fig. 79 for the mechanical oscillator.

The capacitance of a tank circuit corresponds to the elasticity of a spring in a mechanical oscillator. An increase in capacitance, therefore, would be expected to lengthen the period of oscillations in a tank circuit, or equivalently, lower its frequency. This is indeed true. As the capacitance is increased, more charge must be transferred in and out of the capacitor during each cycle. Consequently, the period is lengthened, or the frequency is lowered.

We have pointed out that the greater the inertia element (weight) of a mechanical oscillator, the longer is the period of each oscillation. The self-inductance of a coil in a tank circuit corresponds to mechanical inertia. Accordingly, the greater the inductance of the coil, the greater is its opposition to any **change** in current flow and, hence, the longer is the time required for completion of each cycle.

The greater the value of the inductance, therefore, the longer is the period, or the lower is the frequency of oscillations in the tank circuit.

The frequency of oscillations in a tank circuit is, thus, inversely related to both the inductance and the capacitance. The formula for the **natural frequency** of oscillations is

$$f = \frac{1}{2\pi \sqrt{LC}}$$

where
$\pi = 3.1416$
$f =$ the frequency in cycles per second
$L =$ the inductance in henrys
$C =$ the capacitance in farads

This expression shows that for any given frequency there is an infinite number of combinations of L and C. The same frequency of oscillations can be produced with a large inductor L and a small capacitor C, or with a small L and a large C. In practice, however, the choice of the L/C ratio for a particular frequency is restricted because the ultimate performance of the tank circuit depends to a large extent on this ratio. The impedance of the tank circuit varies directly in proportion to the L/C ratio. Since the desired impedance of the tank circuit in an electron tube oscillator is determined to some extent by the tube operating conditions (i.e., the plate voltage and the plate current), the choice of L/C ratios is limited.

TICKLER FEEDBACK CIRCUIT

Historically, one of the earliest electron tube oscillator circuits is the so-called **tickler feedback oscillator,** devised by Major Armstrong (Fig. 82).

In Fig. 82, the oscillatory tank circuit is made up of L1 and C1, the source of d-c energy is the plate supply voltage, *E,* and the feedback circuit consists of the "tickler" coil L, which is coupled to L1. Capacitor C1 is made variable to adjust the frequency of the oscillations. The bypass capacitor is placed across the battery to provide a low-reactance path for the alternating component of the plate current. The combination of R and C in the grid circuit is called the **grid leak,** and its function is to furnish a self-adjusting negative bias to the tube. We shall discuss its operation later on. The key serves to interrupt the oscillations in accordance with a code, for use in telegraphy transmission.

Once oscillations have been started in the tank circuit L1-C1, they will appear in amplified form in the plate circuit of the tube. Part of the energy is fed back from tickler coil L to tank coil L1 by mutual induction, thus overcoming losses and sustaining the oscillations. You can easily ascertain that the phase of the feedback is correct. The tube itself introduces a phaseshift of 180° between grid and

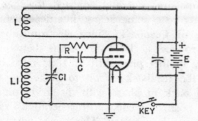

Fig. 82. *Tickler Feedback Oscillator Circuit*

plate circuit, as you will remember. The combination of L and L1 constitutes a transformer, and consequently, another phaseshift of 180° takes place between the plate and the grid circuit. (All transformers introduce a phaseshift of 180° between primary and secondary.) As a result, the voltage fed back is **in phase** with the voltage in the grid circuit, and regenerative or positive feedback takes place.

To understand the physical action you must remember that only a **changing** magnetic field is capable of inducing a voltage in a nearby coil. An **expanding** magnetic field induces a voltage in a certain direction in the coil, while a **contracting** magnetic field induces a voltage of the opposite polarity.

Assume the cathode of the triode tube is heated and electrons are being emitted. The moment the key is closed, plate current begins to flow through the tube and through the external circuit, consisting of the tickler coil L and the battery. This current sets up an **expanding** magnetic field around the tickler coil *L* and so induces a voltage in the tank coil L1. Assume that the initial induced voltage is positive so that the upper end of the L1-C1 tank circuit is of positive polarity. This positive voltage charges the capacitor C1 and places a positive charge on the grid of the tube which is connected to the upper end of L1-C1. Since the grid has initially no bias, this positive charge increases the plate current and thus further builds up the magnetic field around the tickler coil L. As a result, a still larger positive voltage is induced in L1 and placed on the grid, and C1 is further charged. Again the plate current keeps rising and the field of L expands, placing a still greater positive voltage on C1 and the grid. Theoretically, this process continues until the plate current reaches its saturation point and tapers off. At that instant the field about coil L stops expanding and becomes static.

As the magnetic field about coil L stops expanding, the voltage induced in L1 begins to drop and finally reaches zero. Now, however, capacitor C1, which has been charged to the maximum positive voltage, begins to discharge in the *opposite* direction, making the upper portion of L1-C1 negative. As the potential on the upper end of L1-C1 is reduced from its positive value to zero and then becomes negative, the voltage on the grid is equally

reduced, thus *lowering* the plate current. As the plate current decreases, the field about L starts *contracting* and induces a *negative* voltage in L1. This leads to a further discharge of capacitor C1 and a still greater negative grid voltage. As the negative charge on the grid increases, the plate current drops more and more, until finally it is cut off. At this instant the field about L is completely collapsed, the voltage induced in L1 disappears, and the grid voltage rises to zero. Since the grid is now less negative (or more positive) than the value it has at cutoff, the plate current increases again, and the entire cycle is repeated.

It must be understood that the entire process takes place very quickly, and is repeated thousands of times in an extremely brief period of time, at a rate determined by L1 and C1. You must further remember that oscillations will not take place if the amplification of the tube or the energy fed back from the tickler coil L to the tank circuit becomes insufficient to overcome circuit losses. The amount of energy fed back depends on the **mutual** inductance between L and L1. As the coils are spaced farther apart, the mutual coupling between the coils decreases, and at a certain critical value of the mutual inductance, the coupling becomes too loose to sustain oscillations.

Grid-Leak Biasing. The grid-leak resistor R and grid capacitor C in Fig. 82 furnish the negative bias required for the tube's operation. Grid-leak bias, rather than a resistor in the cathode circuit (cathode bias), is universally used in triode oscillators to insure stable operation and make the oscillator self-starting. Capacitor C is large enough to provide a free (low-reactance) path to the grid for the excitation signal, and thus bypasses the high-resistance grid leak R. In operation the grid is driven positive during positive half-cycles of the oscillations and, therefore, draws grid current. The electron current flows from the cathode to the grid and then through the external circuit consisting of L1 and R, thus developing a voltage drop across R. The end of R connected to the grid is made more negative than the other end, and the grid is thus biased negatively by an amount equal to the voltage drop across R. The voltage present across R during grid-current flow charges capacitor C. If R and C are sufficiently large, the charge on C will leak off only very slowly during negative half-cycles, when no grid-current flows. Hence, for all practical purposes, the voltage across C remains constant throughout a complete cycle, and maintains a steady bias on the tube.

The use of grid-leak bias also makes the oscillator **self-starting,** since the grid bias is initially zero, making the plate current and the amplification large. Any initial impulse, such as the closing of the key, will thus be considerably amplified and start the building up of oscillations in the manner described before. Small random variations are always present in the circuit to start the oscillations.

HARTLEY OSCILLATOR

A modified version of the tickler feedback circuit using a tapped coil common to both the plate and the grid circuit, rather than two separate coils is known as the **Hartley oscillator** (Fig. 83). Except for minor modifications in the manner in which coupling is obtained between the plate and grid circuit, the operation of the Hartley oscillator is

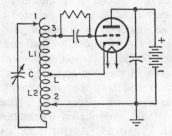

Fig. 83. *Hartley Oscillator*

identical with that of the tickler feedback oscillator just discussed.

The lower portion of the tank coil, L2, is inductively coupled to the upper portion, L1, the combination functioning as an **autotransformer.** Since the variable tank capacitor C is connected across both coil sections, capacitive coupling is present in addition to the mutual inductance. Adjustable taps are customarily provided to control the operating conditions.

Tap 1 together with the variable capacitor adjusts the frequency of oscillations. The inductance L of the tank can be decreased by moving the tap down.

The frequency of the oscillator is $\dfrac{1}{2\pi\sqrt{LC}}$, where

L is the total inductance connected across C.

Tap 2 adjusts the effective impedance of the plate-circuit coil L2, and has only a slight effect on frequency. The more turns are included between cathode and plate, the higher the plate-circuit impedance.

Tap 3 adjusts the grid excitation voltage to the proper value for maximum output. Sufficient excitation must be provided to overcome all losses in the tank and associated plate circuit and hence make the effective circuit resistance zero, as previously explained. At the same time the excitation must not be so large as to overdrive the oscillator and thus distort the waveform of the output. The excitation is increased by moving tap 3 upward.

All the taps are generally adjusted together, since their setting is to some degree interdependent.

Phasing of Feedback. Since electron tubes normally introduce a 180° phaseshift between grid and plate, another phaseshift of 180° must be provided in the feedback circuit to make the voltage being fed back in phase with the grid voltage. In this way positive feedback (regeneration) required for oscillation is obtained. This phase correction is obtained in the tank coil. You can readily confirm that the two opposite ends of the auto-transformer L1-L2 are actually 180° out of phase, i.e., whenever one end of the transformer winding is positive the other end is negative. When the upper end of L1 (connected to the grid) is positive, for instance, the lower end is at a minimum potential. However, the tap connected to the cathode is at an intermediate voltage, and so is negative with respect to the upper end and positive with respect to the lower end. Or, viewed from the tap, the upper end of the coil is positive and the lower end is negative. Since the grid and the plate are connected to opposite ends of the coil, they are of opposite polarity or phase. In this way, the feedback is properly phased to sustain oscillations.

COLPITTS OSCILLATOR

The Colpitts oscillator (Fig. 84) is similar to the Hartley circuit just discussed, except that two variable capacitors, C1 and C2, are used in the tank circuit instead of the tapped coil. The grid voltage is adjusted by capacitor C1, instead of the tap used in the Hartley circuit. The tank is again common to both the plate and grid circuits, but the feedback is obtained by the relative voltage drops across the two capacitors. In Fig. 84, the plate is said to be **shunt-fed,** in contrast to the **series-feed** used in Fig. 83. Blocking capacitor C_b prevents the d-c plate current from reaching the tank circuit while the radio-frequency choke keeps the r-f currents out of the power supply. Grid-leak bias is used, but the

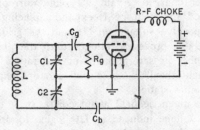

Fig. 84. *Colpitts Oscillator*

grid leak R_g is connected in parallel with the grid circuit, rather than in series with it. This is necessary to provide a d-c return path for the grid current, since otherwise the grid circuit would be completely isolated from the cathode for d.c. Except for these minor modifications, the operation of the Colpitts oscillator is the same as that of the Hartley circuit.

Feedback. The two tank-circuit capacitors C1 and C2 act as a simple a-c voltage divider. The tap between them fulfills the same purpose as the tapped coil in the Hartley circuit. It assures both the proper amount and the correct phase of the feedback voltage. As seen from the cathode-connected tap, whenever the top of C1 is positive, the plate-connected end of C2 is negative, or vice versa. Because of this polarity reversal, the correct 180° phaseshift is obtained.

Frequency. The frequency of oscillations is determined by the coil L and the total capacitance C_T in the tank circuit. As before, the formula is

$$f = \frac{1}{2\pi\sqrt{LC_T}}$$

By a well-known formula, the capacitance of two capacitors in series is

$$C_T = \frac{C1 \times C2}{C1 + C2} \qquad \text{(See Review Section)}$$

Hence we can write

$$F = \frac{1}{2\pi}\sqrt{\frac{C1 + C2}{L \times C1 \times C2}}$$

TUNED PLATE-TUNED GRID OSCILLATOR

The tuned-plate tuned-grid oscillator (abbreviated TPTG) uses a tuned tank circuit in both the grid and the plate circuits, as shown in Fig. 85. The two coils L1 and L2 are *not* inductively coupled. Feedback takes place entirely through the grid-to-plate capacitance (C_{gp}) of the tube. In order to obtain oscillations of the correct frequency, both the grid and plate tank circuits must be tuned to a frequency somewhat higher than the desired operating frequency.

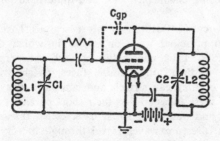

Fig. 85. *Tuned-Plate Tuned-Grid Oscillator*

ELECTRON-COUPLED OSCILLATOR

To minimize the "detuning" effect of a load on the frequency of oscillations, **electron-coupled os-**cillators are frequently employed. These substitute a common electron stream to couple the load to the plate circuit in place of inductive or capacitive output coupling. Two commonly-used types of electron-coupled oscillators are shown in Fig. 86. In each case, a single pentode fulfills both the functions of a triode oscillator and an isolating or **buffer** amplifier. The cathode, control grid, and screen grid (acting as the plate) form a triode oscillator of the conventional type. In Fig. 86a, this is a modi-

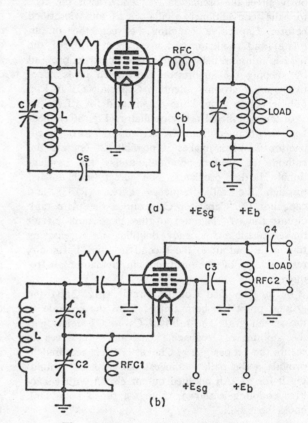

Fig. 86. *Electron-Coupled Oscillators*

fied Hartley; in (b) it is a modified Colpitts oscillator. The load is connected in the plate circuit of the pentode. The circuit derives its name from the fact that the oscillator and plate circuits are coupled solely by the stream of electrons between screen grid and plate.

Since the screen grid is at a positive voltage, electrons flow from cathode to screen grid, and oscillations are generated in the same manner as in the case of the conventional triode oscillator. The frequency is determined by the value of L and C in the grid tank circuit. Only a small number of electrons are intercepted by the screen, but they are sufficient to maintain oscillations in the tank circuit. The remaining electrons, which represent most of the space current, go on to the plate and through the load impedance connected in series with it. Since the screen-grid current varies in intensity with the oscillations, the intensity of the electron stream through the screen grid to the plate will be varied accordingly. Thus the plate current is **modulated** at the oscillator frequency by the action of the screen grid and control grid. In effect, the screen grid and plate act as a triode whose control electrode voltage is varied at the oscillator frequency. Changes in the plate loading do not seriously affect the oscillation frequency, as is the case for the conventional oscillator circuits described before. Capacitive coupling between the output (plate) and oscillator sections (screen grid) of the tube is minimized by the pentode construction, and by keeping the suppressor and screen grids effectively at r-f ground potential. Capacitor C3 in Fig. 86b serves to ground the screen grid for r.f.

In the modified Hartley oscillator (Fig. 86a) feedback is obtained from the screen grid through capacitor C_s. This makes it possible to operate the cathode at ground potential, unlike the conventional Hartley oscillator. The screen is shunt-fed through the radio-frequency choke (RFC) and capacitor C_b. Capacitor C_t grounds the plate tank circuit for r.f. The use of the plate tank circuit makes possible frequency doubling or tripling by tuning the output to the second or third harmonic, respectively, of the grid circuit fundamental frequency.

In the modified Colpitts circuit (Fig. 86b) the ground point has been shifted from the cathode to the screen grid. The ratio of C1 to C2 determines the amount of feedback, while the tap serves to secure the proper phase. Choke RFC1 is required to provide a d-c path to the cathode without grounding it for r.f. An untuned output circuit with capacitive coupling is shown, though a tuned plate tank could be used.

CRYSTAL OSCILLATORS

The most satisfactory method of stabilizing the frequency of radio-frequency oscillators is by the use of quartz crystals. Oscillators of this type are called **crystal-controlled.** They are used in the great majority of commercial and military radio telephone and telegraph transmitters.

Piezoelectric Effect. The control of frequency by means of crystals is based upon the **piezoelectric effect.** When some crystals are compressed or stretched in certain directions, electric charges appear on the surfaces of the crystal that are perpendicular to the axis of mechanical strain. Conversely, when such crystals are placed between two metallic surfaces between which a difference of potential exists, the crystals expand or contract.

Thus, if a slice of a crystal is compressed along its width, or stretched along its length, so that it bulges inward, as in Fig. 87a, opposite electrical charges will appear across its faces, and a **difference of potential is generated.** If the crystal is squeezed or compressed lengthwise, so that it bulges outward, as in (b) of Fig. 87, the charges across its faces reverse. If alternately stretched and squeezed, a crystal slice becomes a source of alternating voltage. Conversely, if an alternating voltage is applied across the faces of a crystal wafer, it vibrates mechanically. The amplitude of these vibrations becomes very vigorous, if the frequency of the a-c voltage equals the natural mechanical frequency of vibration of the crystal, that is, if **resonance** takes place. If all mechanical losses are overcome, the vibrations at this natural frequency will sustain themselves and generate electrical oscillations of constant frequency. Accordingly, a crystal can be substituted for the tuned tank circuit in an electron tube oscillator.

Fig. 87. *Expansion and Contraction of a Crystal*

Types of Crystals. Practically all crystals exhibit the piezoelectric effect, but only a few are suitable as the equivalent of tuned circuits for frequency-control purposes. Among these are quartz, Rochelle salts (sodium potassium tartrate), and tourmaline. Rochelle salt is the **most active** piezoelectric substance—that is, it generates the greatest amount of voltage for a given mechanical strain. Rochelle salts are physically and electrically instable, however, and therefore not suitable for frequency control. They have found applications in microphones and phonograph pickups. Quartz, although being much less active than Rochelle salt, is used universally for frequency control of oscillators because it is cheap, mechanically rugged, and expands very little with heat. Quartz is among the most permanent materials known, being chemically inert and quite hard physically.

Frequency of Oscillation. Most crystals have at least two principal ways or **modes** in which they can vibrate. They can bulge in and out perpendicularly to their long parallel faces, as shown in Fig. 87, or they can stretch and contract along the width of these faces so that their short parallel faces bulge in and out. In the first case, **called thickness vibration**, pressure waves travel through the crystal from one long face to the opposite face and reflected back from there to the first face and are reflected back again. At a particular thickness of the crystal, the reflected waves are in phase with the direct waves, and reinforce each other. As a result **standing waves** are created between the two long faces of the crystal. The fundamental natural frequency of oscillation occurs at that particular thickness, where at least one complete wavelength (or cycle) can exist between the two long faces. However, for the same thickness, two or more shorter complete wavelengths can also exist between the two faces. Due to the inverse relationship between frequency and wavelength this means that the crystal also can vibrate at the second, third, or higher harmonic of this fundamental frequency. But for a given thickness, the fundamental frequency of the first principal mode is fixed.

The second principal mode of vibration (the short faces bulging in and out) is determined by the width of the plate, as measured along the long parallel faces. It is known as **width vibration.** Again, standing waves occur at the natural frequency of oscillation, and harmonics of this fundamental frequency are possible. But the fundamental frequency is determined by the width. Besides these principal modes and their harmonics, additional modes of vibration are possible, produced by various flexural and torsional tensions. Quartz crystals are produced commercially for frequencies from about 50 kilocycles to as high as 50 megacycles.

Crystal Mounting. Crystals become practical circuit elements when they are associated with a crystal holder. In a holder (Fig. 88a), the crystal is placed between two metallic electrodes and forms a capacitor, the crystal itself being the dielectric. The crystal holder is arranged to add as little damping of the vibrations as possible, and yet it must hold the crystal rigidly in position. This is done in various ways. In some holders the crystal plate is firmly clamped between the metal electrodes, while others permit an **air gap** between the crystal plate and one or both electrodes. The size of the air gap, the pressure on the crystal, and the size of the contact plates affect the operating frequency to some degree. The use of a holder with an **adjustable air gap** permits slight adjustments of frequency (about 1/10%) to be made. For the control of appreciable amounts of power, however, a holder which firmly clamps the plate is generally preferred.

Equivalent Circuit. At its resonant frequency a crystal behaves exactly like a tuned circuit, so far as the electrical circuits associated with it are concerned. The crystal and its holder can be replaced therefore, by an **equivalent electrical circuit,** as shown in Fig. 88b. Here, C_m represents the capacitance of the mounting with the crystal in place between the electrodes, but not vibrating. C_g is the effective series capacitance introduced by the air gap when the contact plates do not touch the crystal. The series combination L, R, and C repre-

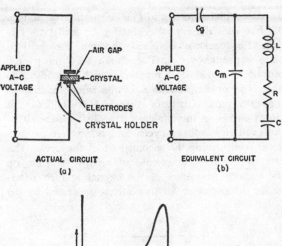

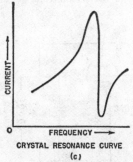

Fig. 88. *Crystal with Holder (a) Actual Circuit (b) Equivalent Circuit (c) Crystal Resonance Curve*

sents the electrical equivalent of the vibrational characteristics of the quartz plate. The inductance L is the electrical equivalent of the crystal mass effective in the vibration, C is the electrical equivalent of the **mechanical compliance** (elasticity), and R represents the electrical equivalent of the mechanical friction during vibration.

The frequency at which L and C are in series resonance is also the frequency of mechanical crystal resonance. Because of the presence of C_m, the circuit will also have a **parallel-resonant** frequency slightly *above* series resonance, when the series branch has an inductive reactance equal to the capacitive reactance of C_m. (A series L-C-R circuit is inductive above resonance.) Since C_m has a very low reactance (high C), only a small inductive reactance is required in the L-C-R branch to produce parallel resonance with C_m. The series and parallel resonant frequencies, therefore, are very close together. The presence of both resonant frequencies is clearly revealed by crystal resonance curve (Fig. 88c). This curve is extremely sharp, which means that the crystal will vibrate only within a very narrow band of frequencies. As you will learn later, this provides high **selectivity,** or discrimination against unwanted frequencies.

CRYSTAL OSCILLATOR CIRCUIT

A typical crystal oscillator is shown in Fig. 89. You can easily verify that this is the equivalent of the tuned plate-tuned grid (TPTG) oscillator of Fig. 85, but with the crystal replacing the grid tank circuit. Feedback is obtained through the grid-to-plate capacitance C_{gp}. The choke RFC keeps r.f. out of the grid leak resistor, which provides the bias. The crystal functions in the same manner as the grid tank circuit of the TPTG oscillator. It stores energy in mechanical form during one-half of the excitation voltage cycle, and releases it in electrical form during the second half of the cycle. The rate of storage and release of energy depends on the natural frequency of the crystal and so determines the frequency of oscillations generated by the

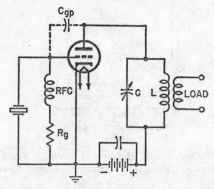

Fig. 89. *Typical Crystal Oscillator Circuit*

circuit. The losses in the crystal are overcome by the energy fed back through C_{gp}.

Frequency. The resonance curve of the crystal (Fig. 88c) is obtained by tuning the plate tank from a frequency below crystal resonance to one above crystal resonance. As the frequency is increased, series resonance of the crystal is reached first as indicated by the high current peak in the resonance curve. (The impedance is a minimum, and hence the current is a maximum for series resonance.) When the frequency of the plate tank is increased slightly above this value, **parallel** resonance of the crystal is reached—that is, the inductive reactance of the crystal proper equals the capacitive reactance of the crystal holder capacitance. This is indicated by the sharp drop of the current in the resonance curve (Fig. 88c) and consequent high crystal impedance. Oscillations commence when the frequency of the plate tank is again slightly increased to make the plate circuit inductive. The presence of oscillations can be detected by a sharp drop in plate circuit as the frequency of the plate tank is raised.

Since the crystal can oscillate practically only at its resonant frequency, the frequency of oscillation remains constant over a wide range of adjustment of the tuning capacitor C. The power output changes, however, substantially, when C is changed.

Changes in plate voltage, filament voltage, and replacement of tubes have only a slight effect on the frequency of oscillations.

Advantages and Limitations. The outstanding characteristic of the crystal is the extreme sharpness of its resonance curve. Because of this characteristic, the crystal can oscillate only over a very narrow frequency range and hence the frequency stability of a crystal oscillator is extremely high. This is taken advantage of in military communications and commercial broadcasting, where the frequency tolerances are very close. In addition to the use of crystal oscillators in fixing the frequency of transmitters, they are used extensively as **frequency standards** for measurement purposes. If a low-frequency crystal is used in a circuit whose output is not tuned (a simple choke) a large number of harmonic frequencies (multiples of the basic frequency) are created. A great number of frequency calibration points can be obtained in this way with a single quartz crystal.

A crystal-controlled oscillator is a **fixed-frequency** oscillator. Its consequent disadvantage is that a different crystal must be used for each desired frequency. In many applications, especially military, it often is required to change the frequency of the transmitter rapidly and continuously. For these applications the ordinary variable-frequency oscillator is preferred, since it may be operated at any frequency within a band at the turn of a dial. Another limitation of the crystal oscillator is its low power output.

MICROWAVE OSCILLATORS

We have seen in Chapter 6 that conventional electron tubes become progressively less effective as amplifiers and oscillators, as the operating frequency is raised above 100 megacycles. You will remember from our discussion there that the reason for this falling off in efficiency is essentially three-fold:

1. The **internal capacitances and inductances** of conventional tubes become low reactances at very high frequencies, thus shunting out an appreciable part of the voltages applied to the tube's electrodes. Moreover, at a certain critical frequency, these internal reactances produce **self-resonance**, regardless of the tuned circuits connected externally to the tube. Tubes cannot be operated above this resonant frequency.

2. The **transit time** of the electrons passing between the electrodes of the tube becomes an appreciable part of the cycle at high operating frequencies (between 300 and 3000 megacycles). As we have seen this leads to a considerable radio-frequency power loss.

3. At very high frequencies the dimensions of the

tube elements and associated circuit become comparable to the wavelength of the signal (wavelength = velocity/frequency) and, hence, considerable radio-frequency losses and power losses due to **direct radiation** from the tube and associated circuit take place.

We have further seen (in Chapter 6) that specially designed electron tubes, such as the acorn and lighthouse tubes, are capable of functioning well in the UHF region between 300 and 2000 megacycles. The generation of frequencies in the **microwave** region between 2000 and 30,000 megacycles (30 billion cycles), however, demands radically different techniques than have heretofore been described. These techniques make use of the very limitations in tube dimensions and electron transit time in achieving their normal operation. Among the most important tubes using these novel principles to generate microwave frequencies are the so-called **Klystrons** and **magnetrons.**

KLYSTRON OSCILLATOR

Klystrons make use of the transit time of the electrons to obtain a transfer of energy from the moving electrons to an electric field produced by an a-c voltage. Oscillations are produced by making most of the electrons pass through the electric field at such times that they **deliver energy** to the source of the alternating electric field. (Oscillations can occur only when more energy is delivered to the field than is taken from it.) This is accomplished by forming the electrons emitted from the cathode of the tube into compact **bunches** or groups, which deliver energy to the electric field on the grid at just the right time to synchronize with the alternations of the field. This principle is called **velocity modulation.** Klystrons are one type of velocity-modulated tube.

As in conventional oscillators, the frequency of the oscillations excited by a Klystron is determined by a resonant circuit. However, the ordinary L-C tank circuit is replaced in microwave tubes by a **resonant cavity** or chamber whose physical dimensions determine the frequency of its vibrations when it is excited by electron oscillations. A pulsating stream of electrons can excite oscillations in a resonant cavity in the same way as pulses of plate current excite a plate tank circuit to its resonant frequency.

Klystron Tubes. A tube which performs this action is shown schematically in Fig. 90a. At one end of the Klystron tube is an **electron gun,** similar to that used in the cathode-ray tube, consisting essentially of a cathode and accelerator grid. The electrons are emitted from the heated cathode and are attracted toward the **positive** accelerator grid. Most of the electrons pass through the grid wires to form

a beam of electrons, all traveling at the same speed. The beam of electrons is then passed through a pair of closely spaced grids, called the **buncher,** each of which is connected to one side of a tuned (resonant) circuit. An alternating voltage exists across the resonant circuit, which speeds up or slows down the electrons, depending on the instant they enter the space between the buncher grids. An electron that passes the center of the buncher at the instant when the a-c voltage passes through zero, leaves the buncher with unchanged velocity. Electrons which pass through the buncher a little earlier, when the a-c voltage is still negative, will be slowed down, while electrons which pass through the buncher a little later, when the a-c voltage is positive, will be speeded up. This action causes the electrons to bunch together at some point beyond the buncher grids. Thus, the electron stream consists of bunches of electrons separated by regions in which there are few electrons.

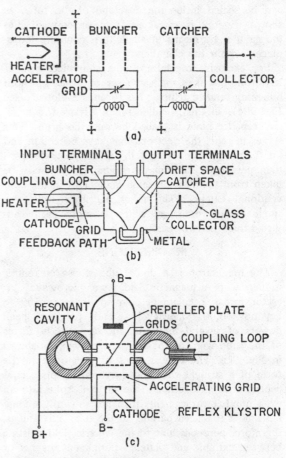

Fig. 90. *Klystron Oscillators*

These bunches of electrons pass through a second set of grids, called the **catcher,** which also are coupled to a resonant circuit. The polarity of the a-c field on the catcher grids is such, that the electrons are slowed down, and thus give up some of their kinetic to the electric field and, consequently, to the tuned circuit. The spent electrons are removed from the circuit by a positive collector plate. The bunches pass through the catcher at the resonant frequency of the tuned circuit (i.e., once each cycle), and so maintain continuous oscillations in the tuned circuit. Some of the energy is fed back from the catcher to the resonant circuit of the buncher (see Fig. 90b) in the proper phase, thus producing self-sustained oscillations. This type of Klystron can be used also as an amplifier or mixer.

As mentioned before, the tuned circuits of the buncher and catcher take the form of hollow metal chambers, called **cavity resonators,** with one of the grids attached to each side of the cavity (Fig. 90b). These cavities possess all the properties of conventional tank circuits, but are much more efficient at microwave frequencies. They are so tiny at these extremely high frequencies, that the entire cavity can be sealed inside the envelope of the tube, as shown. Energy is extracted from or coupled into the cavities by means of single-turn coupling loops, placed within the chambers.

For exclusive oscillator use, a simplified type of Klystron that uses the same set of grids for both bunching and catching, has been developed (see Fig. 90c). In this so-called **reflex Klystron,** a negative **repeller plate** is placed beyond the grids. This serves to repel the electrons that have been bunched on their first trip through the grids. The electrons then pass back through the grids, where energy is taken from them in the same manner as in the conventional catcher. The reflex Klystron utilizes a single cavity resonator, which is more easily adjusted than the two-resonator Klystron.

MAGNETRONS

The magnetron is a diode tube whose current is influenced by a magnetic field. As a microwave generator it can oscillate at frequencies from 300 to beyond 30,000 megacycles, and produce peak powers of several thousand kilowatts. Early forms of the tube were of the split-anode type, as shown in Fig. 91a. The tube consists of a cathode in the form of a straight wire and a cylindrical plate concentric with the cathode; the plate is split into an even number of segments (two are shown). A magnetic field is placed parallel to the tube axis and is, therefore, **perpendicular** to the electric field existing between cathode and plate. A permanent magnet or an electromagnet can be used to provide the magnetic field.

In operation, the segments of the plate are kept at some positive potential with respect to the cathode. In the absence of a magnetic field, the electrons travel to the plate in a straight path. However,

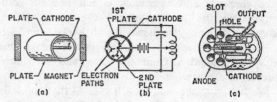

Fig. 91. *Magnetrons* (*a*), (*b*) *and* (*c*)

as a magnetic field is applied and increased, the electron path becomes more and more curved, and finally circular. (It is shown in electron physics that magnetic and electric fields at right angles tend to bend the electron paths into circles.) At a certain critical value of the field strength, the electrons will miss the plate entirely and return to the cathode in a circular orbit, as shown in (b) of the figure. This is the point of **plate-current cutoff.** At higher values of the field strength the radii of the circular orbits become smaller.

When the plate segments are made part of a resonant circuit (Fig. 91b) and the magnetic field is adjusted to plate-current cutoff, so that the electrons just fail to reach the plate, ultra-high-frequency oscillations can be produced. The action depends upon the **transit time** of the electrons. An alternating voltage is superimposed on the constant voltage present on the plate segments, causing the plate voltage to vary about its d-c value. If the period of the a-c voltage is made equal to the transit time of an electron for a complete circular rotation, some electrons will be slowed down by the alternating field at the plate, and lose energy to it; others will be speeded up and, thus, gain energy from the a-c electric field. The tube can be adjusted so that energy is extracted *by* the electric field from the majority of electrons grazing the plates. This energy sustains powerful oscillations in the associated resonant circuit. The frequency of oscillations is determined primarily by the transit time of the electrons. The magnetron also can be operated as a **negative-resistance** oscillator at frequencies that are low compared to the transit-time frequency. In this case, the frequency can be varied continuously by changing the constants of the external resonant circuit.

Modern super-high-frequency magnetrons are transit-time types made in the form of a multianode cavity resonator. The basic structure is shown in Fig. 91c. The assembly is a solid block of copper which assists in heat dissipation. The cathode is a cylinder of appreciable diameter located in a cavity in the center of the structure. The plates or anodes are cut out of the block in the form of large circular holes and are divided into 4 to 16 or more segments. Radial slots lead out from the common cathode region to smaller circular holes, which act as resonant cavities. Each slot and terminating hole is electrically equivalent to a tuned resonant circuit, the slot having predominantly capacitive action, and the terminating hole being primarily inductive. Under proper conditions of voltage and magnetic field strength, energy is transferred from the swarm of gyrating electrons to the resonant cavities and powerful oscillations are sustained. The oscillations are coupled to the output by means of a wire loop, placed into one of the small circular holes. A cutaway view of a multicavity magnetron is shown in Fig. 92.

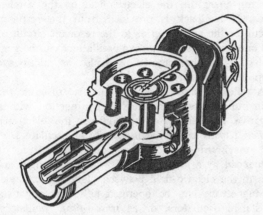

Fig. 92. *Cutaway View of Multicavity Magnetron*

TRAVELING-WAVE TUBES

The traveling-wave tube (TWT) makes use of the interaction between an electron beam and a traveling electromagnetic wave to obtain highly efficient microwave amplification. Continuous-wave powers of hundreds of watts in the frequency range from 1000 to 10,000 megacycles are easily obtained, with voltage gains of 50 db or better over a 1000-mc bandwidth being typical.

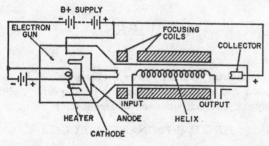

Fig. 93. *Schematic Drawing of Traveling-wave Tube*

Figure 93 illustrates the physical construction of a traveling-wave amplifier tube. One end of the tube is an electron gun that produces a pencil-like beam of electrons at the gun anode. This beam is shot through a long, loosely wound helix and strikes the collector electrode, which is at the anode potential. The electromagnetic wave to be amplified is applied

through a wave-guide section at the input end of the helix and travels along the turns of the helix. As it does so, it produces an electric field at the center of the helix which is directed along the axis of the helix. When the velocity of the electron beam approximates the rate of advance of the axial field of the wave traveling along the helix, an interaction takes place in which the electrons impart energy to the traveling wave. As a result, a substantially amplified wave emerges from the output end of the helix. This wave is then coupled to another wave guide.

In practice, two electromagnetic focusing coils surround the glass envelope of the traveling-wave tube. One focuses the electron beam at the exit of the electron gun, while the other—arranged along the length of the tube—prevents the beam from spreading and guides it along the center of the helix. Although the traveling-wave tube can be used only as amplifier, its essential operating principle has also been applied to an oscillator, known as **backward-wave oscillator.**

SUMMARY

One complete oscillation is a **cycle; frequency** (f) is the number of cycles completed in one second; **period** is the time required to complete one cycle and is equal to 1/f; **amplitude** is the maximum displacement from the origin.

Damped oscillations die down because of internal losses. If losses are overcome, **undamped,** continuous oscillations take place.

Every **oscillating system** has two elements—inertia and elasticity—capable of storing and releasing energy from one to the other at a natural frequency determined by the dimensions of the elements.

Energy in a tank circuit is alternately stored in the electric field of the capacitor and the magnetic field around the coil.

The frequency of natural oscillations in a tank circuit is

$$f = \frac{1}{2\pi \sqrt{LC}}$$

An electrical oscillator acts as an energy converter which changes direct-current energy into alternating-current energy.

The essential parts of a triode oscillator are:

(1) An oscillatory **tank circuit,** containing L and C, to determine the frequency of oscillations.

(2) A **source of (d-c) energy** to replenish losses in the tank circuit.

(3) A **feedback circuit** for supplying energy from the source in the right phase to aid the oscillations (**regenerative** or positive feedback).

The amplification factor of the tube and the

amount of energy fed back must be sufficient to overcome all circuit losses.

In the **tickler feedback oscillator,** positive feedback occurs because of mutual induction between the tickler and tank coils.

The **Hartley oscillator** is a modified version of the tickler feedback circuit, using a tapped coil, rather than two separate coils.

The **Colpitts oscillator** is similar to the Hartley circuit, except that it uses a split capacitor in place of the tapped coil.

The **tuned plate-tuned grid oscillator** uses a tuned tank circuit in both the grid and the plate circuits. Feedback in the TPTG oscillator takes place entirely through the grid-to-plate capacitance (C_{gp}) of the tube.

In an **electron-coupled** oscillator a common electron stream couples the load to the plate circuit; this minimizes the effect of the load on the frequency of oscillation.

The control of frequency by means of crystals is based upon the **piezoelectric effect.** Quartz is the most suitable piezoelectric crystal because it is cheap, mechanically rugged, chemically inert, and expands very little with heat.

Most crystals can vibrate in a **thickness mode** at a high frequency and in a **width mode** at a substantially lower frequency.

At its frequency of mechanical resonance, a quartz crystal behaves exactly like an electrical tuned circuit and can, therefore, be represented by an **equivalent** electrical circuit.

The basic crystal oscillator circuit is the equivalent of the tuned-plate tuned-grid oscillator, with the crystal replacing the grid tank circuit; its frequency of operation is the frequency of mechanical resonance of the crystal.

The advantages of crystal oscillators are extreme sharpness of resonance, high frequency stability, and possible use as frequency standards.

The limitations of crystal oscillators are their fixed frequency and relatively low power output.

The **Klystron** is based on the principle of **velocity modulation;** Klystrons form the electrons emitted by the cathode into compact bunches, capable of delivering energy to the electric field on the **catcher grids.** If feedback is provided from the resonant circuit of the catcher grids to the resonant circuit of the buncher grids, microwave oscillations can be generated. The tube may be used also as microwave amplifier or mixer.

A **reflex** Klystron uses the same set of grids for both bunching and catching, in conjunction with a negative **repeller plate** that reflects the electrons. Reflex Klystrons can be used only as oscillators.

A **magnetron** is a diode whose plate current is influenced by a magnetic field, which is at right angles to the electric field between plate and cathode.

Magnetrons can be operated either as negative-resistance oscillators, or as transit-time oscillators. In the first case, the frequency of oscillation is adjustable by means of the resonant circuit; in the second, the frequency is determined by the transit time.

A traveling-wave tube (TWT) makes use of the interaction between a focused electron beam and a traveling electromagnetic wave to amplify microwave signals of thousands of megacycles over a large bandwidth and with high gain and low noise.

Chapter Twelve

TUNED RADIO FREQUENCY AMPLIFIERS

Most of the "audio" amplifiers we have discussed in earlier chapters will also work at radio frequencies —that is, from about 50 kilocycles up. However, the ordinary untuned, resistance-coupled amplifier becomes less efficient as the frequency goes up and the amplification falls off rapidly. Moreover, the audio amplifier amplifies a wide band of frequencies equally well and does not permit the selection of a particular desired frequency while discriminating against others. The ability to select a desired frequency (or band of frequencies) is quite important, however, inasmuch as radio and television transmissions are "carried" on a specific radio frequency assigned to the station. The use of tuned circuits in conjunction with electron tube amplifiers makes possible the selection and efficient amplification of a specific radio frequency, or narrow band of frequencies.

TUNED (RESONANT) CIRCUITS

The tank circuit we discussed in the chapter on oscillators is a tuned or resonant circuit. We have seen that such an L-C tank circuit is capable of functioning as an oscillating system when d-c pulses of energy are fed to it with the right timing to excite its "natural" frequency of oscillations. In this case a tank circuit functions as an energy converter from direct to alternating current. Oscillations in a tank circuit may also be excited by alternating current, provided the frequency of the alternations is the same as the natural frequency of tank circuit oscillations. In the latter case the tank circuit is said to be **in resonance** at the a-c frequency.

The principle of resonance is best illustrated by ordinary physical objects. Every object has its own natural frequency of vibrations, depending on its size and mass. For example, when a certain note is struck on a piano, a nearby vase may begin to vibrate. This means that the natural frequency of oscillation of the vase has been excited by a piano tone of the same frequency and energy is being transferred to the vase to sustain the vibrations. Similarly, soldiers marching across a bridge in step may cause it to vibrate at its natural frequency. If the constant small impulses from the marching soldiers take place at the same frequency as the natural frequency of bridge oscillations, resonance takes place and the bridge starts to vibrate. This effect is cumulative and the amplitude of vibrations can become so large that the bridge may be destroyed.

Resonance in a tank circuit is analogous to mechanical resonance. When an a-c voltage is impressed across a tank circuit, electrical resonance occurs at that a-c frequency at which the tank circuit breaks into natural oscillations. The circuit then draws just enough energy from the a-c supply to overcome its internal resistance losses. This, indeed, is the fundamental meaning of resonance. At the resonant frequency, the external supply releases just enough energy with the proper timing to sustain the natural self-oscillations of the tank circuit. With these ideas in mind, let us now look more closely at the familiar parallel-resonant or tank circuit, which forms the basis of all tuned circuits.

PARALLEL-RESONANT (TANK) CIRCUIT

As shown in Fig. 94, a parallel-resonant circuit consists of two branches in parallel, one branch containing capacitance (C), the other inductance (L) and resistance (R). A source of a-c voltage (E) of a certain frequency (f) is applied to the two branches. The resistance of the inductive branch is always present, although sometimes not shown, and it represents that of the coil and associated conductors.

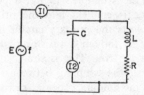

Fig. 94. *Parallel Resonant Circuit*

You will remember from basic electricity that the current in a capacitor **leads** the applied voltage by a

quarter cycle, or 90 degrees in phase, while the current through an inductor (coil) **lags** the applied voltage by a quarter cycle, or 90 degrees. When both capacitance and inductance are present in a circuit, therefore, the current through the capacitor leads that through the inductor by one-half cycle, or 180 degrees. In effect, the two currents are in **phase-opposition** and tend to cancel each other. Similarly, the voltage developed across the inductor by the inductive current is in phase opposition to that developed across the capacitor. You will further recall that the voltage developed across an inductance (E_L) is the product of the current (I) and the inductive reactance (X_L), where the latter is defined as

$$\text{inductive reactance } X_L = 2\pi \times f \times L$$

Here f is the applied frequency and L is the inductance in henrys. Further, the voltage developed across a capacitor (E_c) is the product of the current (I) and the capacitive reactance (X_c), where the capacitive reactance

$$X_c = \frac{1}{2\pi \times f \times C}$$

(Here C stands for the value of capacitance in **farads.**) Of course, the voltage developed across a resistor (E_R) is by Ohm's law simply the product of the current (I) and the resistance (R), and this voltage is always **in phase** with the applied voltage.

By comparing the above relations you will note that the inductive reactance, X_L, increases **directly** with the applied frequency, while the capacitive reactance is **inversely** proportional to the frequency. It stands to reason that there must be some frequency, where the inductive and capacitive reactances are **equal in value** and, hence, present equal opposition to current flow. At that frequency the currents through the two branches in Fig. 94 must be about equal in magnitude (except for the small current through the resistance), but they are, of course, in phase opposition in respect to each other. As far as the external line supplying the two branches is concerned, the branch currents are equal and opposite at this frequency and, hence, cancel each other out. Under these conditions only a tiny current necessary to supply the losses in the resistance flows. The line current is then a **minimum,** as shown in Fig. 95.

Now, **parallel resonance is precisely defined as the frequency where the inductive and capacitive reactances of the two branches are equal and the line current is a minimum.** It is interesting to note that these two conditions take place at the same frequency only if the resistance is **small,** which is generally the case. From the definition, we can easily determine what this frequency is:

At resonance, by definition, the inductive reactance equals the capacitive reactance.

$$X_L = X_c$$

hence, $2\pi fL = \dfrac{1}{2\pi fC}$

solving for the resonant frequency,

$$f = \frac{1}{2\pi\sqrt{LC}}$$

This turns out to be the same as the natural frequency of oscillations of a tank circuit, discussed in the last chapter. You should not be surprised by this, since at resonance a tank circuit breaks out into natural or self oscillation.

Although the line current (I1 in Fig. 94) is very small at resonance, the current **circulating** in the two branches (I2 in Fig. 94) can be very large and may, in fact, be several hundred times the value of the line current. As evident from Fig. 95, the line

Fig. 95. *Line Current vs. Frequency in a Parallel-Resonant Circuit*

current is a minimum at resonance and increases above and below resonance. The **impedance** presented by the parallel branches is simply the ratio of the applied voltage to the line current (E/I1), and since the line current is small at resonance, the impedance is correspondingly large. This is graphically shown in Fig. 96 by the impedance resonance curve. The impedance rises to a steep peak at resonance.

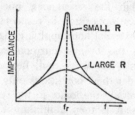

Fig. 96. *Impedance Curve of Parallel-Resonant Circuit*

"Q"-Factor. As is apparent from Fig. 96 the impedance diminishes rapidly when the frequency is varied in either direction from resonance. Note also that the impedance peak is much less pronounced when the resistance in the parallel-resonant circuit is large than when it is small. This is due to the fact that a large resistance consumes a considerable amount of power and draws a relatively large line current. In general it is desirable to have the resonance curve as sharp as possible, in order to provide the necessary **selectivity** to discriminate between different radio frequencies. The sharpness of the resonance curve (Figs. 95 and 96) is determined by a quality factor called "Q." The **Q is defined as the ratio of the reactance of either the coil or the capacitor at the resonant frequency to the total resistance of the circuit.** Mathematically,

$$Q = \frac{X_L}{R} = \frac{2\pi fL}{R}, \text{ or } Q = \frac{X_c}{R} = \frac{1}{2\pi fCR}$$

Since $X_L = X_c$ at resonance, both definitions result in the same value for Q.

Q is also a measure of the ratio of the reactive power stored in the tank circuit to the actual power dissipated in the resistance. The higher the Q, the greater the amount of energy stored in the circuit compared with the energy lost in the resistance during each cycle. Consequently, the higher the Q, the greater is the efficiency.

The quantitative performance of a parallel-resonant circuit is easily evaluated by means of the Q-factor. The higher the Q of a parallel-resonant circuit, the greater is its resonant impedance and circulating current (I2 in Fig. 94), and the smaller is the line current. (The circulating current is approximately Q times the line current.)

COUPLED CIRCUITS

Fig. 97a shows a circuit frequently used for coupling a radio-frequency amplifier to a resistive load. Here the radio-frequency energy from the tank circuit (in a transmitter, for instance) is coupled to the load by means of an air-core transformer, consisting of coils L1 and L2. By changing the mutual inductance between the coils, the impedance of the tank circuit can be matched to the load resistance. The impedances must be made comparable in value to obtain the greatest possible energy transfer from the tank circuit to the load. The easiest way to change the mutual inductance and, thus, obtain the required impedance match is to vary the coupling between the coils by changing the distance between them.

When coil L2 is coupled to L1, a portion of the load resistance is **coupled into** the primary (tank) circuit and affects the primary circuit in exactly the same manner as though a resistor had been added **in series** with the coil. This is shown dotted in Fig. 97a. The closer the coupling between the two coils, the greater is the amount of series resistance coupled into the primary (tank) circuit. You may consider this apparent series resistance coupled into the tank circuit as being **reflected** from the load (secondary) circuit into the tank (primary) circuit. Increasing the

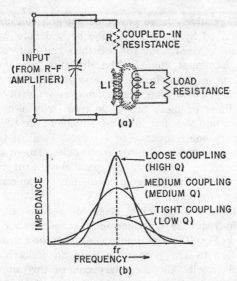

Fig. 97. *Effect of Variable Load Coupling on Tank Circuit*

coupling by bringing the coils closer together increases the reflected series resistance in the tank circuit and, hence, lowers the Q in the same manner as shown in Fig. 96 for a primary resistance.

In Fig. 97b you can see the effect of various degrees of coupling on the shape of the impedance resonance curve. When the coils are loosely coupled (large distance between them), the reflected resistance is small and, hence, the Q is high and the resonance curve is sharp. When the coupling is increased to an intermediate value, the coupled-in resistance is larger, the Q is lowered and the resonance curve is broader. Finally, when the coupling between the coils is very tight (coils close together), the reflected resistance is large, the Q is low, and the resonance curve is very broad, as shown by the bottom curve in Fig. 97b.

Coupled Resonant Circuits. Two resonant tank circuits are frequently used to couple the output of one stage of a radio-frequency amplifier to the input of the following stage. Fig. 98 shows such a circuit used primarily in the intermediate-frequency amplifier stages of superheterodyne receivers.

The primary tank circuit in Fig. 98 is connected

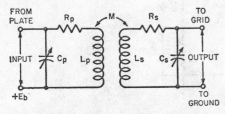

Fig. 98. *Coupled Tuned Circuits in Intermediate-Frequency Amplifier*

between the positive terminal of the plate supply voltage and the plate (output) of the tube, while the secondary tank circuit is connected between ground and the grid (input) of the following stage. The primary and secondary circuit resistances, R_p and R_s respectively, are chiefly the resistances associated with the coils L_p and L_s. Both the primary and secondary tank circuits are tuned to resonance at the same frequency (possibly the frequency of a radio signal). The combined resonant response of such a coupled circuit depends primarily on the degree of coupling—that is, the amount of mutual inductance (M) between the primary and secondary circuits.

Fig. 99 illustrates some typical resonance curves for two coupled resonant circuits for various degrees of coupling. These curves were obtained by plotting the current in the secondary circuit against frequency for a constant input voltage applied to the primary circuit. The resonance frequency of both circuits is designated by f_r. When the coupling between the primary and secondary is quite loose, the

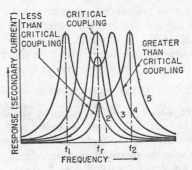

Fig. 99. *Resonance Curves of Coupled Tuned Circuits for Various Degrees of Coupling*

secondary current is small, but the resonance curve is sharply peaked (curve 1). As the coupling is increased slightly, the secondary peak becomes larger and the resonance curve becomes somewhat broader (curve 2). This tendency continues with increased coupling until the secondary current becomes a maximum for a **critical** degree of coupling (curve 3).

Critical coupling is said to occur when the resistance **reflected back** into the primary from the secondary is equal to the primary resistance (R_p). The coefficient of critical coupling, k, for this condition is then found to be

$$\text{critical } k = \frac{1}{Q}$$

provided both the primary and the secondary Q-factors are the same. Since the Q is usually high (above 100), the coefficient of critical coupling is generally very small.

When the coupling is increased beyond the critical value defined above, the secondary resonance curve begins to show two resonance humps, a condition known as **double-peaking** (curve 4). As the coupling is increased still further, the two peaks begin to spread apart in frequency and the valley or dip between the peaks becomes more pronounced (curve 5). For extremely tight coupling between the two circuits, the response between peaks may go almost to zero.

The double-peaked resonance characteristic shown in Fig. 99 is taken advantage of in the transformers of intermediate-frequency amplifiers to obtain the required **bandpass** characteristic. (I-f amplifiers in radio sets must pass a band of frequencies of about 10 kilocycles on both sides of the desired radio frequency.) By slightly **overcoupling** the tuned circuits of an i-f transformer beyond the critical k, the secondary current will be approximately constant near resonance over a range of frequencies between the two peaks. Beyond the peaks, however, the response falls off rapidly, so that the discrimination or **selectivity** against adjacent unwanted channels is excellent. Various resistance-loading methods are used to smooth out the dips between peaks at the extreme frequencies of the pass band. The greater the required bandwidth, the tighter must be the coupling to spread the two peaks sufficiently far apart.

TUNED AMPLIFIERS

Radio-frequency amplifiers can be divided into two main types: those used in **radio and TV receivers,** where small voltages are to be amplified, and those used in **transmitters,** where large voltages are to be amplified. The first type is always a Class A voltage amplifier, which means (as you will remember) that plate current flows during the entire a-c cycle of the input voltage. The second type is generally a Class C voltage or power amplifier; in Class C relatively large power output and efficiency are obtained by biasing the tube beyond cutoff, so that plate current flows for less than one half-cycle of the a-c input voltage. The distortion attendant with Class C amplification is largely eliminated by use of a resonant plate tank circuit, as we shall see later. Either of these two basic types may also be operated untuned, but most r-f amplifiers are tuned to amplify one frequency or a narrow band of frequencies and reject all other frequencies. This is accomplished by means of the parallel-resonant circuits, which—as we have seen—have frequency-selective properties.

A circuit for the initial stages of a tuned Class A radio-frequency amplifier in a radio receiver is shown in Fig. 100. The input signal from the antenna is applied to the primary coil of an inductively coupled circuit. The secondary coil of this circuit is tuned to resonance at the frequency of the desired radio station. The voltage across the tuned secondary is then applied to the grid of the first r-f amplifier. The plate circuit of this tube is coupled to the grid of the second stage by a similar coupled circuit. Both of these tuned coupled circuits serve to select a voltage of the desired frequency and reject all others. Maximum input voltage to each stage is developed across the tuned secondary of the coupling transformer when its impedance is a maximum. As you can see from the resonance curve of a parallel-resonant circuit (Figs. 96 and 97), the impedance is a maximum at the resonant frequency and falls off sharply on either side of resonance. The tuned coupled circuits are, therefore, highly effective in selecting the desired frequency and rejecting all others.

The maximum voltage gain of a tuned radio-frequency amplifier is obtained when the secondary of the coupling transformer is tuned to the frequency of the incoming stage and the coupling between

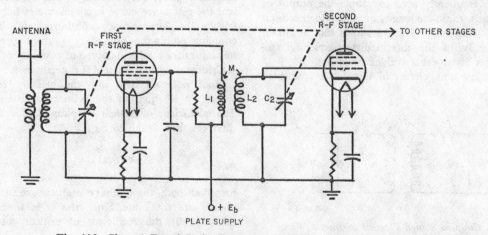

Fig. 100. *Class A Tuned Radio-Frequency Input Stages of Radio Receiver*

primary and secondary is adjusted to the critical value. (See curve 3 in Fig. 99.) The amplification under these conditions is then approximately

Amplification at resonance $= 2\pi\, f_r\, g_m\, M\, Q$

where f_r is the resonant (incoming) frequency

g_m is the transconductance of the amplifier tube

M is the mutual inductance between coils

and Q is the effective Q $(= 2\pi f_r L_2/R)$ of the secondary.

Tuned r-f voltage amplifiers always employ pentode tubes of the types used with resistance-coupled and video amplifiers (Chapters 8 and 9). Triodes are not suitable for use as r-f voltage amplifiers for two reasons. First, triodes have a low plate resistance compared to pentodes. This results in a low value of the mutual inductance (M) for critical coupling and, hence, a low gain per stage. Secondly, triodes have a large grid-to-plate interelectrode capacitance, which results in a feedback of energy from the plate circuit to the grid circuit. Since this feedback is positive, instability and oscillations may result.

Class C Amplifiers. Class C radio-frequency amplifiers are used in radio transmitters for voltage and power amplification. The grid excitation voltage required in transmitters is many times larger than is permissible with Class A amplification and the power output and efficiency in Class C are higher than with either Class A or Class B operation. Aside from the changes in tube operation and bias, the circuit used for Class C may be a single-ended amplifier, like that shown in Fig. 100, or it may be a **push-pull** amplifier, similar to that shown in Fig. 73 (Chapter 10). However, the important point of any Class C amplifier is that a **tuned tank circuit must be used** in the plate circuit of the tube to avoid distortion. As was explained in Chapter 10, the plate current in a Class C amplifier is zero during most of the a-c input cycle and flows only in short pulses during the intervals when the grid voltage is near the positive peak of its cycle. Because of this the plate current is far too distorted to be used directly. However, by feeding a plate current pulse once during each cycle to the plate tank circuit, continuous sinewave oscillations can be maintained in the tank circuit and large amounts of undistorted power may be withdrawn.

INTERMEDIATE-FREQUENCY AMPLIFIERS

Let us now consider the use of two coupled resonant circuits in an intermediate-frequency (i-f) amplifier. These amplifiers are high-gain circuits used in superheterodyne receivers. As will be explained in a later chapter, i-f amplifiers are permanently tuned to a fixed frequency, which is the difference between the incoming radio signal and a locally generated frequency. By using a fixed intermediate frequency, the tuned circuits of an i-f amplifier may be adjusted for maximum amplification and selectivity. The i-f transformers used in these amplifiers consist of pairs of coupled resonant circuits, as shown in Fig. 101. They are tuned either by varying the capacitance in the circuit, or by moving a powdered iron core in or out of the coils to change the inductance of the tuned circuit.

Fig. 101 illustrates the circuit of a single-stage i-f amplifier with two i-f transformers (T1 and T2). Transformer T1, the input transformer, has its primary winding connected to the plate circuit of the previous stage and is tuned to the selected intermediate frequency (usually 455 kilocycles). The secondary circuit, L2-C2, is inductively coupled to the primary and is in series with the grid of the amplifier pentode. Transformer T2 in the output circuit is identical with transformer T1. For use in wideband and high-fidelity receivers, the transformers are usually **overcoupled** to produce a resonance

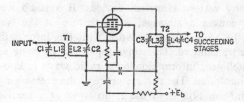

Fig. 101. *Single-Stage Intermediate-Frequency Amplifier*

curve similar to curve 4 of Fig. 99. This provides a small band of (audio) frequencies in addition to the radio "carrier" frequency.

SUMMARY

Electrical **resonance** occurs in a tank circuit at the frequency of natural oscillations. The circuit then draws just enough energy to overcome its internal resistance losses.

Parallel resonance in a tank circuit takes place at the frequency where the inductive and capacitive reactances of the two branches are equal and the line current is a **minimum.** The impedance is then a **maximum.** The resonant frequency

$$f_r = 1/2\pi\sqrt{LC}$$

The **quality factor Q** is the ratio of the reactance of the coil or the capacitor at the resonant frequency to the total resistance. Q is also a measure of the ratio of the reactive power stored in the tank circuit to the actual power dissipated in the resistance.

When a tuned resonant circuit is coupled to a load resistance, an apparent series resistance is reflected from the load (secondary) circuit into the tank (primary) circuit. This reflected load resistance lowers the Q and thus decreases the sharpness of resonance.

When two resonant tank circuits are coupled to each other, the resonant response depends on the degree of coupling between primary and secondary. As the coupling is increased, the secondary current peak becomes larger and the resonance curve becomes broader. Maximum response takes place at **critical coupling** ($k = 1/Q$). When the coupling is increased beyond the critical value, the secondary resonance curve shows two humps, which move apart with tighter coupling. This characteristic is used in wideband r-f and i-f amplifiers.

Radio-frequency amplifiers make use of resonant circuits to amplify a desired frequency and reject all others. Class A voltage amplifiers are used for receivers, while Class C power amplifiers are used in transmitters.

I-f amplifiers with two coupled resonant circuits are used in superheterodyne receivers.

Chapter Thirteen

MODULATION AND DETECTION

Radio frequencies generated by oscillators and sent out as radio waves from transmitting antennas are by themselves mute messengers. You can neither see nor hear them, although you may detect their presence by various electrical devices. If a radio wave is to convey a message, some feature of the wave must be varied in accordance with the information to be transmitted. This superimposition of some sort of intelligent information on a radio wave is known as **modulation.** The reverse process, the extraction of this information from the radio wave at a receiver, is called demodulation or **detection.**

There are various ways for conveying information by means of radio waves. You can turn a radio transmitter on or off in accordance with some pre-arranged code, which may be the dots and dashes of the telegraph (Morse) code. This system of radio-telegraphy by means of **continuous waves** (c.w.) permits the transmission of written messages, but it can neither reveal the sound of a voice nor the appearance of a face. The latter, more complicated task can be performed in several ways. The amplitudes of the radio-frequency waves can be varied in accordance with the pressure changes of sound waves (after conversion to electrical **audio** frequencies) or they may be varied in accordance with the light intensity of a portion of a picture to be transmitted. This process is called **amplitude modulation** of a radio wave. It is used in **radio-telephony** for the transmission of sounds, in **facsimile** for the transmission of still pictures and in **television** for the transmission of moving pictures or actual scenes.

A second way of modulating a radio wave is to vary its **frequency** (number of alternations per second) in accordance with the pressure of a sound wave, the light intensity of a picture, or some other form of intelligence. This is called **frequency modulation.** Fig. 102 illustrates the difference between am-plitude and frequency modulation—the two most important methods of modulation. The illustration shows the modulation of a radio-frequency "carrier" wave (so called because it carriers the modulation) by a single audio tone.

Fig. 102a shows the electrical equivalent of a single musical tone of a particular pitch. This is seen to be a simple a-c sinewave of the corresponding frequency. A radio-frequency "carrier" wave of constant frequency and amplitude is shown in (b). Since the carrier completes ten complete cycles during the time of one audio cycle, its frequency must be ten times that of the audio tone. In (c) of Fig.

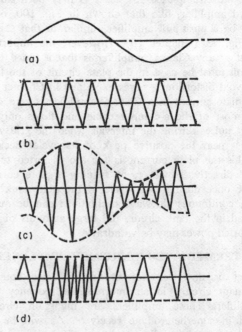

Fig. 102. *Amplitude and Frequency Modulation of Radio-Frequency Carrier by an Audio-Frequency Wave*

102 the radio-frequency carrier is being amplitude-modulated by the audio-frequency wave. Note that the amplitudes of both the positive and negative half-cycles of the r-f wave vary in accordance with the audio signal. The detector of a radio receiver simply eliminates the radio-frequency carrier, while retaining its audio-frequency amplitude variations. Finally, in (d) the **frequency** of the carrier is varied in accordance with the **amplitude** of the audio signal. The greater the positive amplitude of the audio wave, the higher is the frequency of the radio-frequency carrier.

There are other methods of modulating a radio wave in addition to amplitude and frequency modulation. For instance, the radio signals can be sent out as a series of sharp discontinuous pulses and the timing or spacing of these pulses may be varied in accordance with some information to be conveyed. This is known as **pulse-time modulation** and it has the advantage of permitting many different messages to be sent out over the same radio-frequency channel. Still other ways of conveying information via radio waves exist, but we shall be concerned primarily with the modulation and detection of radio waves, which have been modified either in their amplitude or frequency in accordance with the desired information.

Need for a Radio-Frequency Carrier. You may well wonder why we cannot simply talk into a microphone and then transmit the electrical equivalents (audio frequencies) of the sound waves directly from a radio transmitter. As a matter of fact, this is not impossible, but quite impractical. As you will recall, the audio frequencies corresponding to sound waves range from about 15 to 20,000 cycles per second. The wavelengths of these frequencies range from approximately 12,400 miles to 9.3 miles (wavelength = velocity/frequency; the velocity of radio waves is about 186,284 miles per second). To transmit a radio wave, the antenna of the transmitter must be approximately of the same size as the waves to be radiated. It is obviously impossible to construct antennas thousands of miles long. The wavelength of a 1000 kilocycle radio wave, on the other hand, is only about 328 yards and antennas of this order of size are easily constructed. The only practical solution, therefore, is to modulate a radio-frequency carrier with audio (sound) or video (picture) signals.

AMPLITUDE MODULATION

We have defined amplitude modulation as the process of modifying the amplitudes of a radio-frequency carrier wave in accordance with the strength of an audio (sound) or video (picture) signal. For a pure tone, or single audio-frequency, amplitude modulation looks as illustrated again in Fig. 103. Assume, for example, that (a) represents a 1000-kc

carrier and (b) a pure 1-kc note. The effect of impressing the carrier with the audio tone simultaneously across a **resistor** is shown in (c). Note that the amplitude of the carrier does not vary at all, but the instantaneous polarity of the radio-frequency cycles varies continuously. This, obviously, is *not* amplitude modulation. A radio receiver has no means for examining the momentary polarity of this signal and, hence, for extracting the audio signal.

It has been found that the desired amplitude modulation, shown in (d), can be obtained only when the audio signal and carrier are impressed on a circuit where the current is *not* directly proportional to the applied voltage; that is, a circuit that does *not* obey Ohm's law. Because the graph of current vs. voltage in such a circuit is a curve, it is called **non-linear**. The necessity for this can be proved mathematically, but you may simply accept it as a fact that an amplitude modulator must be a **non-linear circuit**.

It will be somewhat easier to appreciate the significance of this fact if you recall our discussion on amplitude distortion in the chapter on power amplifiers (Chapter 10). We demonstrated there that the non-linearity of electron tube characteristics results in amplitude distortion. In a sense, you may consider amplitude modulation as a form of **intentional** amplitude distortion. To produce this distortion, or modulation, a non-linear characteristic is

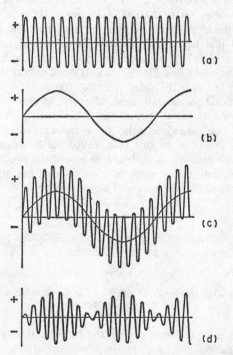

Fig. 103. *Impressing an R-F Carrier (a) and Audio Tone (b) Simultaneously Across a Resistor Results in Waveform (c), Which Is Not Amplitude-Modulated; to Obtain Amplitude Modulation (d) a Non-Linear Device Must Be Used*

required. Furthermore, electron tubes make ideal modulators, when they are operated in the non-linear portions of their characteristic.

Sidebands. We further pointed out in Chapter 11 that amplitude distortion of a waveform results in the production of **new frequencies not present in the original waveform.** The same thing happens with amplitude modulation. Although we intended to change only the amplitude of the waveform in Fig. 103, the process of amplitude modulation has also resulted in the production of some additional frequencies. These new frequencies are equal to the **sum and the difference** of the carrier frequency and the modulating frequency. In the example of Fig. 103 only two new frequencies result. One is 1001-kc frequency equal to the *sum* of the 1000-kc carrier wave and the 1-kc audio signal; this is called the **upper side frequency.** The other frequency is 999 kc, or the *difference* between the carrier and audio tone; this is called the **lower side frequency.**

If it is desired—as in radio broadcasting—to modulate a carrier with the greater part of all audible frequencies, say up to 10,000 cycles, each of these audio frequencies will produce upper and lower side frequencies during modulation, resulting in **upper and lower sidebands.** Thus, for the 1000-kc carrier the upper sideband will extend to 1010 kc (for a 10,000-cycle modulating frequency), while the lower sideband will reach to 990 kc. To broadcast audio frequencies up to 10,000 cycles on a 1000-kc carrier, therefore, a transmitting channel must be provided that has a bandwidth of 20 kilocycles (from 990 to 1010 kc), or *twice* the highest audio modulating frequency (see Fig. 104). This is not only true for audio but also for video (picture) transmission, and in general the overall bandwidth of a channel must be *twice* the band of frequencies included in the information (sound or sight) to be transmitted. There is an exception to this rule, called **single sideband transmission,** which utilizes only one of the two sidebands, but this system is chiefly used for transoceanic radio-telephony and amateur radio.

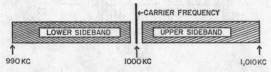

Fig. 104. *Sidebands of 1000-kc Carrier Produced by 10-kc Amplitude Modulation*

We now understand why the tuned amplifiers in transmitters and radio receivers must be able to pass a whole band of frequencies, rather than only the frequency of the radio carrier wave. In order to extract the information contained in the amplitude modulation of a carrier, it is clearly necessary that all tuned transmitter and receiver circuits pass the bandwidth of the entire channel with its upper and lower sidebands. While passing the desired channel, the tuned circuits should nevertheless be sufficiently selective to discriminate against unwanted adjacent channels on either side.

Depth of Modulation. So far we have described the principle of modulating the amplitude of a carrier, but we have not said how much. The degree of modulation is a very important consideration, since it determines the strength and quality of the transmitted signal. Fig. 105 illustrates a radio-frequency carrier modulated to various degrees by an audio-frequency signal. The audio signal is shown in (a) of the figure and the unmodulated carrier in (b). The waveform of the modulated carrier for a low degree or **depth of modulation** is shown in (c). While the carrier varies faithfully in accordance with the audio signal (a), the amount of variation is rather small. Now, as we shall explain later, the detector in a receiver responds only to the **variations** in the amplitude of the carrier and not its absolute magnitude. When the carrier is modulated only to the small degree shown in (c), the audio signal will not be very strong and it may possibly be drowned out by extraneous noise. The greater the **depth** of the modulation, the stronger and clearer will be the audio signal.

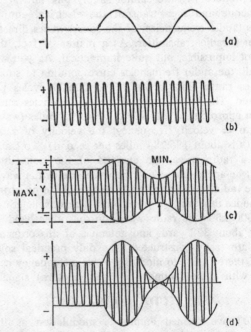

Fig. 105. *Radio-Frequency Carrier Modulated to Various Degrees by an Audio-Frequency Signal*

In (d) of Fig. 105 the r-f carrier has been modulated to the maximum possible extent, namely, to twice its normal amplitude during the positive peak of the modulating signal and to zero amplitude on the negative peak of the modulating signal. This is known as **100 percent modulation.** Any further increase in the amplitude of the modulating signal will result in distortion during reception.

The depth of modulation of a carrier is conveniently expressed as a **percentage** of the normal (unmodulated) carrier amplitude. Knowing the relative amplitudes of the modulating signal and the carrier, as indicated in Fig. 105, the percentage of modulation is easily computed from the following relation: (See Fig. 105.)

$$\text{percentage modulation} =$$
$$\frac{\text{MAX.} - Y}{Y} \times 100, \text{ or } \frac{Y - \text{MIN.}}{Y} \times 100$$

where Y represents *twice* the carrier amplitude

If the modulating signal is a pure sinewave, as in (a), both relations will result in the same percentage; if the modulating signal is distorted, however, this is not the case and the larger of the two values is generally used.

For the example of 100% modulation, shown in (d) of Fig. 105, the MAX. value is twice the value of Y, or 2Y, while the MIN. value is zero. Thus, substituting in the above relation, we verify:

$$\text{percentage of modulation} =$$
$$\frac{2Y - Y}{Y} \times 100 = 100\%$$
$$\text{or}$$
$$\frac{Y - 0}{Y} \times 100 = 100\%$$

TYPES OF MODULATORS

To obtain amplitude modulation of a radio carrier, the modulating signal must be injected in some way into the radio-frequency power amplifier stages of a transmitter. Depending upon where the audio signal is inserted, different types of modulating circuits result. If the modulating signal is inserted in series with the d-c plate voltage supply of a transmitter, **plate modulation** results; if it is injected into the control grid of a transmitter stage, **control-grid modulation** occurs; if injected into the screen or suppressor grids, **screen grid** or **suppressor grid modulation,** respectively, result. Because of its efficiency and ease of adjustment, plate modulation is the most widely used modulating method.

Plate Modulation. A basic form of plate modulation circuit is illustrated in Fig. 106. Here the r-f carrier signal is applied to the grid of a tuned r-f amplifier stage and the audio-frequency modulating signal from the output of an audio amplifier is inserted in series with the plate circuit.

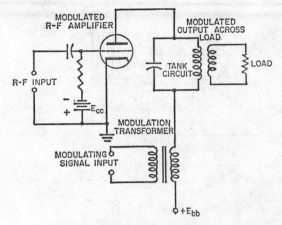

Fig. 106. *Basic Plate Modulation Circuit*

In the absence of the modulating signal, the r-f stage amplifies the carrier signal and the output voltage appears across the plate tank circuit, from which it is transformer-coupled either to the next stage or to the antenna. When the modulating signal is applied through the **modulation transformer,** the amplitude of the plate circuit signal is made to vary in accordance with the audio signal, and the modulated carrier appears across the secondary of the r-f output transformer. You can easily see how this happens. During one half-cycle of the audio signal, a positive voltage is induced in the secondary of the modulation transformer that adds to the plate supply voltage and, thus, causes an *increase* in the **r-f voltage** across the tank. During the next, negative half-cycle of the audio signal, the voltage induced in the secondary of the modulation transformer subtracts from the plate supply voltage, thus causing a *decrease* in the r-f voltage across the tank circuit. Since the tube and associated circuit contains nonlinear elements, true amplitude modulation takes place.

To attain maximum (100%) modulation the peak amplitude of the audio modulating signal must be made equal to the plate supply voltage. If it is smaller than this value, the percentage of modulation will be less than 100%; if it is larger, the plate current will be cut off during part of the audio cycle and the tank output will be zero during this portion of the cycle. This so-called **overmodulation** results in excessive distortion of the transmitted modulated carrier signal.

A very popular circuit for plate modulation is shown in Fig. 107. Here the final r-f stage of a transmitter is plate-modulated by a Class B push-pull audio amplifier. This is known as **high-level** modulation. When the modulating signal is injected into one of the earlier stages of the transmitter, where the r-f power is small, the system is called **low-level**

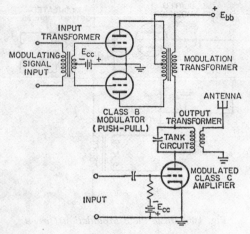

Fig. 107. *Class C R-F Amplifier Plate-Modulated by a Class B Push-Pull Audio Amplifier*

modulation. In Fig. 107 the r-f carrier signal is applied to the grid of a Class C amplifier and the output voltage is developed across the plate tank circuit. The modulating signal is obtained through a push-pull Class B power amplifier stage (the modulator). The output of the modulator is applied in series with the r-f plate tank circuit through the secondary of the modulation transformer. The d-c plate power for both modulator and r-f amplifier is obtained, in this case, from the same plate supply ($+E_{bb}$), but this is not necessarily so.

You may have wondered how the power of a modulated wave compares with that of an unmodulated r-f carrier. Well, with 100 percent modulation each sideband is one-half the amplitude of the carrier wave, as we have seen (Fig. 105). Since power is proportional to the **square** of the amplitude, the power associated with each of the two sidebands is **one-quarter** of the carrier power. The total audio-modulating power for two sidebands, therefore, is **one-half** the carrier power and, hence, the average output of the transmitter **increases by 50 percent** with 100 percent modulation compared to the power of an unmodulated carrier. This additional power in the modulated wave *must be supplied by the modulator* in a plate-modulated system. The modulator must, therefore, be designed to furnish 50 percent of the desired transmitter output power without overloading and resulting distortion. Stated in another way, the maximum transmitter power that can be utilized with 100 percent modulation is *twice* the power available from the modulator. Since this places quite a burden of expense on the modulator, other methods of amplitude modulation are occasionally employed.

Control Grid Modulation. If it is desired to eliminate the expense of a high-power modulator, the modulating signal can be injected into the grid of the Class C r-f power output stage of a transmitter and control-grid modulation results. However, regardless of how the modulation is applied, the transmitter output power is still required to increase by 50 percent with 100 percent modulation, and this power must come from somewhere. In control-grid modulation, the modulator power requirements are very light and the required additional output power is generated *within* the r-f amplifier stage itself. Grid modulation differs from plate modulation primarily in that the **plate supply voltage remains constant,** and the extra output power is obtained by causing the plate current and plate efficiency of the r-f amplifier to vary in accordance with the modulating signal. The plate current is made to double at the peak of the modulation swing, which produces the required extra amount of output power.

Fig. 108 illustrates a typical control-grid modulated r-f power amplifier circuit. The r-f carrier signal is applied to the grid of the Class C amplifier, which is biased well beyond cutoff by a fixed d-c bias source. The audio modulating signal is applied in series with the bias lead through the secondary of the modulation transformer, thus superimposing the a-f signal on the d-c bias. Consequently, the plate current of the r-f stage also varies in accordance with the audio modulating signal.

As you will remember, the grid of the r-f stage is driven positive in Class C operation and grid current flows. Since the grid current varies with the

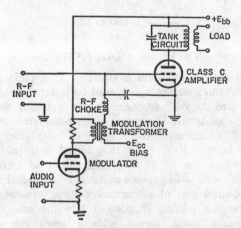

Fig. 108. *Control-Grid Modulated Class C Power Amplifier*

amplitude of the input voltage to the grid, a variable load is presented to the modulator output, which can cause distortion. To reduce the load variation and resulting distortion, a resistor is connected as constant load across the output circuit of the modulator.

FREQUENCY MODULATION

Historically, frequency modulation (FM, for short) arose from the search for something better than amplitude modulation. While theoretically highly effective, amplitude modulation (AM) suffers from two practical defects. The first is noise. Practically all the natural and man-made noises, such as atmospheric static, razors, electrical machines, etc., consist of electrical **amplitude** disturbances. Since a radio receiver cannot distinguish between amplitude variations that represent noise and those that contain the desired sound (i.e., the modulation), AM reception is generally noisy anywhere, except within close proximity of the radio transmitter. Raising the power of the transmitter improves the "signal-to-noise ratio," but this brute-force method is costly and becomes less effective the greater the distance from the transmitter.

The second practical defect of AM is the lack of fidelity or audio quality. To attain high-fidelity reception, all audio frequencies within the span of human hearing up to about 15,000 cycles must be reproduced. This necessitates a channel bandwidth of 30,000 cycles, or 30 kc, since both sidebands must be reproduced. But AM broadcasting stations are assigned channels only 20 kc wide and most of them use only 15 kc to avoid interference with adjacent channels. This means, in practice, that the highest audio modulating frequency can be only 7500 cycles, which is hardly enough to reproduce music realistically. The reason for the narrow width of the AM channels is that at the time of construction of the early broadcast stations, which were exclusively AM, only the medium frequencies from about 500 kc to 1600 kc were suitable for the existing electronic equipment. To accommodate as many broadcast stations as possible within this narrow broadcast band of 1100 kc, the width of each channel was deliberately restricted. Hi-Fi was a remote consideration at the time these channel assignments were made. Outside of this arbitrary restriction, there is nothing in the nature of AM that would not permit full high-fidelity reproduction of speech and music.

Frequency modulation, which came into practical use just before World War II, dramatically removes these limitations of amplitude modulation. By impressing the modulation on the carrier through a variation in its frequency, while keeping the carrier amplitude *constant*, all the amplitude-sensitive noises are immediately eliminated, since variations in amplitude (i.e., noise) are simply not reproduced. This means that the signal-to-noise ratio for noisefree FM reception can be much lower than for AM and, hence, the FM transmitter power may be lower for the same quality of reception. Furthermore, FM *is* high fidelity, since the FCC requires the transmission of the full audio band from 20 to 15,000 cycles.

As we shall see later, this requires a channel bandwidth even wider than that required for AM. By the time FM was developed, it became feasible to transmit in the very wide VHF frequency band from 30,000 kc to 300,000 kc (30 mc to 300 mc). FM broadcasting was assigned a band from 88 mc to 108 mc (i.e., 20,000 kc wide) and each station was allotted a channel width of 200 kc; this permits simultaneous operation of 100 stations in the same area.

BASIC PRINCIPLES

The appearance of a frequency-modulated carrier wave is illustrated in Fig. 109. A pure sinewave audio-modulating tone is shown in (a) and an unmodulated carrier wave in (b). When this tone is frequency-modulated onto the carrier wave, the result is as shown in (c) of the figure.

As you will note, the **amplitude** of the carrier in Fig. 109c remains constant, but its frequency is continuously **varied** in accordance with the instantaneous amplitude (strength) of the audio signal. During the times when the audio voltage is zero, as at *A, C, E, G,* and *I,* the carrier frequency is not modulated and, hence, remains the same as in (b). When the audio signal approaches its positive peaks, however, as in *B* and *F,* the frequency of the carrier is increased toward its maximum positive **deviation,** as shown by the closely spaced cycles in (c). When the audio signal approaches its negative peaks, on the other hand, as in *D* and *H,* the car-

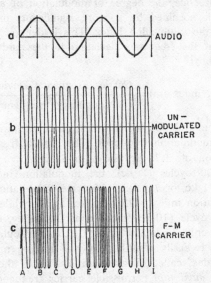

Fig. 109. *Frequency Modulation of an R-F Carrier by an A-F Wave*

rier frequency is deviated toward its minimum frequency, as shown by the wider than normal spacing of the cycles in (c).

Assume, for example, that the frequency of the

carrier without modulation, called the **center frequency,** is 100,000 kc (100 mc) and the audio-modulating frequency is 10,000 cycles, or 10 kc. Say that the **frequency deviation** of the carrier is 50 kc during the peaks of the audio signal. The carrier is then deviated to 100.05 mc during the **positive** peaks of the audio signal, and to 99.95 mc during the **negative** peaks of the audio signal. The total range of the FM carrier frequency, called **frequency swing,** is then 100 kc (from 99.95 mc to 100.05 mc), or *twice* the deviation in either direction.

Modulation Index. The depth of modulation in AM depends on the change in carrier amplitude caused by the modulating signal, 100 percent modulation taking place when the carrier amplitude goes to zero during the negative peak of the modulating signal. Since the amplitude of an FM carrier is constant, the depth of modulation depends on the amount of frequency deviation during the peaks of the audio signal. Theoretically, maximum or "100%" modulation could be attained if the carrier *frequency goes to zero* during the negative peak of the audio-modulating signal. This is obviously not practical, since a zero-frequency signal could not be radiated and in any case would require a prohibitive channel bandwidth. Maximum or 100 percent modulation is, therefore, arbitrarily defined by the FCC as the maximum **permissible frequency swing.** This is 150 kc ($\pm$75 kc deviation) for FM broadcast stations and varies from 15 to 25 kc for various FM communications services.

In practice, the degree of modulation of an FM wave is specified by a more significant quantity, called the **modulation index.** This is defined as the **ratio of the frequency deviation to the modulating frequency:**

$$\text{Mod. index (m)} = \frac{\text{Carrier Frequency Deviation}}{\text{Modulating Frequency}}$$

The modulation index varies, of course, with the audio-modulating frequency. Thus for a frequency deviation of $\pm$75 kc and a modulating frequency of 1000 cycles (1 kc), the modulation index is 75 kc/1 kc, or 75; for 5000 cycles modulation, the modulation index is 75 kc/5 kc, or 15; while for a 10,000-cycle (10 kc) modulating frequency, the index is only 75 kc/10 kc, or 7.5. A particular FM system is easily identified by the limiting modulation index, called **deviation ratio.** This is the ratio of the *maximum* permissible carrier frequency deviation to the *highest* audio-modulating frequency. For FM broadcasting the maximum deviation is 75 kc and the highest audio frequency is 15,000 cycles, or 15 kc; hence the deviation ratio is 75 kc/15 kc = 5.

FM Sidebands. The significance of the modulation index arises out of the curious fact that it determines the FM bandwidth. You might think that the FM bandwidth is simply the frequency deviation on both sides, that is, the total frequency swing of the carrier. But this is not so. An inspection of Fig. 109c reveals that a frequency-modulated carrier consists of **distorted** sinewaves. Distortion, however, results in new upper and lower side frequencies, as you will remember. Fact is that frequency modulation by a *single* audio tone sets up a whole *group* of upper and lower side frequencies, rather than just one pair, as for AM. These side frequencies are spaced at intervals equal to the modulating frequency, and their **number and intensity** depends directly on the modulation index. A useful rule of thumb states that **the total bandwidth of an FM channel is approximately equal to the total frequency swing plus twice the highest audio modulating frequency.**

REACTANCE TUBE MODULATOR

To produce FM the frequency of an r-f oscillator must be varied in accordance with the amplitude of an audio signal. The best way to do this is by means of a **reactance tube modulator.** The basic circuit of such a modulator together with the input circuit of a conventional Hartley oscillator is shown in Fig. 110.

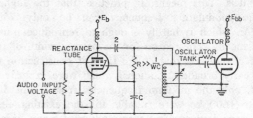

Fig. 110. *Frequency Modulation of Oscillator by Reactance Tube*

An a-c voltage divider R-C is connected across the oscillator tank circuit between plate and cathode of the pentode tube. The tap of the voltage divider is connected to the control grid of the tube. The resistance R is made large compared to the reactance of C by choosing a large resistor and a large capacitor. (Capacitive reactance varies inversely with capacitance.) With the reactance almost negligible compared to the resistance R, the current through the voltage divider is practically **in phase** with the voltage across the tank circuit. However, the current through a capacitor always **leads** the voltage across it by 90 degrees, or equivalently, the voltage **lags** the capacitive current by 90 degrees. The voltage across C, which is the input voltage to the control grid of the tube, thus **lags** the current through C by 90 degrees. Since the plate current is *in phase* with the control-grid voltage, the r-f plate current of the tube also must *lag* behind the current through C by 90 degrees, or equivalently, it *lags the oscillator tank voltage by 90 degrees*. The effect is the same as if an *additional inductance were connected across the tank*. This shunt inductance *decreases* the total tank circuit inductance and, consequently, *raises* the oscillator frequency.

To accomplish frequency modulation, we must raise the tank circuit frequency in proportion to the amplitude of the modulating frequency. Now, the larger the lagging plate current of the reactance tube, the greater will be the apparent shunt inductance of the tank circuit and, hence, the higher will be the oscillator frequency. The amplitude of the reactance tube plate current is controlled by the audio input voltage on the screen grid. Evidently, then, the amplitude of the lagging plate current varies directly with the voltage on the screen grid and, hence, the tank circuit frequency is proportional to the audio modulating voltage applied to the screen grid. True frequency modulation is thus accomplished.

The frequency-modulated oscillator is usually operated at a relatively low frequency to attain stability of the carrier frequency. The oscillator frequency is then raised to the desired carrier frequency by a number of r-f frequency multipliers. You will remember that the output frequency of an r-f amplifier may be doubled or tripled by tuning the plate tank circuit to twice or three times the input frequency, respectively. When the output frequency of a frequency-modulated oscillator is thus multiplied, the **deviation increases by the same factor** as the frequency is raised. For example, if a 12.5 mc oscillator frequency is raised to a 100-mc carrier frequency, an original frequency deviation of, say, 2 kc will be increased to 16 kc deviation at the output. This fact is sometimes taken advantage of by connecting a reactance tube modulator to a relatively low-frequency crystal oscillator, thus obtaining exceptional frequency stability. The reactance tube cannot vary the crystal frequency by more than a small amount, which may be *less than one cycle* (360 degrees). When the variation is less than a cycle, it is in effect only a **phase shift,** and the modulation is then called **phase modulation,** rather than frequency modulation. The two types of modulation are essentially the same. The phase modulation of a crystal oscillator can easily be converted to frequency modulation by multiplying the crystal frequency and, hence, the frequency deviation to the desired value through a series of frequency doublers or triplers.

DETECTION

The process of **demodulation** or **detection** is the inverse of modulation, since it consists of the extraction of the information contained in the modulated wave. As we shall see later in more detail, the detection of an amplitude-modulated wave involves the rectification of the r-f carrier and the filtering out of the audio-frequency modulation. Detection of an FM wave is somewhat more complicated and usually first requires the limiting of the carrier amplitude to a constant value to eliminate noise and then the conversion of the carrier frequency variations to corresponding audio-frequency amplitude variations.

AM DETECTORS

The modulation of an AM wave consists of amplitude variations of the carrier by the audio signal. An AM detector, consequently, must extract these a-f amplitude variations of the carrier, while eliminating the r-f carrier itself. The first step in this process is **rectification** of the carrier wave to eliminate the negative half-cycles. This is necessary, since the negative and positive peaks of the carrier tend to cancel each other. The second step in the detection process is to remove the carrier-frequency variation so as to leave only the audio modulation. This is accomplished by a **filter** circuit. We shall consider three popular types of AM detectors out of the multiplicity of existing circuits. These are the **diode detector,** the **grid-leak detector,** and the **infinite impedance detector.**

Diode Detector. The basic circuit and waveforms of the diode detector is illustrated in Fig. 111. Although an electron tube is shown in the figure,

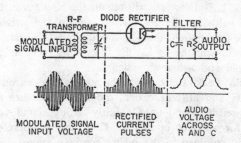

Fig. 111. *Basic Action of Diode Detector*

crystal diodes made of silicon or germanium may equally well be used.

The action of the diode detector is essentially that of a half-wave rectifier. (See Chapter 14.) The modulated r-f signal is coupled to the plate of the diode through the tuned r-f input transformer. As you know the tube conducts only when its plate is positive in respect to the cathode. Consequently, a current pulse flows through the tube and load resistance, R, whenever the signal voltage at the diode plate is positive. No current can flow during the negative half-cycles of the carrier signal. The output of the diode, thus, consists of a series of positive half-cycles of the carrier wave, as shown.

The rectified current pulses are still much too rapid to be made audible in a loudspeaker. The remaining r-f carrier variations are, therefore, smoothed out by a filter, consisting of filter capacitor C and load resistor R. The value of the capacitor must be sufficiently large to present a low reactance (opposition) to the r-f current pulses, while presenting a relatively high reactance to audio frequencies. Under these conditions, the remaining r-f carrier frequencies are bypassed around R, while the audio-frequency currents (the modulation) develop a voltage across R.

Let us consider the filtering action in more detail. When the tube conducts during the positive half-cycles, the capacitor quickly charges up to the **peak value** of each voltage pulse. The values of the capacitor and resistance are chosen so that the capacitor is able to discharge only very slowly through the resistor. The capacitor thus tends to hold its charge between successive positive peaks, when the applied signal voltage drops to zero and cuts off the plate current. As a result the voltage across the load resistance R does not drop to zero between pulses, but decreases slowly as the charge on the capacitor leaks off. During each successive r-f pulse the capacitor charges up again to the peak value of the new pulse and the voltage across the load rises to the new peak value. The upshot is that the load voltage is unable to follow the rapid r-f carrier variations, but follows only the peak values of the applied r-f signal voltage, as shown in Fig. 111. But, as you will recall, the changes in the peak value of the r-f carrier contain the original audio modulation, that is, the desired information. By following the peaks, the filter thus recovers the original audio-frequency signal.

Grid Leak Detector. Although having excellent fidelity and little distortion, diode detectors are not very sensitive and need a fairly strong signal voltage for efficient operation. Their use is therefore chiefly confined to the highly sensitive **superheterodyne** receivers, where large amplification takes place *before* demodulation. When extreme sensitivity for weak signal reception is required, the **grid leak detector** is generally used. A typical triode grid leak detector circuit with its associated waveforms is shown in Fig. 112. A pentode circuit could be used to obtain added amplification.

The circuit is highly sensitive because it provides amplification, as well as detection of the modulated wave. Detection of the modulated signal takes place in the control-grid-to-cathode portion of the tube, while the grid-to-plate portion amplifies the signal. The grid and cathode of the tube are operated like a diode rectifier. Resistor R_g (the grid leak) repre-

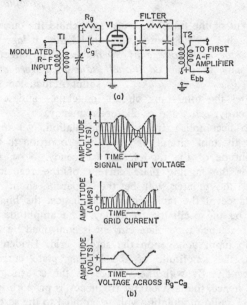

Fig. 112. *Grid Leak Detector* (*a*) *and Waveforms* (*b*)

sents the load for the rectifier, while grid capacitor C_g acts as bypass for radio frequencies. The modulated r-f signal is coupled to the grid of the tube through transformer T1.

The tube has initially zero grid bias. Whenever the signal voltage at the grid becomes positive, grid current flows through R_g and the secondary of T1 back to the cathode. This grid current produces a negative voltage drop across R_g and C_g, which serves to bias the tube negatively. A large enough capacitor is chosen so that it holds its charge, and hence the bias, during the negative half-cycles of the r-f input signal. Since the bias builds up during the positive half-cycles, the tube is eventually driven to cutoff and plate current flows only during positive half-cycles. This results in **rectified** r-f grid current pulses, as shown in Fig. 112b.

As in the diode detector, the rectified r-f current pulses develop a load voltage across the combination of R_g and C_g that **varies at the audio-modulation** frequency rather than at the frequency of the r-f carrier. Again capacitor C_g is chosen large enough so that its charge can leak off only slowly and, hence, the voltage across it can follow only the **peaks** of the r-f carrier, just as in the diode detector. The peaks of the r-f carrier contain the desired audio modulation, as shown in the bottom waveform in Fig. 112b. But in contrast to the diode detector, the r-f component of the grid voltage also is amplified by the tube and appears at the plate. The audio modulation component of the plate current, therefore, must be separated from this r-f component by a suitable filter. The filter may consist of a resistance-capacitance combination, as shown, or an inductance-capacitance filter may be used with even greater effectiveness. In either case the r-f component is bypassed to ground and the audio component of the plate current is coupled through transformer T2 to the input of an audio-frequency amplifier.

For the detection of weak signals, the grid resistor R_g is usually between 1 and 5 megohms and C_g has a value between 100 and 300 micromicrofarads. Since a weak input signal operates the tube near the bottom, curved portion of the characteristic, the detection is **non-linear** and considerable distortion takes place. Furthermore, with a strong input signal the positive grid current overloads the tube and additional distortion results. Because the curved portion of the tube's characteristic follows a square law (that is, the plate current increases as the *square* of the grid voltage), this type of detector is sometimes referred to as a **square-law detector.**

Infinite-Impedance Detector. The infinite-impedance detector shown in Fig. 113 is used frequently in high-fidelity AM receivers because of its ability to handle relatively large signal voltages at low distortion and with excellent fidelity and selectivity. No amplification takes place, however, since the output is taken from the cathode of the tube; the sensitivity of the circuit is therefore low. This is of no importance, however, in high-gain receivers with plenty of r-f amplification.

The circuit of Fig. 113 derives its name from the fact that its grid input has theoretically infinite impedance. The grid cannot be driven positive by the input signal and, hence, no grid current can flow.

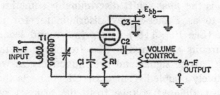

Fig. 113. *Infinite-Impedance Detector*

The cathode-load resistor, R1, serves several purposes. First, it provides automatic grid bias near the plate-current cutoff of the tube. Because of the cutoff bias, plate current flows only during the positive half-cycle of the r-f signal, the negative half-cycles being cut off. The rectified current pulses appear across R1, which thus acts also as a load for the rectified audio signal. Capacitor C1 filters out the remaining r-f carrier component, just as in the diode detector. The value of the capacitor is chosen so that it bypasses the r-f fluctuations to ground, but does not shunt out the audio-frequency signal appearing across load resistor R1. This audio output voltage is then coupled through capacitor C2 and a volume control (a potentiometer) to the grid of the first audio amplifier stage. Since the load resistor R1 is common to both the plate and grid

circuits of the tube, it has still another effect. It feeds a portion of the audio output signal back to the grid input circuit. As you can easily verify, this feedback is *negative* and, hence, results in further reduction of distortion and improvement of linearity. It also prevents any gain from being realized.

This detector operates somewhat differently from those discussed before, since it does *not respond to the peak values* of the r-f carrier wave. What happens is this: the series of rectified plate current pulses result in a certain **average value** of the plate current, which would be the current measured by a d-c ammeter. When the amplitude of the signal at the grid increases, the rectified current pulses become larger and the average value of these pulses also increases. The *average* value of the plate current, rather than the peak value, will thus faithfully follow changes in the carrier signal amplitude and, in this way, reproduce the original modulation. Furthermore, when the average plate current increases for a large input signal, the bias developed across R1 also increases, and prevents the grid from being driven positive and drawing current.

FM DETECTORS

The function of an FM detector is to recover the audio modulation of the r-f carrier wave, just like an AM detector. But because of the nature of the FM signal, the process of FM detection differs quite substantially from AM detection. We can divide FM detection conveniently into two main tasks. The first is to smooth out or limit any variations in the amplitude of the carrier wave, to obtain a signal that varies in frequency only, as shown in Fig. 109c. This is necessary because the actual FM signal arriving at the antenna is likely to vary quite a bit in amplitude owing to noise voltages superimposed on it and the effects of signal fading, reflections and absorption by various objects. If these extraneous amplitude variations were allowed to pass through the detector, noisy and distorted reception would result. The task of limiting the FM signal to constant amplitude is carried out by an FM limiter circuit. As we shall see, however, some FM detectors do not require a limiter.

The second task to be performed in FM detection is to obtain a detector output voltage that varies instantaneously as the frequency of the carrier wave. In other words, the detector output voltage should be zero or a fixed d-c value for the unmodulated carrier, and it should rise and fall proportionally with increases and decreases in the r-f carrier frequency. This task is performed either by a **discriminator** or **ratio detector** circuit.

LIMITERS

An amplitude limiter is a kind of electronic "gate" that restricts the positive and negative amplitude excursions of the carrier to a predetermined value, thus removing any amplitude variations present. To obtain limiting in an ordinary triode or pentode, advantage is taken of the fact that the characteristic curves of tubes show a leveling off or **saturation** when the input voltage at the grid exceeds a certain value. You may further remember that plate current saturation sets in for relatively small values of the grid input voltage if the tube's plate and screen potentials are kept low. Such early saturation effectively limits the **positive** swings of an r-f input signal. To limit the **negative** signal swings the tube is simply biased in the lower portion of the characteristic, so that negative amplitudes beyond a set level will produce plate-current cutoff.

Plate Circuit Limiter. A typical limiter circuit using a sharp cutoff pentode is illustrated in Fig. 114. Resistors R2 and R3 are chosen large enough to drop the plate and screen voltages to a low value, in the order of 30 volts. The grid transfer characteristic is thereby flattened out in the manner shown in (b) of Fig. 114. Because of the early saturation,

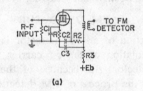

(a)

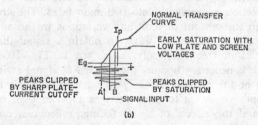

(b)

Fig. 114. *Plate Circuit Limiter* (a) *and Grid Transfer Characteristic* (b)

the amplitudes of relatively small input voltages are cut off above a certain positive value. Thus, all positive amplitude peaks exceeding line *B* in the figure are clipped off. At the same time the sharp plate-current cutoff of the pentode eliminates all negative signal peaks that exceed line *A*. The plate circuit limiter achieves effective limiting action, but very little amplification is obtained because of the low plate and screen grid voltages.

Combined Plate and Grid Circuit Limiter. Fig. 115 shows a circuit that combines both plate and grid circuit limiting. By adding the grid-limiting action, the plate and screen voltages may be raised somewhat, thus producing higher gain without a loss in limiting action. The values are typical of a practical receiver circuit.

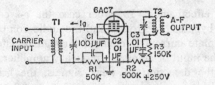

Fig. 115. *Combined Plate and Grid Circuit Limiter*

The input transformer T1 couples the FM carrier wave to the grid of the pentode tube. Plate circuit limiting is achieved in the same way as in the circuit of Fig. 114, except that the plate and screen voltages need not be quite as low. Resistors R2 and R3 drop the screen and plate voltages to the desired value and together with capacitors C2 and C3 provide bypassing action for the radio frequencies.

Grid circuit limiting takes place through the combined action of R1 and C1, which act essentially as a grid leak biasing circuit. The input signal must have sufficient amplitude to drive the grid of the tube positive and, hence, draw grid current. The grid current flow charges capacitor C1 to nearly the peak voltage of the positive signal half-cycle. When the signal falls away from its peak value, the capacitor discharges through resistor R1 and produces a negative bias. Since most of the input voltage appears across R1 rather than at the grid, the flow of grid current tends to flatten the **positive** peaks of the signal, i.e., limiting results.

PHASE-SHIFT DISCRIMINATOR

One of the most popular and effective FM detectors is the **phase-shift discriminator** sometimes referred to as **Foster-Seeley** discriminator for its inventor. (See Fig. 116.) The primary and secondary tank circuits are tuned to the frequency of the incoming carrier signal and are inductively as well as capacitively coupled. Capacitor C1 connects the frequency-modulated voltage from the primary coil to the center of the secondary and also blocks the d-c plate voltage from the diodes V1 and V2. A choke coil *L*, connected from the center of the secondary to the junction of diode load resistors R1 and R2, serves as a return path for the rectified current of the two diodes.

The two equal load resistors R1 and R2 are bypassed for the r-f carrier frequency through capacitors C3 and C2, respectively. When the carrier is at the center frequency in the absence of modulation, the rectified voltages across R1 and R2 are **equal and opposite in polarity;** hence, no audio output voltage appears between point "X" and ground. As the carrier frequency increases or decreases with modulation, however, the voltage across one of the load resistors increases or decreases with respect to the other and a net audio output voltage appears at "X." The variation in carrier frequency is translated

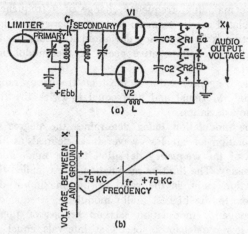

Fig. 116. *Phase-Shift Discriminator (a) and Characteristic (b)*

teristic is seen to be linear between the two resonant peaks at the extremes of the frequency swing.

RATIO DETECTOR

A variation of the phase-shift discriminator, called **ratio detector,** is insensitive to carrier amplitude variations and, hence, permits eliminating the limiter circuit. The basic circuit, shown in Fig. 117, is seen to be very similar to the discriminator. As a matter of fact, the circuit behaves in about the same way as the discriminator, as far as the detection of frequency-modulated waves is concerned. The only physical difference is that one of the diodes, V1, has been reversed in polarity and that the audio output voltage is taken off across one of the capacitors, rather than across both of them. Slight though these changes may be, they add up to a significant difference, as far as possible amplitude variations in the carrier wave are concerned.

into an equivalent voltage change by means of the **changing phase** relationship between the voltages across the primary and secondary tuned circuits. The process is rather complicated and is explained in detail in more advanced texts. Essentially, the voltages across the top and bottom half of the secondary coil add vectorially to the primary voltage. When the secondary tank circuit is resonant to the carrier (center) frequency, the top and bottom secondary coil voltages are exactly 90 degrees out of phase with the primary voltage. The vector addition thus produces equal and opposite voltages across the diodes, and the rectified diode output voltages, E_a and E_b, cancel each other out. When the applied frequency of the input wave is either higher or lower than the resonant frequency of the secondary tank, the secondary voltages are shifted in phase with respect to the voltage across the primary coil. Vector addition will then no longer result in equal and opposite voltages across the top and bottom diodes, but one will be larger and the other smaller, depending on the direction of the frequency deviation. The difference between the unequal, rectified voltages developed across load resistors R1 and R2, appears as audio output voltage between point "X" and ground (the chassis). The output voltage thus varies in accordance with the instantaneous frequency of the input signal.

A typical **discriminator characteristic** is shown in Fig. 116b. The characteristic is a plot of the discriminator output voltage versus frequency. At the resonant frequency (f_r) the output voltage is zero, as explained. For frequencies above resonance, the output voltage increases positively and becomes maximum for a frequency deviation of 75 kc. Below resonance the output voltage becomes negative and reaches a minimum at the maximum negative deviation of −75 kc. The voltage-frequency charac-

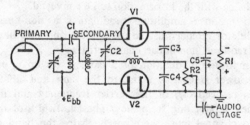

Fig. 117. *Basic Ratio Detector Circuit*

Because of the inversion of V1 the two diodes conduct on the same r-f half-cycle, whenever the bottom half of the transformer secondary is positive. A current then flows in **series** through the two diodes, the secondary coil and resistor R1, and produces a voltage across R1 with the polarity shown. This voltage charges capacitor C5 to the peak value. The capacitor is chosen large enough to hold the voltage across R1 constant during the non-conducting half-cycles. When frequency modulation is present, unequal, rectified diode output voltages are developed across capacitors C3 and C4. However, since C3 and C4 are in *series* across C5, the **sum of their voltages must remain the same as the voltage across C5, regardless of the ratio between them.** Since the sum of the two output voltages across R1 remains a fixed value, amplitude variations cannot occur. The **ratio** between the individual diode output voltages, however, changes continuously in accordance with the instantaneous frequency of the carrier wave. The audio output voltage cannot be taken off across the output resistor (R1), since that voltage remains constant, but must be picked off from one of the capacitors, C3 or C4, as shown. The output voltage-frequency characteristic is practically the same as that shown in Fig. 116b for the discriminator.

SUMMARY

Modulation is the process of superimposing information on a radio carrier wave. **Detection** or **demodulation** is the extraction of the original information from the carrier wave at the receiver.

Amplitude modulation modifies the amplitudes of a radio-frequency carrier wave in accordance with the strength of an audio (sound) or video (picture) signal.

In **frequency modulation,** the **amplitude** of the carrier remains **constant,** but its **frequency** is continuously varied in accordance with the **instantaneous amplitude** of the audio or video signal. The **number** of periodic **frequency changes** of the carrier (per second) **equals the modulating frequency.** By keeping the amplitude constant, frequency modulation eliminates static and electrical interference.

In **pulse-time modulation** the radio signal is transmitted in the form of sharp, discontinuous pulses and the timing or spacing of these pulses is varied in accordance with the information to be conveyed.

To produce amplitude modulation a **non-linear** modulator circuit must be used. Electron tubes have the required non-linear characteristic.

Amplitude modulation of a carrier sets up **upper and lower sidebands** equal to the **sum and difference,** respectively, of the carrier frequency and the highest modulating frequency. The **bandwidth** of the radio channel must therefore be **twice** the band of frequencies included in the modulation.

The greater the depth or **percentage of amplitude modulation,** the stronger and clearer is the audio or video signal. The maximum possible modulation without distortion occurs when the carrier reaches **twice** its normal amplitude on positive swings and **zero** amplitude on negative swings of the modulating signal. This is known as **100 percent modulation. Overmodulation** (more than 100%) results in distortion of the modulated signal.

The **power** of an amplitude-modulated signal increases by **50 percent** with full (100%) modulation, compared to the unmodulated wave.

In **plate modulation** the modulating signal is injected in series with the d-c plate voltage supply of the transmitter; in **control-grid modulation** the modulating signal is injected into the control grid, and in **screen grid** or **suppressor grid modulation** the signal is injected into either the screen or suppressor grid.

The maximum frequency change of a frequency-modulated carrier from its **center frequency** in either direction is its **frequency deviation;** it occurs during the positive and negative peaks of the modulating signal.

The **frequency swing** is the total frequency range of the FM signal and it is equal to **twice** the frequency deviation.

The **modulation index** determines the degree of modulation of an FM wave; it is defined as the **ratio of the frequency deviation to the modulating frequency.** The limiting modulation index, called the **deviation ratio,** is the ratio of the maximum deviation to the highest audio modulating frequency.

Frequency modulation sets up **groups of upper and lower sidebands,** spaced at intervals equal to the modulating frequency; their number and intensity depends on the modulation index.

The **bandwidth** of an FM channel is approximately equal to the **total frequency swing plus twice the highest audio modulating frequency.**

A **reactance tube modulator** produces FM by varying the frequency of an oscillator in accordance with the amplitude of an audio signal.

Detection of an amplitude-modulated wave requires the **rectification** of the r-f carrier and the separation (filtering) of the modulation.

In a **diode detector** the r-f carrier is rectified and an R-C filter develops a load voltage that follows the **peak values** of the rectified pulses; these peak values contain the audio modulation. The detector has good fidelity and low distortion, but is not sensitive.

A **grid leak detector** is highly **sensitive,** but tends to overload for large signals; it also follows the **peak values** of the r-f signal.

An **infinite-impedance detector** handles large input signals with low distortion; it follows the **average values** of the rectified plate current.

A **limiter** confines the amplitudes of an FM carrier to a predetermined, constant value. Limiting is obtained by plate-current **saturation** and **cutoff,** by **grid-current flow,** or by a **combination of both.**

In a **phase-shift discriminator** the variations in carrier frequency are translated into equivalent voltage changes by means of the changing phase relationships between the primary and secondary tank circuit voltages.

A **ratio detector** eliminates the carrier amplitude variations, while demodulating the FM carrier in nearly the same way as a discriminator circuit.

Chapter Fourteen

POWER SUPPLIES

Most power sources supply alternating current because it is easily generated and transmitted over long lines. Electron tubes, on the other hand, require direct current supply for all their electrodes, except the filaments, which may be heated either by a.c. or d.c. The most convenient way to change alternating to direct current is by means of a rectifier. A rectifier is capable of changing alternating current into a **pulsating** form of direct current; to obtain smooth d-c power additional filter circuits are required. A complete power supply also contains a **voltage divider** for providing d.c. at various desired electrode potentials and sometimes a **voltage regulator** to keep the d-c output voltage at a relatively constant value. We shall be primarily concerned with rectifiers and filters, which are the main elements in all power supplies.

RECTIFIERS

Rectifiers change a.c. into pulsating d.c. by eliminating the negative half-cycles or alternations of the a-c voltage. Thus, only a series of sinewave pulsations of positive polarity remain. An ideal rectifier may be thought of as a switch that closes a load circuit whenever the alternating current is positive, and opens the circuit whenever the alternating current is of negative polarity. Such a switch would have in effect **zero resistance** when the circuit is closed during positive a-c half-cycles, and **infinite resistance** for the time when the circuit is open during negative half-cycles. Practical rectifiers do not attain this goal, but come close to it. The resistance during the non-conducting interval (called **back resistance)** of electron-tube rectifiers is extremely high—for practical purposes infinite—but the resistance during the conducting interval (called **forward resistance)** is never zero or even constant. In any case, all rectifiers must provide a substantially **one-way path** for electric current; that is, conduction must take place primarily in *one direction* only. This is called **unilateral conduction,** or a **unidirectional characteristic.** You will remember that diode tubes have such a unidirectional characteristic.

Diode as Rectifier. A diode is any electronic device consisting of two elements, one being an electron **emitter** or **cathode,** the other an electron **collector** or **anode.** Since electrons in a diode can flow in one direction only, from emitter to collector (or cathode to plate), the diode provides the unilateral

conduction necessary for rectification. This is true for all diodes regardless of type. As we have already seen, diodes come in many forms. They may be electron tubes of the vacuum or gas-filled type, crystals or semiconductors made of germanium or silicon, or metallic types, such as copper-oxide and selenium rectifiers. Although the rectifier circuits that follow will illustrate primarily electron tube diodes, any of these other types may equally well be substituted, with the additional advantage that no filment supply is required, since no filments are present. The latter is one of the main reasons that selenium rectifiers and crystal diodes have become increasingly popular in television and radio receivers.

HALF-WAVE RECTIFIER

Since a diode will permit current to flow only during the positive half-cycles of the applied a-c voltage, a single diode is known as a **half-wave rectifier.** A half-wave rectifier circuit with its input and output waveforms is shown in Fig. 118. As you can see this is a very simple circuit. The a-c supply voltage is applied through a power transformer in series with the diode tube and load resistance, R_L. Plate current I_b flows through the tube every other

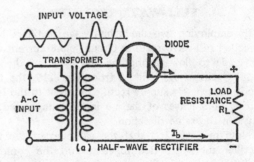

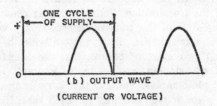

Fig. 118. *Half-Wave Rectifier and Output Waveform*

half-cycle, during positive alternations of the input voltage; it is blocked during the negative half-cycles. The current consequently flows through the load always in the same direction so as to make the cathode-connected end of R_L positive. Although unidirectional, the current is *not* direct (d.c.), because of its continuous changes in amplitude, or pulsations. It can be shown mathematically that these pulsations contain both a d-c component and an alternating (a-c) component, known as **ripple.** The current can be converted into a steady d.c. by filtering out the a-c ripple with a suitable **smoothing filter,** as we shall see later on.

It is evident from Fig. 118 that during the time the plate current flows, its instantaneous amplitude follows exactly the changes in the applied voltage. The shape of the plate-current waveform (Fig. 118b), therefore, is an exact replica of the a-c input voltage waveform during positive half-cycles. The plate current flowing through the load resistance develops a d-c output voltage, whose waveform is exactly the same as that of the plate current and is also represented by Fig. 118(b). But since only the positive half-cycles of the input voltage are reproduced, one-half of the input voltage is in effect lost. The efficiency of the half-wave rectifier is therefore low and is used only for applications requiring a small current drain. Another disadvantage of the half-wave rectifier is that the pulsations of the output current and voltage are at the **same frequency** as the a-c power line. Elaborate filter circuits are required to eliminate this low-frequency ripple (usually 60 cycles) and produce smooth direct current.

FULL-WAVE RECTIFIER

By employing two diode half-wave rectifiers in a so-called **full-wave rectifier** circuit, plate current can be made to flow during the full cycle of the a-c supply voltage. As evident from Fig. 119, the two diodes alternately supply rectified current to the load during both halves of the a-c input voltage, and always in the same direction.

Note in Fig. 119a that the cathodes of the two rectifier tubes are tied together and the common junction is connected to one side of the load resistor, R_L. The other end of R_L is connected to the center tap of the secondary winding of the power transformer. Since each tube is connected between one end of the transformer winding and the center tap, only *one-half* of the transformer secondary voltage appears between the plate and cathode of each diode. This means, of course, that the transformer secondary winding must supply a total voltage (called **plate-to-plate voltage**) that is *twice* the value of the plate voltage required for each tube. To provide sufficient plate and output voltage, therefore,

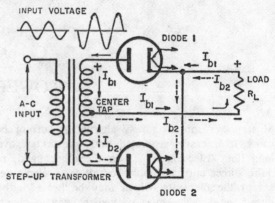

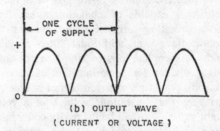

(a) FULL-WAVE RECTIFIER

(b) OUTPUT WAVE
(CURRENT OR VOLTAGE)

Fig. 119. *Full-Wave Rectifier and Output Waveform*

the transformer usually has a considerable **step-up** ratio between the primary and secondary winding. You will remember from elementary electricity, that the voltage across the transformer secondary winding is the primary input voltage multiplied by the turns ratio (or step up) between primary and secondary.

When an a-c voltage is applied to the primary winding of the transformer, a voltage of the same shape, but enlarged in amplitude by the step-up ratio, appears across the secondary winding, as illustrated in Fig. 119a. This secondary voltage is split in half, one-half appearing across diode 1 in series with the load (R_L), the other half appearing across diode 2 in series with the load. Assume that the polarities are such that the top of the transformer secondary winding is initially **positive** during the first half-cycle of the a-c input voltage. The plate of diode 1, therefore, is positive with respect to the cathode junction, and a plate current, I_{b1}, flows from the cathode to the plate of diode 1 through the top-half of the transformer secondary and through the load R_L. The direction of this current, indicated by the solid arrows in Fig. 119, is such as to make the cathode-end of the load **positive.** (You will recall that electron current flows **from minus to plus.**) The current I_{b1} develops a voltage across the load, which reproduces the first half-cycle of the a-c input voltage. Note also that during this first a-c half-cycle the **bottom** of the transformer secondary winding is **negative** with respect to the center tap and, hence, the plate of diode

2 is negative. Consequently, no plate current can flow through diode 2.

During the second half-cycle of the a-c input voltage the polarities reverse, making the **top** of the transformer secondary winding **negative** with respect to the center tap. Hence, the plate of diode 1 is negative in respect to its cathode and no plate current flows. During this same half-cycle, however, the **bottom** of the secondary winding is **positive** and thus the plate of diode 2 is positive with respect to its cathode. Consequently, a plate current, I_{b2}, flows from the cathode to the plate of diode 2, through the bottom half of the transformer secondary and through the load, R_L. As indicated by the dotted arrows, the current flows through the load in the **same direction as the previous half-cycle** so that rectification is obtained. The current I_{b2} flowing across the load develops an output voltage, which reproduces the second half-cycle of the a-c input voltage. Thus, **two** positive half-cycles appear across the load during **one** a-c cycle of the supply voltage, as is illustrated by the output waveform in Fig. 119b.

During successive half-cycles of the a-c input voltage, diodes 1 and 2 will continue to conduct alternately, each permitting current to flow during one half-cycle whenever its plate is positive with respect to the cathode junction. The resulting output current is a series of unidirectional pulses, as shown in Fig. 119b. Since there are two output pulses for each complete cycle of the a-c input voltage, the output frequency is **twice** that of the a-c supply frequency; that is, for a 60-cycle supply, the a-c **ripple** in the output is 120 cycles per second. With the ripple frequency being twice the a-c supply frequency and the current much less discontinuous than that of the half-wave rectifier, the pulsations are easily smoothed out by a suitable filter circuit. Furthermore, inasmuch as both halves of the a-c input cycle are rectified, the efficiency of the full-wave rectifier is far better than that of the half-wave type.

The full-wave rectifier has another important advantage. You may have noted in Fig. 119 that the individual tube currents flow in **opposite** directions through the secondary winding of the power transformer. The rectified current pulses, therefore, cancel each other out and no d-c current flows through the secondary winding. This avoids **d-c magnetization** of the transformer core and the resultant saturation, which is one of the serious defects of the half-wave rectifier. Since saturation of the core **reduces the inductance** of the transformer winding, a transformer that is subject to d-c saturation must be much larger for the same power rating than a unit that has no d.c. flowing through its winding. This advantage—combined with the high

efficiency, high permissible current drain, low ripple at twice the a-c frequency, and relatively high d-c output voltage—makes the full-wave rectifier suitable for a wide variety of applications in electronics. It is the standard circuit for low power applications.

FULL-WAVE BRIDGE RECTIFIER

The need for a center-tapped power transformer is eliminated by the **bridge rectifier** circuit, in which four diodes are used. (See Fig. 120.) The a-c input to the bridge circuit is applied to diagonally opposite corners of the network, while the output to the load is taken from the remaining two corners. The circuit is a full-wave type because both halves of the a-c input cycle are utilized. As we shall see, two tubes in series carry the load current on alternate a-c half-cycles.

Assume as before that the top of the transformer secondary winding is initially **positive** during the first half-cycle of the a-c supply voltage. You may consider the transformer secondary voltage during this half-cycle to be applied across a **series** circuit, consisting of diode 1, the load resistor, and diode 2. Since the cathode of diode 1 is at maximum **negative** potential and the plate of diode 2 at maximum **positive** potential, an electron current flows through diode 1, the load resistor, diode 2, and the transformer secondary, in the direction indicated by the solid arrows. During this first half-cycle the plates of diodes 3 and 4 are more negative than their cathodes and, hence, these tubes do not conduct. This is indicated by the dotted positive half-cycles for diodes 3 and 4, in Fig. 120. Thus, the first half-cycle is reproduced by the conducting tubes (diodes 1 and 2) and across the load resistor, as illustrated by the solid waveforms in Fig. 120.

One half-cycle later the **top** of the transformer secondary is **negative** and the **bottom is positive,** so that diodes 1 and 2 cannot conduct. This is indicated by the dotted negative half-cycles for diodes 1 and 2. The plates of diodes 3 and 4 are now positive, however, with respect to their cathodes and an electron current flows through diode 3, the load resistor, diode 4, and the transformer secondary, in the direction indicated by the dotted arrows. Thus, the second half-cycle of the a-c supply voltage is reproduced by conducting diodes 3 and 4 and also across the load resistor, as shown by the solid waveforms in Fig. 120. Note that the current pulses flow through the load resistor in the *same* direction during *both* a-c half-cycles, each pulse flowing from the plate junction of diodes 1 and 3 through the load resistor to the cathode junction of diodes 2 and 4. This makes the plate-end of the load resistor **negative** and the cathode-end **positive**, since electrons flow from minus to plus. The current pulses

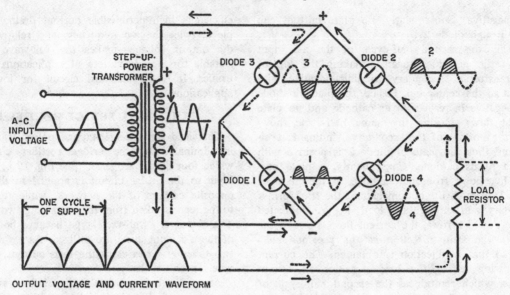

Fig. 120. *Full-Wave Bridge Rectifier Circuit*

are, therefore, unidirectional and their ripple frequency is *twice* that of the a-c supply frequency, just as for the conventional full-wave rectifier.

One advantage of the bridge rectifier over the conventional full-wave rectifier (Fig. 119) is that the bridge circuit produces a voltage output nearly *twice* that of the conventional circuit for a given power transformer. This is so because in the bridge circuit the full voltage of the transformer secondary winding is applied across two conducting tubes during each half-cycle, in contrast to the circuit of Fig. 119, where the secondary voltage is split in two. The bridge circuit, however, has the disadvantage of requiring four tubes, which makes for uneconomical operation. It is not possible to combine two of the tubes in one envelope of a **duodiode,** because the diode cathodes are not at the same potential and hence cannot be connected in parallel. Hence, separate filament transformers are required to heat the cathodes of the individual diodes, which also adds to the expense of the circuit. The last disadvantage does not apply to metallic-oxide rectifiers and crystal diodes, which do not require any cathode heating power. The bridge circuit is therefore widely used with selenium, copper-oxide, and crystal rectifiers.

VOLTAGE DOUBLER

Diode rectifiers can be made to deliver d-c voltages that are twice or several times the peak amplitude of the a-c voltage supplied to the tubes. Moreover, by multiplying the supply voltage to the desired value it becomes possible to omit the expensive step-up power transformer, required in conventional rectifiers. **Voltage-multiplying** circuits are

therefore especially useful in high-voltage circuits, where a transformer with a sufficiently high secondary voltage would be expensive and inconvenient, and in transformerless radio and TV receivers, operated directly from the a-c supply line. The voltage doubler described in the following paragraphs illustrates the principles involved in all voltage multiplier circuits.

The **full-wave voltage doubler** illustrated in Fig. 121 is capable of delivering a d-c output voltage of *twice* the peak value of the applied a-c input voltage. Basically, the circuit combines the output voltages of two half-wave rectifiers (diodes 1 and 2) in **series.** It is a full-wave circuit because each diode passes current to load resistors R1 and R2 on alternate half-cycles of the a-c input voltage.

The full-wave doubler operates as follows. When point *A* of the transformer secondary winding is

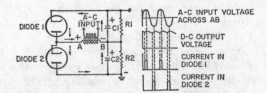

Fig. 121. *Full-Wave Voltage Doubler Circuit and Waveforms*

instantaneously **positive** during positive a-c half-cycles, diode 1 passes a current in the direction of the *solid* arrows, which charges up capacitor C1 so that its upper plate becomes positive. (The transformer could be omitted, as mentioned.) You can see that the upper plate becomes positively charged, since electrons flow *out* from it to the cathode of diode 1, leaving an excess positive charge. The lower plate of C1 becomes negatively charged, since electrons from the plate of diode 1 flow through the transformer secondary *into* the lower plate of C1, thus giving it an excess negative charge. As we have previously explained, a capacitor always charges up to the **peak value** of the voltage applied to it and hence capacitor C1 charges up to the peak value of the transformer secondary voltage. This is shown in the output voltage waveform of Fig. 121. Note also that only a brief charging current pulse flows through diode 1 during the first half-cycle.

During the next a-c half-cycle, point *B* of the transformer secondary becomes *positive* and point *A* becomes *negative,* so that diode 1 cannot conduct. Since the tube is an open circuit, the charge (and hence voltage) on capacitor C1 remains essentially *constant,* except for a small amount that leaks off through R1. This is shown in Fig. 121 by the slow dropping off of the d-c output voltage between conducting half-cycles.

Since point *A* of the transformer secondary is negative during the second half-cycle, the cathode of diode 2 also becomes **negative** with respect to its plate. A current flows, therefore, through diode 2 in the direction of the *dotted* arrows, which charges up the lower plate of capacitor C2 **negatively** to the peak value of the transformer secondary voltage. The waveform of the brief charging pulse flowing through diode 2 is illustrated in Fig. 121. Again, the voltage across capacitor C2 remains essentially constant during the next non-conducting half-cycle since the capacitor tends to hold its charge. A small amount of charge, however, leaks off through resistor R2, as shown by the dropping off in the output voltage waveform between the second and third a-c half-cycles.

Note that capacitors C1 and C2, as well as load resistors R1 and R2, are connected **in series.** The total voltage across the capacitors and resistors in series is therefore the sum of the voltages across each capacitor-resistor combination. Moreover, since the voltage across each combination equals the *peak* value of the transformer secondary voltage, the **total d-c output voltage** across load resistors R1 and R2 is **twice the peak value** of the transformer secondary voltage. This value is attained only when no load current is drawn from the output of the doubler. When a d-c load current is drawn, the capacitors become partially discharged and the output voltage drops considerably. By making the capacitors large (more than 10 microfarads), a fairly large load current can be supplied without losing too much d-c output voltage. Voltage triplers and quadruplers operate on these same basic principles.

FILTER CIRCUITS

Although the rectifier circuits we have discussed deliver an output voltage that always has the same polarity, this voltage is not suitable as d-c supply for electron tubes because of the pulsations in amplitude, or **ripple,** of the output voltage. These pulsations must be smoothed out before the output voltage can be applied to the plate, screen, and grid circuits of electron tubes. The required smoothing action is obtained by **filter networks,** consisting of choke coils and capacitors. (See Figs. 122 and 123.)

The filtering action of coils and capacitors depends on basic electrical principles. You will remember from basic electricity that a capacitor opposes any change in the voltage applied across its terminals by **storing up energy** in an electrostatic field whenever the voltage tends to rise and converting this stored energy back into voltage or current flow whenever the voltage across its terminals tends to fall. Thus, if some of the energy of the rectifier output pulsations could be stored in the electric field of a capacitor, and the capacitor would then be allowed to discharge *between* current pulses, the fluctuations in the output voltage could be considerably reduced. This is exactly what happens when a capacitor is connected in parallel with the rectifier output and load.

An inductor coil (**choke,** for short) opposes any *change* in the magnitude of the current flowing through it by storing up energy in a **magnetic field** when the current through it tends to increase; and by taking energy away from this field to maintain the current flow when the current through the inductor tends to decrease. Hence, by placing a choke coil **in series** with the rectifier output and load, abrupt changes in the magnitude of the load current and voltage are minimized. Another way of looking at the action of a series choke coil is to consider that the coil offers only the low resistance of the winding to the passage of d.c. while offering at the same time a high *impedance* to the passage of fluctuating or alternating currents. Thus, the d-c passes through, while the a-c ripple is largely reduced.

CHOKE-INPUT FILTER

Practical filter circuits are derived by combining the voltage stabilizing action of shunt capacitors with the current smoothing action of series choke coils. If the first component of the filter is a shunt capacitor connected *across* the rectifier, a **capacitor-input filter** results; if the first component of the filter consists of a choke coil connected in *series* with rectifier output, a **choke-input filter** results. A typical choke-input filter is illustrated in Fig. 122. Only one filter section is shown, but several identi-

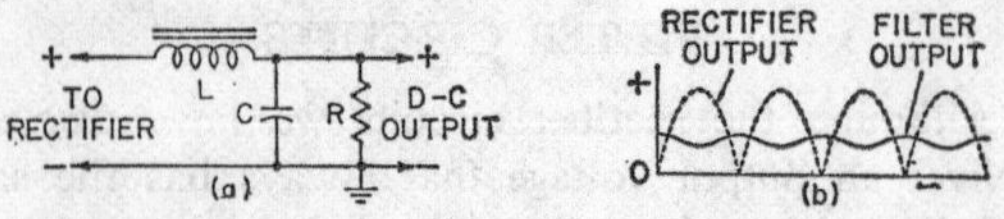

Fig. 122. *Choke Input Filter and Output Waveform*

cal sections are often used to improve the smoothing action.

The choke coil L at the input of the filter readily passes the direct current from the rectifier, but opposes the a-c pulsations, or ripple. Any fluctuations in the current that remain after it passes through the choke are largely bypassed around the load by the shunt capacitor, C, in the output of the filter. However, a small ripple is still present in the filter output, as shown in Fig. 122b. This is considered negligible if it is less than 1 percent of the steady d-c voltage. A filter with a 1 percent ripple may be obtained, for example, with a 10-henry choke and an 8-microfarad capacitor.

Note in Fig. 122b that the d-c output voltage of a choke-input filter is *not* equal to the peak values of the pulsations. This is so because the series choke prevents the capacitor from charging to the peak voltage when a load current is drawn. In the absence of a load current, the d-c output voltage of the filter is nearly equal to the peak value of the applied a-c voltage. As soon as even a small load current is drawn, however, the voltage drops off to some lower value, as shown. Beyond this initial drop, the d-c output voltage of the choke-input filter changes little with changes in load current and the filter is said to have good **voltage regulation.**

You many wonder why a resistor, R, is connected across the output of the filter. This resistor is known as a **bleeder resistor** and its main purpose is to place a **minimum load** across the rectifier during the time when the receiver tubes (or other load) are heating up and do not draw any current. Without the resistor there would be an initial high voltage surge when the rectifier is turned on, which might damage the rectifier and other tubes. The bleeder also helps to maintain a constant output voltage for changing loads by drawing a minimum current at all times. Moreover, the bleeder resistor discharges the capacitors after the rectifier has been turned off and so helps to prevent dangerous shocks. For these reasons a bleeder resistor is usually present in the output of any filter.

CAPACITOR-INPUT FILTER

The action of the capacitor-input filter, illustrated in Fig. 123, is slightly different from the choke-input filter. Here the rectifier output voltage first charges the capacitor to the **peak value** of the

pulsations. The capacitor tends to hold this charge between successive pulses, though discharging slowly through the choke and load. As a result, the filter output voltage drops off slightly between successive pulses, as indicated in the waveforms (Fig. 123b), but remains substantially near the peak value. The output voltage of a capacitor-input filter is therefore higher than that of a choke-input filter for the same a-c input voltage.

The remaining fluctuations of the rectifier output current are opposed by the series choke and bypassed to ground by the output capacitor, $C2$. A

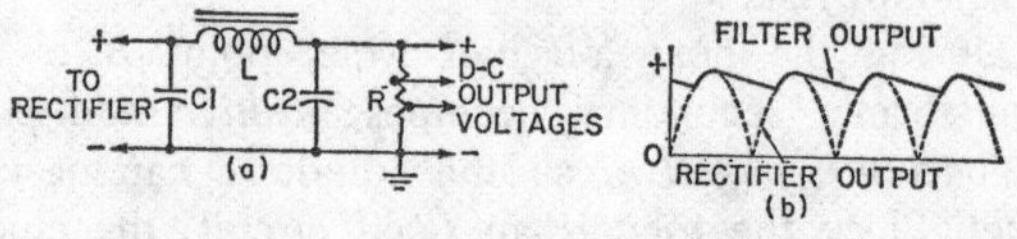

Fig. 123. *Capacitor Input Filter and Output Waveform*

small a-c ripple remains which may be further reduced by adding additional, identical filter sections in series. The output voltage of a capacitor-input filter falls off rapidly with increasing load current and, hence, the voltage regulation of this filter is considerably poorer than that of the choke-input type.

Note that the bleeder resistor in the filter of Fig. 123 has been provided with taps for supplying the proper voltages to various tube electrodes. The resistor, therefore, acts as a **voltage divider,** as well as providing the required bleeder current.

COMPLETE POWER SUPPLY

Now that we have learned something about rectifiers and filters, let us look at a complete d-c power supply, which combines these basic elements. A typical high-quality power supply, such as might be found in a high-fidelity radio receiver or amplifier, is illustrated in Fig. 124. It operates from the 117-volt a-c power line.

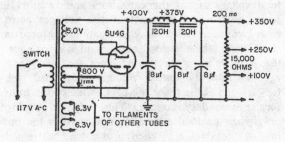

Fig. 124. *Typical Rectifier Power Supply*

Note that the power supply consists of a full-wave rectifier circuit and a two-section capacitor-input filter with a 15,000-ohm bleeder (voltage divider) in the output. The two diodes are housed in a single envelope of a 5U4G full-wave rectifier tube, which supplies a d-c load current of approximately 200 ma maximum. The step-up power transformer has four secondary windings: an 800 volt rms winding to supply the plates of the rectifier, a 5 volt winding for the heater of the rectifier, and two 6.3 volt filament windings for the heaters of the tubes in the set. Since a 800-volt plate-to-plate voltage is used at the input of the rectifier, the first filter capacitor theoretically charges up to a peak voltage of one-half this value, or about 565 volts, *when no load current is drawn*. With the maximum load current of 200 ma being drawn, however, the rectifier output voltage at the input of the filter drops to 400 volts because of the voltage lost in the tube and transformer windings. This value is further reduced to about 350 volts at the output of the filter due to additional voltage drops occurring in the two filter chokes, which have about 125 ohms resistance each. The upshot is that the filter output provides a maximum d-c output voltage of 350 volts at a load current of 200 ma. This output voltage would probably be used for the plates of the final power amplifier tubes in the audio amplifier. Reduced voltages of 250 volts and 100 volts are provided at the taps of the voltage divider for the plates and screens of the earlier amplifier tubes. Sometimes an additional tap is provided to supply a fixed **negative** bias voltage. The voltage divider provides a minimum load current (bleeder current) of about 23 milliamps.

The two-section capacitor-input filter employs large (20-henry) chokes and (8-microfarad) capacitors and, hence, is extremely effective. The ripple voltage at the output of the filter is only about 0.01 percent of the d-c output voltage, or about 0.035 volts a.c. (350 volts × 0.0001 = 0.035 volts).

D-C TO A-C INVERTERS

We have spent a considerable amount of time describing power supplies that furnish smooth d-c power from the a-c lines. In mobile and field applications frequently the opposite need exists—namely, to furnish a ready source of a-c power for lighting and equipment. We are all familiar with mobile, gasoline-driven motor generators which furnish a-c power directly for emergency and military uses. We also take the tube-type automobile radio for granted without ever questioning how the 6–12 volt automobile battery is capable of providing the 100 volts d.c. or so needed to operate the electron tubes. In the age of the transistor radio, this is no longer a problem, since transistors will operate directly and efficiently from the automobile battery. In tube-type auto radios, however, the low-voltage d.c. from the battery first had to be converted to a.c., then transformed to a suitable high voltage, and finally rectified again to direct current. This formidable task used to be accomplished by means of mechanical "vibrator"—to chop up the direct current to alternating current—and the resulting a.c. was then transformed, rectified and filtered by a conventional rectifier power supply. You will probably recall the humming sound made by the vibrator in the old-type automobile radios.

In general, any electrical or electronic device for converting direct current into alternating current is known as an **inverter.** How d-c motors coupled to a-c generators (motor generators) and vibrator power supplies accomplish this feat should be fairly obvious. The majority of high-power inverters are still of the motor-generator type. However, the advent in recent years of transistors and other solid-state components capable of operating from a low-voltage d-c source has made it possible to design all-electronic inverters—without any moving parts. These are called **static inverters.**

Static Inverter. The functional operation of a typical static inverter used in military aircraft is portrayed in the block diagram (Fig. 125). This particular inverter, manufactured by Magnetic Amplifiers,

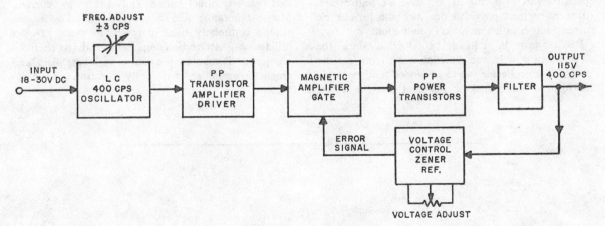

Fig. 125. *Static Inverter*

Inc., will provide 80 volt-amperes of 400-cycle, 115-volt alternating current from the standard 24-volt direct-current primary aircraft power supply. The entire package comes in a compact box that weighs approximately nine pounds.

The conversion from d.c. to 400-cps a.c. is accomplished in the first stage by means of a transistorized 400-cycle L-C oscillator. A push-pull transistor amplifier strengthens the output of the oscillator and drives the succeeding magnetic amplifier. A magnetic amplifier is essentially an a-c inductor (reactor) whose inductance—and hence output—can be varied by means of a control winding that changes the saturation of the iron core. The purpose of the magnetic amplifier is twofold. First, it provides amplification of the 400-cycle square-wave output of the transistor amplifier. In addition, and more important, it regulates the voltage output of the inverter in accordance with a feedback signal derived from a voltage control circuit. The latter consists of a Zener diode, which "samples" the output of the inverter and compares it with a Zener reference voltage. The comparison provides an "error signal," which is used to "gate" the magnetic amplifier to provide more or less amplification to compensate for variations in the output voltage. A "voltage adjust" potentiometer permits manual control of the output.

The magnetic amplifier drives the power output stage, consisting of two push-pull power transistors. The transistors are "gated" (i.e., switched on and off) by the magnetic amplifier to provide the proper output power and voltage. A bandpass filter smoothes the output.

SUMMARY

Rectifiers change alternating current into **pulsating direct** current by eliminating the negative half-cycles of the a-c voltage.

An **ideal** rectifier acts like a **switch** that closes a load circuit during the positive a-c half-cycles and opens the circuit during the negative a-c half-cycles. All rectifiers must provide a **one-way path** for electric current; this is called **unilateral conduction.**

Diodes provide unilateral conduction since the current can flow in one direction only, from **emitter** (cathode) to **collector** (anode). Diodes may come in the form of evacuated or gas-filled electron tubes, germanium or silicon crystals (semiconductors), and copper-oxide or selenium metallic rectifiers. Crystal diodes and metallic rectifiers do not have filaments and hence do not require heating power.

A single diode is called a **half-wave rectifier** because it permits plate current flow only every other half-cycle during positive alternations of the a-c input voltage. Since only positive half-cycles are reproduced, one half of the input voltage is lost and the efficiency is low. Furthermore, a half-wave rectifier has a **large a-c ripple voltage** at the same **frequency** as the a-c power line.

A **full-wave rectifier** rectifies **both** halves of the a-c input cycle by employing two diodes back to back. The conventional full-wave rectifier requires a plate-to-plate voltage **twice** the value of the plate voltage for each tube. The circuit has a **small a-c ripple at twice the a-c supply frequency,** high efficiency and high permissible current drain.

A **full-wave bridge rectifier** uses **four** diodes, but produces a voltage output nearly **twice** that of the conventional full-wave circuit. Separate transformers are required to heat the filaments of the tubes.

Voltage multipliers (doublers, triplers, quadruplers) deliver d-c output voltages that are several times the **peak amplitude** of the applied a-c voltage; this is done by charging series-connected capacitors through diode rectifiers on alternate a-c half-cycles so that the d-c voltages across the capacitors add up in **series.**

Filter circuits smooth out the pulsations in amplitude (a-c ripple) of the rectifier output voltage. Filter **capacitors** oppose any change in the voltage applied **across** them by **storing up energy** in an **electric** field, while **choke coils** oppose any change in the magnitude of a current flowing **through** them by storing up energy in a **magnetic** field.

Choke-input and capacitor-input filters combine choke coils with shunt capacitors to obtain increased filtering action.

An **inverter** is any electrical or electronic device that converts direct current into alternating current. Motor generators and vibrator power supplies are among commonly used inverters. A **static inverter** utilizes only electronic components and has no moving parts. Transistor oscillators and amplifiers and **magnetic amplifiers** frequently constitute static inverters.

Chapter Fifteen

RADIO COMMUNICATION

We have progressed in previous chapters from the first simple ideas about atoms and electrons to the amazing tricks performed by "trained" electrons and ions in electron tubes, and to the variety of tube and transistor circuits, which form the building blocks of the electronic art. Let us now put together these electronic building blocks into various configurations to make up different electronic systems. These systems employ electronic circuits, as well as electromechanical and other components, to carry out specific tasks assigned to them. In the chapters that follow we shall explore some of the electronic systems that make possible the fascinating marvels of the modern world. Among these are radio communication, high-fidelity systems, television, radar, and electronic aids to air navigation.

Radio communication and broadcasting was one of the first important uses electronics was put to. Simple radio transmitters and receivers had existed already long before the application of electronics made them what they are today; the existence of electromagnetic (radio) waves was predicted as early as 1865 in a mathematical treatise by Professor Clerk Maxwell, an English scientist. The German scientist Heinrich Hertz had verified the existence of these mysterious waves in 1887, using an elementary spark transmitter and "detector." A somewhat more sophisticated system, using an elevated antenna, ground, and an improved detector, sent out the first dot-dash radio-telegraphy messages in 1896. The range of this communication system was two miles, just beyond shouting range. A milestone in radio communication occurred on a December day in 1901 atop Signal Hill at St. John in Newfoundland, when the young Italian Guglielmo Marconi and his associates succeeded in spanning the Atlantic with three faint buzzes, the letter "S" of the Morse code. Further attempts, in 1902, to institute a commercial transatlantic radio telegraph service failed, however, because of the seasonal variations in signal strength, which nobody had suspected.

Radio telegraphy and telephony continued to make slow and erratic progress in the following years, but it was not until 1907 that the invention of the crystal detector and Lee De Forest's "audion" (triode) tube finally brought it to fruition. Two years later, in 1909, De Forest broadcast a Metropolitan opera by radio-telephone, a feat that forecast our present system of commercial radio broadcasting.

BASIC ELEMENTS

Modern radio communication and broadcasting may consist of the periodic interruptions of continuous radio waves (c.w.) in accordance with a telegraphic code, called **radio telegraphy,** or it may consist of the modulation of radio waves by speech or music—**radio telephony.** Fig. 126 illustrates the basic elements required to transmit messages, speech or music from one location to another by radio.

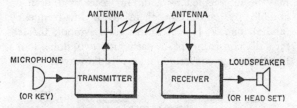

Fig. 126. *Elements of Radio Communication System*

These elements are:

1. A radio transmitter to **generate** the radio frequency waves.

2. A telegraph key or a microphone to **control** the radio waves in accordance with the information to be transmitted.

3. A transmitting antenna to **radiate** the waves into space.

4. A receiving antenna to **intercept** a portion of the radiated waves.

5. A radio receiver to **select** and **amplify** the desired transmitter signal and to **demodulate** the information contained in the radio waves.

6. A loudspeaker, or headphones, to **convert** the demodulated electrical waves into sound and, thus, **reproduce** the original information.

This picture is rather crude and we shall have to fill in many important details to gain a significant understanding of the process of radio communication. Let's take the elements one by one.

RADIO TRANSMITTERS

Radio transmitters generate r-f energy at a definite frequency and convey this energy to the transmitting antenna for radiation. To obtain a useful radiated signal, information must be superimposed on the radio waves. In the continuous-wave (c-w) transmitters used for radio telegraphy, the desired information is added by interrupting the radio-frequency oscillations in accordance with a telegraphic code. In radio-telephone (modulated) transmitters the information is added by **modulating** either the amplitude or the frequency of the radio-frequency **carrier** wave with the speech or music to be transmitted. Let us first turn to c-w transmitters.

CONTINUOUS-WAVE (C-W) TRANSMITTERS

Any of the oscillators described in Chapter 11 are capable of generating radio frequencies and, when connected to a suitable antenna, can radiate a useful signal. The radio-frequency output of such a simple one-stage transmitter may be interrupted in accordance with a coded telegraphic message by turning the plate-supply power on and off with a switch, or **key.** The output of the transmitter then consists of interrupted continuous r-f waves, which may be intercepted by a distant receiver and made audible by converting them to a low-frequency (audio) tone. A typical Morse code symbol, the letter "d" consisting of a dash and two dots, looks like this when converted into c-w radio waves.

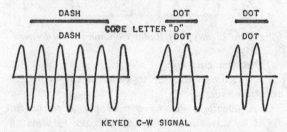

Fig. 127. *C-W Carrier Wave Keyed by Code Letter "D"*

Master-Oscillator Power-Amplifier. While a simple one-stage transmitter illustrates the principles involved, most practical transmitters use several stages to obtain additional r-f amplification or to multiply the oscillator frequency. A popular combination that provides more power than is possible with a simple oscillator is known as the **master-oscillator power-amplifier (mopa)** transmitter. As its name implies, the mopa consists of an oscillator stage, which may be crystal-controlled, and one or more r-f power amplifier stages coupled to an antenna. The block diagram of a mopa transmitter and a typical schematic circuit diagram are shown in Fig. 128.

Note that in the block diagram (Fig. 128a) the key controls the power to the oscillator stage, while in the circuit diagram (Fig. 128b) the key has been inserted in the plate-supply circuit of the power amplifier stage. (Batteries have been shown for convenience.) Either method of keying is practical, but **plate-circuit keying** of the power amplifier is preferred. In addition to providing increased power, the mopa transmitter has the advantage of **isolating the**

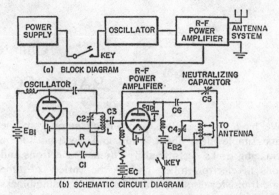

Fig. 128. *Master-Oscillator Power-Amplifier* (*Mopa*) *Transmitter,* (*a*) *Block Diagram and* (*b*) *Schematic Circuit Diagram*

antenna from the oscillator stage through the intervening r-f power amplifier. An antenna may change its effective resistance and capacitance as a result of the swaying in the wind and other influences. If coupled directly to the oscillator, these changes in antenna impedance cause **detuning** of the oscillator tank circuit and, hence, frequency instability.

The typical mopa transmitter illustrated in Fig. 128b consists of a conventional Hartley oscillator and a Class *C* power amplifier, both being **shunt-fed** through r-f chokes and capacitors. The output of the oscillator is coupled to the grid of the power amplifier through capacitor C3, which also prevents the bias voltage, E_c, of the amplifier from being short-circuited. The plate-tank circuit of the amplifier is tapped, as shown, and coupled to an antenna.

Neutralizing capacitor C5 in the plate-to-grid circuit of the power amplifier is used to balance out or neutralize the effects of the triode's grid-to-plate interelectrode capacitance, C_{gp}, shown dotted in (b). You will recall that the grid-to-plate capacitance of a triode feeds back energy from plate to grid with a resulting loss in power output. By feeding back a voltage in **opposite phase** from the bottom of the plate tank circuit to the grid of the tube, through capacitor C5, the interelectrode feedback voltage may be cancelled out and its detrimental effects are overcome. When this is done through a neutralizing capacitor in the plate circuit, as shown in (b), the process is called **plate neutralization.** When the neutralizing voltage is developed in the grid circuit of the tube, it is known as **grid neutralization.** In either case, the capacitance of the neutralizing capacitor (C5) will be approximately equal to the grid-to-plate capacitance of the tube, when the stage is properly neutralized.

Buffer Amplifier. A two-stage mopa transmitter can supply only a limited amount of output power. When high output power is required, several intermediate amplifiers are inserted between the oscillator and the final power amplifier that feeds the antenna.

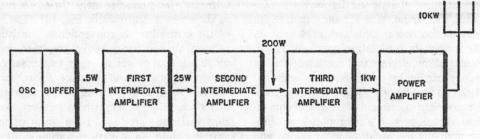

Fig. 129. *Block Diagram of Medium-Frequency 10 KW C-W Transmitter*

A sufficient number of intermediate stages must be inserted so that the output of the last intermediate amplifier is capable of driving the final power amplifier to the required transmitter power. Furthermore, an additional **buffer amplifier** is generally inserted between the oscillator and the first intermediate amplifier to isolate the oscillator from the following stages and minimize changes in the oscillator frequency due to variations in coupling and antenna loading. The buffer stage is a **voltage amplifier** operated Class *A;* it therefore draws no grid current, requires no input power and does not load down the oscillator stage. Fig. 129 illustrates the block diagram of a typical medium-frequency (300 kc to 3 mc) 10-kilowatt c-w transmitter, which employs a buffer stage and three intermediate amplifiers. The combined oscillator-buffer stage has the correct transmitter frequency, but supplies only one-half watt output. This is multiplied to 1 kilowatt by three intermediate r-f power amplifiers, operating Class *C,* and to 10 kw by the final power amplifier stage.

High-Frequency and VHF Transmitters. When the transmitter must operate in the high-frequency band between 3 mc and 30 mc, or in the VHF band between 300 mc and 3000 mc, **frequency multipliers** must be used to obtain the required output frequency. Generally, crystals cannot be obtained that will oscillate well in the VHF band and other oscillators are far too unstable for direct frequency control in these bands. The obvious way around this difficulty is, of course, to employ a suitable low-frequency oscillator and multiply its output frequency to the required value by a series of multiplier stages whose output tank circuits are tuned to multiples of the oscillator frequency. When the plate-tank circuit of the multiplier is tuned to *twice* the input (oscillator) frequency, the stage is called a **doubler;** when tuned to *three times* the input frequency, a **tripler;** and when tuned to *four times* the input frequency, the stage is called a **quadrupler.** Frequency multiplications of several dozen may be obtained in this way. Since a multiplier is an r-f power amplifier, the power output is, of course, also boosted by each additional stage, though not as much as by a conventional r-f power amplifier. In high-power VHF transmitters, one or more intermediate r-f amplifiers may be required, therefore, in addition to the multiplier stages.

The block diagram of a typical VHF transmitter, which is continuously tunable between 256 and 288 mc, is shown in Fig. 130. Four frequency multiplier stages are used here to increase the oscillator frequency by a factor of 64. The oscillator itself provides an output frequency that may be varied between 4 mc and 4.5 mc. This is multiplied successively by four, four, two, and two by two quadruplers and two doublers. The transmitter output frequency, thus, is variable between 256 mc and 288 mc. A final power amplifier stage boosts the multiplier output to the required transmitter power and couples it to the antenna to be radiated.

AMPLITUDE-MODULATED TRANSMITTERS

To broadcast speech or music with a radio-telephone transmitter requires either frequency or am-

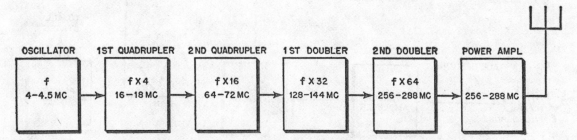

Fig. 130. *Block Diagram of VHF Transmitter with Four Multipliers*

plitude modulation of the r-f carrier waves in accordance with the sound to be transmitted. Both types of modulation were explained at length in Chapter 13. Amplitude modulation, you will recall, is the process of modifying the **amplitudes** of a radio-frequency carrier in correspondence with the **strength** of the audio (sound) signal. This is accomplished by converting the sound waves into equivalent electrical variations with a microphone or similar device, amplifying the resulting audio signal, and finally superimposing it on the plate or grid voltage of an r-f power amplifier. The generation, amplification and frequency multiplication of the radio-frequency signal are done in exactly the same way as in c-w transmitters and the same types of transmitter arrangements are used. The only new components present are the microphone and audio amplifier (modulator), which accomplish the amplitude modulation of the r-f carrier wave.

A basic block diagram of a typical amplitude-modulated radio-telephone transmitter is shown in Fig. 131. The electrical waveforms, when a pure audio tone (such as from a tuning fork) strikes the microphone, are shown above each stage. The tone is converted by the microphone into an audio signal, which is then amplified by a conventional audio amplifier. Power amplification to the required value of one-half the transmitter power (for plate modulation) takes place in the modulator stage, which—you will recall—is an ordinary audio power amplifier. The output of the modulator is applied through the modulation transformer either to the final r-f power amplifier (high-level modulation) or to an in-

termediate r-f amplifier stage (low-level modulation), as shown by the arrows in Fig. 131. The r-f portion of the transmitter is conventional, consisting in this case of an oscillator, buffer amplifier, intermediate amplifier, final power stage, and antenna. A common power supply serves all stages of the transmitter.

Microphones. We haven't said anything as yet about the microphones which convert the air pressure variations produced by a voice or musical instrument into an electrical variation of the same frequency and corresponding amplitude. There are many types of microphones operating on various principles, but their chief practical distinguishing marks are their **output** (voltage or current) and range of **frequency response.** Usually the better the frequency response, the lower is the output voltage (or current), and vice versa. The type of microphone used depends on its application. A microphone with a frequency response from 75 to 4500 cycles per second (cps) may be just the right thing for commercial radio communication, where speech intelligibility is of paramount importance, but it will never do for broadcast purposes. A frequency response of at least 30 to 10,000 cps is required for AM broadcasting and the whole audio range from about 20 to 15,000 cps is necessary for FM broadcasting or high-fidelity work. Three types of "mikes," in ascending order of audio quality, are the **carbon microphone,** the **crystal microphone,** and the **dynamic** or **moving coil microphone.**

Carbon Microphone. A typical **carbon microphone,** similar to that used in the telephone, is illustrated in Fig. 132. A loose pile of carbon gran-

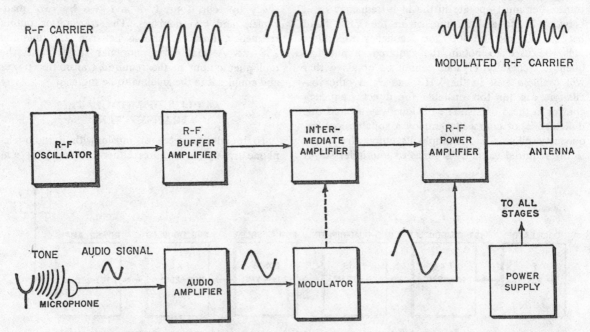

Fig. 131. *Block Diagram of Amplitude-Modulated Radiotelephone Transmitter and Waveforms*

ules is contained in an insulated cup, called **button,** which is in constant contact with a thin steel diaphragm. When sound waves strike the diaphragm, the resulting vibrations vary the pressure on the button and, thus, on the pile of carbon granules. These variations in pressure change the d-c resistance of the loosely piled granules in accordance with the vibrations of the diaphragm and, hence, in accordance with the sound waves. As you can see, the pile of carbon granules is in **series** with a battery and the primary of a step-up microphone transformer. As a consequence, the changing resistance of the carbon granules produce corresponding changes in the current through the circuit, resulting in a **pulsating** direct current. The current pulsations in the primary of the transformer induce an alternating (a-c) voltage in the secondary of the transformer, which is coupled to a suitable audio amplifier. The microphone transformer actually serves two purposes: It *matches* the relatively low resistance (a few hundred ohms) of the microphone to the high impedance of the grid circuit of the first audio amplifier tube and it also *steps up* the voltage. With a battery from 1.5 to 6 volts and microphone currents between 20 and 100 ma, a good carbon mike will develop up to 25 volts peak in the secondary of the transformer.

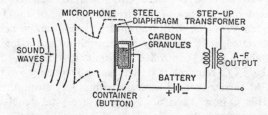

Fig. 132. *Carbon Microphone and Associated Circuit*

The trouble with the carbon microphone is that it has too many troubles. Because of random changes in the resistance of individual carbon granules, there is a constant background noise or hiss, which drowns out weak sounds. The carbon granules may stick together, or **pack,** resulting in a loss in sensitivity and serious distortion. Finally, and most serious, the frequency response of a carbon microphone is limited to a few thousand cycles on the upper end, which makes it unsuitable for anything but speech communication. There are various tricks to extend the response, such as stretching the diaphragm, but compared to other types of microphones, it's not worth the effort.

Crystal Microphone. The operation of a crystal microphone is based on the **piezoelectric effect,** which we discussed in Chapter 11 (oscillators). You will recall that many crystalline materials, such as quartz and Rochelle salt, generate a voltage when mechanical stress is applied to one of their faces.

Rochelle salt is used for crystal microphones, since it has greater voltage output than other piezoelectric materials. A basic crystal microphone is shown in Fig. 133.

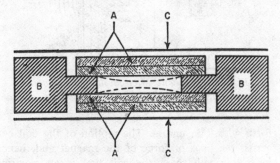

Fig. 133. *Construction of Crystal Microphone*

The basic unit, or **cell,** is made by cementing a metal foil to the surfaces of two thin crystal slabs and clamping the two slabs together. A sound diaphragm is coupled to the cell and causes a twisting of the crystal faces, when sound waves strike the diaphragm. This in turn generates a voltage between the metal foil on the crystal face, which is proportional to the sound pressure.

The unit shown in Fig. 133 is known as a **sound cell** or **grill** type microphone, since it consists of a number of basic crystal units or cells, connected in series or parallel for greater sensitivity. This superior type of crystal microphone dispenses with the diaphragm, sound striking the crystal plate directly. The crystal cells (A) are mounted in a bakelite framework (B) and the assembly is covered with a flexible, airtight cover (C), which permits the cells to vibrate freely when sound strikes them. Crystal microphones have a high output impedance and can therefore be connected directly into the grid circuit of the first audio amplifier tube. Because their output voltage is low (a few **millivolts),** crystal microphones require several stages of high-gain audio amplification. However, their frequency response is relatively uniform from about 50 to 10,000 cps, which is sufficient for good quality reproduction of speech and music.

Dynamic Microphone. One of the highest-quality microphones commercially available is the dynamic or moving coil microphone, illustrated in Fig. 134. Its operation is based on the fundamental electric principle that a voltage is induced in a wire or coil which moves across a magnetic field.

As shown in the illustration, a coil of fine wire is suspended in the field of a strong permanent magnet and is fastened on one end rigidly to the back of a diaphragm. When sound waves strike the diaphragm, it vibrates, and the coil moves back and

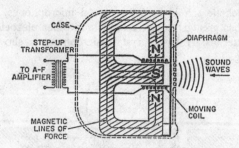

Fig. 134. *Dynamic Microphone with Matching Transformer*

forth with it in unison. The motion of the coil cuts across the lines of force of the magnet and, hence, induces a voltage in the coil that is an electrical representation of the original sound.

The dynamic microphone is practically ideal for the most demanding purposes. It requires no external voltage, is light in weight, rugged, and practically immune to the effects of mechanical vibration, temperature and moisture. It is highly sensitive and has a fair voltage output, a little lower than that of the crystal mike. Most important, the dynamic microphone has an essentially uniform output over the entire audio-frequency range from about 30 to 15,000 cps. The output impedance of the moving coil, however, is only about 30 to 500 ohms maximum and, hence, a step-up transformer is required to match the microphone to the input stage of an audio amplifier. This transformer is usually built right into the microphone.

FREQUENCY-MODULATED TRANSMITTERS

As described in Chapter 13, in frequency modulation the **amplitude** of an r-f carrier remains *constant,* but its **frequency** is continuously *varied* in accordance with the instantaneous amplitude (strength) of an audio signal. Thus, the original

sound waves are converted into a **frequency deviation** of the r-f carrier frequency that is proportional to the intensity of the sound waves. The number of carrier frequency swings per second is equal to the frequency of the sound, you will recall. We have already become acquainted with the **reactance tube modulator** (in Chapter 13), which changes the amplitude variations of an audio signal into corresponding phase or frequency variations. To construct an FM transmitter on paper, therefore, we need only fit together the pieces.

Fig. 135 shows the block diagram of a typical FM transmitter. The upper row of the diagram practically explains itself. It consists of a microphone, an audio amplifier, a **reactance modulator** (for converting the audio amplitude changes into frequency changes), a 10-mc transmitter oscillator, a series of frequency multipliers that multiply the oscillator frequency 10 times to the required operating frequency of 100 mc, a final r-f power amplifier and an antenna. Since we have already explained the operation of the reactance modulator, there is nothing unfamiliar in the upper row of Fig. 135.

Now what about the complex circuitry indicated by the blocks of the lower row in Fig. 135? The basic purpose of these circuits is to **stabilize the center frequency** of the transmitter. Since the frequency of the FM transmitter is constantly being deviated, it is obviously not possible to use a highly stable, crystal-controlled transmitter oscillator. Yet the FCC requires that the **center frequency** of the FM carrier be held constant within very close limits. This is how it's done. A separate crystal oscillator generates a 9-mc signal, in this case. A series of doublers multiply the oscillator frequency to 99 mc, that is, 1 mc below the carrier center frequency. The output of these multipliers is applied to a **mixer stage** together with a portion of the 100-mc output of the frequency multipliers in the upper (main

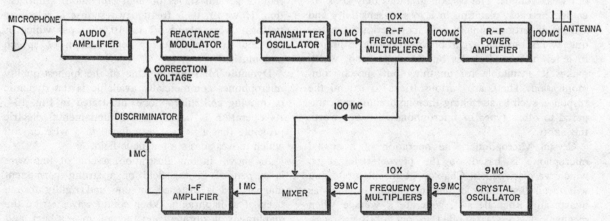

Fig. 135. *Block Diagram of FM Transmitter with Automatic Frequency Control*

transmitter) row. These two signals beat together in the mixer to generate a 1-mc difference or **intermediate frequency** (i-f) in a manner which we shall explain in connection with the superheterodyne receiver. The intermediate-frequency signal is boosted in amplitude by an i-f amplifier tuned to 1 mc. The 1-mc output of the i-f amplifier, finally, is applied to a phaseshift **discriminator** circuit, identical with that discussed in Chapter 13. The discriminator, you will remember, produces a fixed d-c output voltage as long as the frequency of its input signal to which it is tuned remains constant. As soon as the input frequency differs from the resonant value, however, the d-c output voltage of the discriminator changes. This d-c output voltage is fed to the grid of the reactance tube and serves as a reference potential for establishing the proper operating frequency of the transmitter.

You can easily see what will happen if the center frequency of the transmitter oscillator should momentarily rise, let's say. As soon as this happens, the mixer output frequency will also rise, since the difference between the transmitter and crystal oscillator frequency has increased. The frequency at the input of the discriminator, consequently, will increase and its d-c output voltage will change. In effect, a correction voltage is now being applied to the grid of the reactance tube modulator. This correction voltage changes the phaseshift of the reactance tube in a direction that will automatically restore the correct transmitter frequency. The system will work, of course, similarly for a momentary lowering of the transmitter center frequency. The same **automatic frequency control** (AFC) system is used in television receivers.

ANTENNAS

We have investigated the behavior of electrons within the atom, inside evacuated tubes, in conductors and circuits, and we have seen them oscillating rapidly back and forth between the capacitor and inductor coil of a tank circuit. But thus far the dance of electrons has always been confined within the boundaries of a conductor, tube, or circuit and we have not explained the central point of radio, the radiation of waves into space. How indeed do the electrons oscillating at a high frequency in the plate tank circuit of a transmitter set up "electromagnetic" (radio) waves to be radiated throughout space? Part of the answer is childishly simple: they do so automatically without any outside help. It is one of the consequences of Maxwell's famous theory that **accelerated electric charges (i.e., electrons) always radiate energy in the form of electromagnetic waves.** Emphasis belongs on the word "accelerated." A charged glass rod or capacitor will not radiate; the charges must be

accelerated, they cannot be static. We cannot go into Maxwell's equations here, but fact is that the simple L-C tank circuit with its accelerated electrons (i.e., the electron oscillations) radiates energy in the form of electromagnetic waves. As a matter of fact, all the conductors in a high-frequency oscillator or amplifier radiate energy to some extent. The radiation from a tank or oscillating circuit, however, is extremely weak. To radiate or receive electromagnetic waves effectively we need an **antenna.** An antenna is a system of elevated conductors, which couples or **matches the transmitter or receiver to space. A transmitting antenna,** connected to the transmitter by a transmission line or **lead-in,** forces radio waves into space by "grasping it with long fingers," as one radio engineer expressed it. Similarly, a **receiving antenna,** connected to a radio receiver, grasps or intercepts a portion of the electromagnetic energy traveling through space.

It is the **size** of the antenna conductors that makes them effective radiators and absorbers of electromagnetic waves. The antenna must have the proper size to **match, in effect, the impedance of the transmitter or receiver to that of space.** (Space actually has a certain **impedance,** or opposition to the passage of electromagnetic waves.) You can visualize the effect of antenna size as follows: If you stir up the water in a pond with your hands or stone pebbles, small water wavelets will be produced. Using a paddle will create far larger waves with the same expenditure of energy.

BASIC ACTION OF ANTENNA

Radiation of electromagnetic waves from an antenna is a complex process, which can be completely understood only by means of the mathematical theory. But you can form a fairly accurate picture of the action by going back to the description of tank circuit oscillations in Chapter 11. You will recall that the electrons in a tank circuit rapidly flow from one plate of a capacitor to the other through the tank coil, thus charging and discharging the capacitor with alternating polarity. While the electrons rush from one plate to the other, they build up an intense **magnetic field about the coil.** When they have fully charged the capacitor the electron current stops, but now an intense **electric field** exists between the plates of the capacitor. Thus, the sequence of charge and discharge results in **accelerated** electron motion, or an oscillating current, with the energy alternately (each quarter-cycle) being stored in the electric field of the capacitor and in the magnetic field of the inductance coil.

Let us now attach a wire to each side of a tank circuit and observe what happens in this simple

Hertz antenna during alternate quarter-cycles. The arrangement is illustrated in Fig. 136 and is similar to one used by Hertz during his early experiments.

Fig. 136a shows the original tank circuit at the moment of maximum current flow, when an intense **magnetic field** has built up around the coil. In (b), which portrays the same instant, two antenna wires have been attached to the tank circuit. The electron current now rushes to the end of each wire and back to the tank circuit, setting up a magnetic field along the whole length and **perpendicular** to the wires. As we shall see later, the length of the wires has to be just right so that the current can complete its round trip to the ends of the wires and back to the tank in just one half-cycle and is thus **in phase** with the current at the tank. Note that the magnetic field is strongest near the center point of the antenna, where it is fed with energy, as indicated by the concentrated lines of force. This field extends far into space, although it grows progressively weaker with increasing distance from the antenna.

Fig. 136c portrays the tank circuit a quarter-cycle later, when the current has stopped, but the charge on the capacitor is a maximum and consequently a strong **electric field** exists between its plates. The action at the same moment in the Hertz antenna is illustrated in (d). The current has mo-

mentarily stopped after reaching the ends of the antenna rods, thus charging the ends of the conductors with opposite polarity. The charged conductors are now surrounded by an electric field, with lines of force extending from the positive end to the negatively charged end parallel to the antenna. This field, too, projects far into the space surrounding the antenna.

During the following quarter-cycle the current rushes back from the ends of the antenna wires to the center, setting up a magnetic field of opposite polarity (in respect to Fig. 136b). After reaching the center-connected ends of the antenna, the current again dies down momentarily and the wires are charged up with opposite polarity. An electric field parallel to the antenna, but with opposite polarity from that shown in (d), now surrounds the antenna conductors. The same sequence is repeated during the next cycle of the current oscillations.

It is evident that the antenna rods are surrounded alternately by electric and magnetic fields which are in step with the field alternations (every quarter-cycle) in the tank circuit. Since one field is building up, while the other is dying down, *both* the electric and magnetic fields effectively surround the antenna at all times. The resulting complex **electromagnetic field** about the antenna is illustrated in Fig. 137. You have to imagine that the

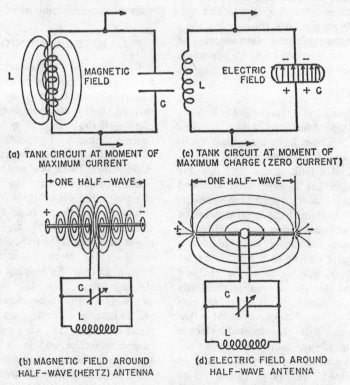

(a) TANK CIRCUIT AT MOMENT OF MAXIMUM CURRENT

(c) TANK CIRCUIT AT MOMENT OF MAXIMUM CHARGE (ZERO CURRENT)

(b) MAGNETIC FIELD AROUND HALF-WAVE (HERTZ) ANTENNA

(d) ELECTRIC FIELD AROUND HALF-WAVE ANTENNA

Fig. 136. *Action of Hertz Type of Transmitting Antenna*

outermost field lines detach themselves from the antenna and extend far out into space as traveling **electromagnetic waves.**

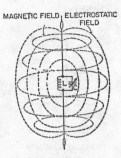

Fig. 137. *Complex Electromagnetic Field Surrounding Antenna*

Antenna Length and Resonance. It is correct to consider the antenna conductors as being in **resonance** or **tuned** to the frequency of the tank circuit oscillations. You will recall that the oscillating current flowing through the tank circuit coil reverses every **half-cycle** in polarity or phase. (It takes a quarter-cycle to charge the capacitor and another quarter-cycle to discharge it.) For the oscillations in the antenna rods to be in step or in phase with the tank circuit oscillations, it must take just as long for the current to rush to the tips of the antenna rods and back as it takes to charge and discharge the capacitor. Since it takes a quarter-cycle to charge the capacitor and another quarter-cycle to discharge it, it should also take a quarter-cycle for the current to reach the ends of the two antenna rods and another quarter-cycle for it to return to the center, so that the timing is just right. Furthermore, since the current flows in the opposite direction when returning from the antenna tips than when flowing out, its polarity or phase on the return trip has been automatically **reversed.** The returning current wave is thus **in phase** with the next half-cycle of the tank current oscillation, provided the length is right. Under these conditions the antenna is **in resonance** with the frequency of the tank circuit oscillations.

For the current to travel out to the ends of the antenna in a quarter-cycle, each antenna rod must be a quarter-cycle or **quarter-wave** long, or the entire antenna must be a **half-wave** long from tip to tip. The **electrical** length of a Hertz antenna is therefore a **half wavelength** and it is referred to as a **half-wave antenna.** What this amounts to in *physical* length depends, of course, on the time allotted to one cycle (the period) and how far the waves travel during this time (the velocity). A wavelength may therefore be determined by multiplying the velocity of the waves by the period, or equivalently, by **dividing the velocity by the fre-**

quency of the oscillations, since frequency is the reciprocal of the period. The velocity of electromagnetic waves is the same as that of light (which is an electromagnetic radiation), or about 186,284 miles per second. We have, therefore,

$$\text{wavelength (in miles)} = \frac{186,284}{\text{frequency (cps)}}$$

or more conveniently, converting to *feet* and *megacycles,*

$$\text{wavelength (feet)} = \frac{984}{\text{freq. (mc)}}$$

The wavelength of a 1500 kc (1.5 mc) broadcast station, for example, is 984/1.5 = 655 feet. A Hertz antenna for this transmitter should be one half-wavelength, or about 327 feet long. Each antenna rod must therefore be one-quarter wave, or 163.5 feet long. Actually, electromagnetic waves do not travel quite as fast in a conductor (such as an antenna) as in free space, so that a small correction must be applied. A convenient formula for the length of a half-wave or Hertz antenna, good to about 30 mc, is:

$$\text{Length of Half-Wave Antenna (feet)} = \frac{468}{\text{freq. (mc)}}$$

Substituting in the latter formula, we see that the half-wave antenna for the 1500 kc broadcast station should be only

$$\frac{468}{1.5} = 312 \text{ feet long, instead of 327 feet}$$

Each conductor must therefore be 156 feet long.

Polarization of Electromagnetic Wave. We have stated that the **electric** field of a (Hertz) antenna is always **parallel** to the plane of the antenna, while the **magnetic** field is at **right angles** to it. The magnetic and electric fields of an electromagnetic wave, thus, always travel at right angles with respect to each other. The orientation of the **electric** field with respect to earth is known as the **polarization** of the wave. If the plane of the electric field is horizontal in respect to earth, the wave is said to be **horizontally polarized;** if the electric field is oriented vertically, the wave is said to be **vertically polarized.** Fig. 138 is a graphic presentation of an electromagnetic wave emanating from a half-wave antenna. Since the intensity of the fields varies sinusoidally with the amplitude of the current oscillations, the fields are represented by traveling sine-waves at right angles to each other. As you can see the antenna rods are placed vertically with respect to earth and hence the parallel electric field is also vertically oriented. The wave is therefore

vertically polarized. If the antenna rods were turned to a horizontal position, the electric field would turn parallel with it, and the wave would be horizontally polarized. Evidently, therefore, the wave polarization is primarily a function of the antenna orientation. However, the plane of polarization of an electromagnetic wave may actually be turned somewhat from its original direction during its travels through space.

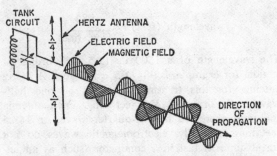

Fig. 138. *Vertical Polarization of Electromagnetic Wave*

Antenna Radiation Pattern. Note in Fig. 136 that the lines of force of the **magnetic** field are strongest near the **center** of the antenna and **weakest** near the tips of the antenna. **Maximum** radiation of the magnetic field therefore takes place **broadside** to the plane of the antenna and **minimum** radiation occurs from the antenna tips. The effectiveness of the spread of the **magnetic** field is measured by the **radiation pattern** of the antenna. Fig. 139a shows the radiation pattern of a half-wave antenna in **one plane** (the plane of the paper). As you can see, it is an 8-shaped pattern, extending from the front and rear of the antenna. The length of a line drawn from the center to the periphery of the pattern is a measure of the amount of radiation in the direction of the line. The longest line that can be drawn is at right angles (broadside) to the antenna, representing maximum radiation. No line can be drawn along the length of the antenna, where the pattern is tangent to it, and hence the radiation is a minimum or zero in this direction.

The actual **free-space** radiation pattern of the half-wave antenna extends in all directions around the antenna and is **doughnut-shaped,** as illustrated in Fig. 139b. Only half of the doughnut pattern is shown. The complete radiation pattern is obtained by rotating the figure-8 pattern of Fig. 139a through a circle around the axis of the antenna. This pattern may be considered **attached** to the antenna, since it rotates with it when the antenna is turned. Thus, when the antenna is vertical in respect to the earth, the doughnut-pattern will be horizontal, as shown in Fig. 140a, and maximum radiation takes place in all horizontal directions

around the antenna. But if the antenna is turned horizontally the doughnut is vertical, and maximum radiation occurs in a vertical plane perpendicular to the antenna (Fig. 140b).

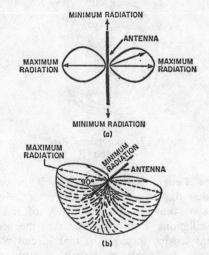

Fig. 139. *Radiation Pattern of Half-Wave Antenna, in Plane of Paper (a), and Doughnut Pattern in Space (b)*

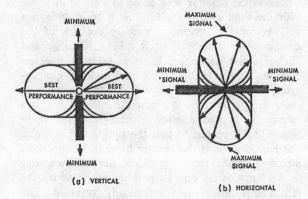

Fig. 140. *Doughnut Radiation Pattern of Vertical (a) and Horizontal (b) Half-Wave Antenna*

Marconi or Grounded Antenna. At low frequencies the length of a half-wave (Hertz) antenna becomes excessively large, as is apparent from our formula for antenna length. Marconi found in his experiments that the length of a Hertz antenna could be cut in half by grounding one end of the antenna. The resulting **Marconi antenna,** illustrated schematically in Fig. 141, is a **vertical** grounded mast equal in length to one-quarter wavelength of the signal it is designed to radiate or receive. The theory of the Marconi antenna is similar to the ungrounded half-wave antenna, but with the earth serving to provide a "mirror image" of the antenna, one quarter-wave in length (see Fig. 141). Since the antenna is vertical, the radiation emitted from it is **vertically polarized.**

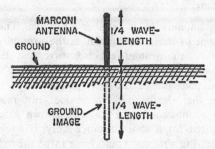

Fig. 141. *Marconi Antenna with Ground Image*

Other Transmitting and Receiving Antennas. Although little has been said about receiving antennas, in learning about the radiating properties of a transmitting antenna, we have also found out the criteria for a good receiving antenna. In general, a **good radiator is a good absorber** and, hence, a good transmitting antenna is an equally good receiving antenna. The action of a receiving antenna is, of course, the reverse of that occurring in a transmitting antenna. A receiving antenna intercepts or absorbs a tiny portion of the electromagnetic wave energy traveling past it and the resulting electron current along the conductor induces a small voltage of the same waveform as the radio waves. The voltage is passed on to the receiver, where it is amplified and demodulated. In principle, the design and shapes of receiving antennas are the same as those of transmitting antennas, but in practice the receiving antenna often consists only of a convenient length or loop of wire because of space restrictions. In areas close to the transmitter sufficient voltage pickup is obtained by these makeshift antennas, but

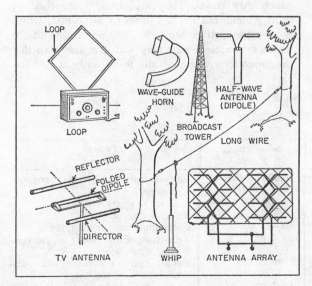

Fig. 142. *Various Types of Transmitting and Receiving Antennas*

farther away from the transmitter in the "fringe areas" of reception, the design of receiving antennas becomes almost as elaborate as that of the transmitting antenna.

There are many additional types of transmitting and receiving antennas, in additional to the half-wave (Hertz) antenna and the quarter-wave (Marconi) antenna we have discussed. The size and shape of antennas depends on many factors, such as the frequency of the signal, the required "directivity" of the radiation pattern, the use to which the antenna is to be put, etc. Fig. 142 illustrates a variety of typical transmitting and receiving antennas.

The simplest type of antenna is a long wire, insulated from the ground and supported by trees or poles. The wire is tapped near one end and connected by a lead-in wire to the transmitter or receiver. Vertical quarter-wave grounded (Marconi) antennas may vary in size from a large broadcast tower to a small mobile "whip" mounted on a vehicle. Half-wave (Hertz) antennas, called **dipoles,** also come in many varieties from the simple two-conductor, center-fed type we have studied to sophisticated structures. Television antennas may consist of a "folded" dipole, a **director** and a **reflector,** which serve to improve the directivity and, hence, pickup from a particular station. Sometimes "arrays" of half-wave antennas are stacked vertically or horizontally to obtain a specific radiation pattern and improve the "gain" of the combination over that of a single antenna. An array may take the form of a metal horn, fed by a "waveguide," to beam an ultra-high-frequency signal in a specific direction like a search light. An antenna may be bent into the shape of a **loop** (or looped coil) to obtain highly directional characteristics at low or medium frequencies. Maximum response is obtained when the plane of the loop faces the transmitter, and practically zero response (called a "null") is obtained when the loop is broadside to the transmitter. This characteristic of loops is exploited in **direction finders** for determining the horizontal compass bearing (azimuth) between the direction finder and a transmitter. Two or more such bearings permit locating the position of the direction finder or that of an unknown transmitter on a map.

WAVE PROPAGATION

We have launched radio waves from an antenna and they are traveling outward in all directions at a velocity of over 186,000 miles per second. What happens to them out in space? We know that radio waves are little affected by the surrounding atmosphere, rain, snow, etc., and are able to penetrate non-metallic objects easily, but are stopped dead by metals or fine meshed wire screens. They seem to travel best in free space, a vacuum, and do *not* need a medium of propagation, as do water waves, for example.

Beyond these generalizations, however, there is little we can say about the propagation of radio waves without specifying their wavelength or frequency. You must remember that radio waves are only a part of the tremendous **spectrum** of electromagnetic wave motion, a spectrum that includes such familiar radiations as heat and light, as well as such mysterious ones as cosmic radiation and X-rays. The electromagnetic spectrum roughly extends from the slow oscillations of commercial power generators (60 cycles), through the **audio** frequencies (20 to 15,000 cycles), and **radio** frequencies (about 15 kc to 300,000 *mega*cycles), the **infrared** or radiant heat region (from about 400,000 mc to about *400 million mc*), the visible light region (400–800 million mc, approx.), the **ultraviolet** region (to several *billion* mc), and X-rays, gamma rays, and cosmic rays, oscillating at frequencies of quadrillions of megacycles, or equivalently, wavelengths of fractions of an Angstrom (1 Angstrom = 10^{-8} cm). All these radiations are electromagnetic waves, traveling at 186,000 miles per second, different though their effects may be. In general, as the frequency goes up, the radiations tend to travel more nearly in **straight lines,** as light does, the effect already being noticeable at the higher radio (microwave) frequencies. The **quasi-optical** behavior of these high-frequency radio waves makes it relatively easy to reflect, refract, and focus them in beams by appropriate microwave "mirrors" and "lenses."

Since we are going to talk about radio waves from now on, let us further subdivide the radio portion of the electromagnetic spectrum into its accepted sub-classifications. (See Table No. 2.)

TYPES OF WAVE PROPAGATION

Depending primarily on its frequency, a radio wave may travel from the transmitting to the re-

ceiving antenna in a number of ways. The most important of these are known as the **ground or surface wave,** the **ground-reflected wave,** the **direct** or **line-of-sight wave,** and the **sky wave.** Fig. 143 illustrates these main radio wave paths.

Note that the ground wave does not reach the receiving antenna in the location shown since it is propagated for a short distance only. To receive it the receiving antenna would have to be moved much closer to the transmitter. You can also see, in Fig.

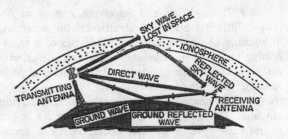

Fig. 143. *Main Types of Radio Wave Propagation*

143, that a portion of the sky wave traveling at a high angle is lost into outer space, while the portion traveling at a lower angle with respect to the horizontal is **reflected back** to the receiving antenna by the electrically charged layers of the **ionosphere** surrounding the earth. Let us take these main types of wave propagation one by one.

Ground (Surface) Wave. In ground wave propagation, the electromagnetic waves are actually guided by the earth and follow its curved surface from transmitter to receiver. Because the waves hug the earth so closely, they are profoundly influenced by the electrical characteristics of the ground over which they travel. They are strongly absorbed by solid ground, but much less so by salt water, which is very conductive. Moreover, the absorption of the waves increases considerably with frequency, so that the ground wave is generally useful only at low fre-

TABLE No. 2

THE RADIO SPECTRUM

Frequency Band	Frequency Range	Wavelength Range	Typical Uses
Very low frequency (VLF)	10–30 kc	30,000–10,000 m	Long-distance point-to-point communication
Low frequency (LF)	30–300 kc	10,000–1000 m	Marine, navigational aids
Medium frequency (MF)	300–3000 kc	1000–100 m	Broadcasting, marine
High frequency (HF)	3–30 mc	100–10 m	Communication of all types
Very high frequency (VHF)	30–300 mc	10–1 m	Television, FM, radar, air navigation, short wave
Ultra-high frequency (UHF)	300–3000 mc	1 m–10 cm	Radar, microwave relays, short-distance communication
Super-high frequency (SHF)	3000–30,000 mc	10–1 cm	Radar, radio relay, navigation, experimental
Extremely high frequency (EHF)	30,000–300,000 mc	1–0.1 cm	Experimental

quencies. Where it is present, however, reception is extremely reliable and not subject to the seasons or atmospheric conditions. Below 500 kc, reliable communication can be obtained over distances up to 1000 miles by the ground wave alone. Amplitude-modulated radio broadcasts in the medium-frequency band are transmitted primarily via the ground wave. But at the higher frequencies used by frequency modulation and television the ground wave is so reduced in strength by absorption as to virtually become useless beyond an area of a few miles around the transmitter.

Ground-Reflected Wave. As shown in Fig. 143, the ground-reflected wave travels from the transmitting antenna to the earth (which may be a mountain, for instance) and is reflected to the receiving antenna. Since it is not subject to continuous absorption by the earth, this wave travels considerably farther than the ground wave. The reflected wave exists even at very high frequencies and provides together with the direct wave the main means of transmitting FM and television programs. This type of wave is reversed in phase (by 180 degrees) at the point of reflection, which occasionally leads to undesirable results. If, for example, the ground-reflected and direct waves have traveled over the same length of path, they will arrive at the receiving antenna 180 degrees out of phase with each other and, thus almost completely cancel out. In contrast, if the path lengths of the two waves differs by one half wavelength (180°), the waves will reinforce each other, resulting in increased signal strength. By raising the height of the receiving antenna, the respective path lengths of the waves can be altered, and alternate points of maximum (reinforcement) and minimum (cancellation) signal strengths are encountered; this phenomenon is called **selective fading.** A similar phenomenon takes place by the interaction of the sky and ground waves.

Direct (Line-of-Sight) Wave. As its name implies, this wave travels directly in an almost straight line from transmitter to receiver. The path is not completely straight, because of a certain amount of refraction or bending occurring in the earth's atmosphere. For this reason the direct wave travels somewhat farther than the optical line-of-sight path. We therefore speak of a "radio horizon" over which these waves may be "seen," to distinguish it from the optical horizon. This is illustrated in Fig. 144. Note that the width of the horizon depends on the relative heights of the transmitting and receiving antennas.

The direct wave becomes increasingly important as the frequency goes up and the radio waves tend to travel more and more in straight lines. Direct line-of-sight transmission commences with the VHF band from 30 megacycles up, approximately. FM and television signals are transmitted in this way,

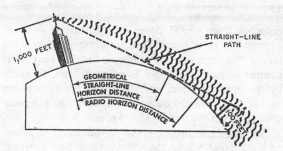

Fig. 144. *Line-of-Sight and Radio Horizons*

with contributions from the ground-reflected wave. UHF transmission takes place exclusively by the direct wave. Thus, radar, microwave relays, air navigation aids, and many other services in the VHF and UHF bands rely solely on direct line-of-sight transmission.

The Sky Wave. Transmission by means of sky waves is an elusive and tricky business. It is subject to **fading** and erratic changes with the seasons, night and day, and atmospheric conditions. Yet it is the primary means of around-the-world short-wave communication and, hence, quite important. Let us see why the sky waves are so unreliable.

The fact that makes long distance transmission via sky waves possible is the existence of a series of electrically ionized layers in the earth's upper atmosphere, called collectively the **ionosphere.** These act as a "radio mirror" to bounce the sky waves back to earth. (See Fig. 143.) The layers of ionized particles are created by various strong radiations from the sun and are therefore subject to its disturbances, as well as atmospheric conditions.

Although there are at least four distinct ionized layers in the ionosphere, the most important of these are the **E** and **F** layers. The **E** layer, sometimes called **Kennelly-Heaviside** layer for its discoverers, extends from about 55 to 90 miles above the earth's surface, its height varying somewhat with the season of the year. The E layer makes possible transmission of signals above 20 mc (approximately) over distances of up to 1500 miles. During the night the E layer disappears almost completely, making it useless for high-frequency (short-wave) radio communication. The highly ionized **F** or **Appleton** layer exists at heights from about 90 to 250 miles above the earth's surface and is the most useful for long distance transmission via sky waves. Depending on season and time of day, frequencies from 2 to 30 mc may be transmitted around the world through the F layer.

The actual mechanism of sky-wave transmission is illustrated in Fig. 145. Here the transmitting antenna is seen to radiate sky waves over a wide range of vertical angles with respect to the earth. Sky wave *1,* radiated at a small vertical angle, is reflected from the F layer and returns to earth at a great distance from the transmitter. Sky wave *2* leaves the transmitting antenna at a somewhat greater vertical angle and is reflected back to earth by the lower E layer. Since it enters the ionosphere sooner, it is reflected back more quickly, at a shorter distance from the transmitting antenna.

The distance between the transmitting antenna and the point where the sky wave is first received after returning to earth is called the **skip distance.** Right near the transmitter, the signal is heard, of course, via the ground wave. As you leave the ground-wave range, however, you encounter a zone of silence, called **skip zone,** where no signal at all can be picked up. If the sky-wave signal is sufficiently strong after returning to earth it may be reflected by the earth and again by the ionosphere, thus returning to earth in a series of hops, separated by the skip distance. The signal may be transmitted in this way one or more times around the world. Since it takes about 1/7th second for a radio signal to travel around the earth, you would then hear a series of **echoes** each spaced 1/7th of a second, depending on how many trips around the earth the signal completes.

As the vertical angle of the sky wave with respect to earth is increased, the ionsopheric layers are no longer capable of reflecting the energy back to earth, as shown by sky waves 3, 4, and 5. The angle above which the sky wave no longer returns to earth, but travels outward into space, is known as the **critical angle of radiation.** Sky waves at or above this angle may be **refracted** (bent) by the ionosphere but they are *not* reflected back to earth. The critical angle depends primarily on the **density of ionization** and on the **frequency** of the signal. As the frequency increases, the ionosphere becomes progressively less effective and lower and lower

angles of radiation must be used to reflect the signal back to earth. At frequencies above about 30 mc the signal is no longer returned even for the smallest vertical angle, and the ionosphere has little effect upon it. The exact frequency at which this occurs depends on the density of ionization, which is affected, in turn, by sun spot cycles, daily and seasonal variations.

If you realize that the actual number of ionized layers, their heights and density of ionization vary from hour to hour, day to day, month to month, season to season, and from year to year, you will no longer wonder why sky-wave transmission is less than reliable and why sudden "fadeouts" and disappearances of signals occur.

Scatter Propagation. Though we have stated that signals in the **VHF** and **UHF** bands (beyond 30 mc) penetrate the ionosphere and can be received only via the direct wave over a line-of-sight distance, exceptions to this have been reported many times in recent years. Continuing research after World War II revealed that consistently strong reception of VHF and UHF signals far beyond the radio horizon **(trans-horizon)** is possible. Although not too much is known about this phenomenon, it is generally explained by the presence of minute disturbances in the electrical properties of the atmosphere. These disturbances result in a **forward scattering** of high-frequency radio waves, and since disturbances are always present, consistent transhorizon reception is possible at all times. A new form of high-power signal transmission, based on this **scatter propagation,** is now in use for long-distance **VHF** and **UHF** communication.

RADIO RECEIVERS

We have followed the radio signal from the transmitting antenna through the radio path to a receiv-

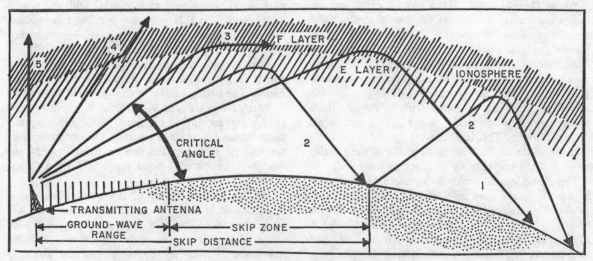

Fig. 145. *Various Sky-Wave Transmission Paths*

ing antenna somewhere within radio range of the transmitter. The receiver with its antenna must intercept a portion of the electromagnetic wave energy and recover the information contained in it. Now regardless of the type of signal or modulation, every radio receiver must perform the following six functions, illustrated in Fig. 146:

1. The **receiving antenna** must **intercept** a portion of the passing radio waves.

2. **Tuned (resonant) L-C circuits must select the desired signal** from the mass of radio signals intercepted by the antenna.

3. The weak signals from the antenna must usually be **amplified** by one or more **tuned radio-frequency amplifiers,** before they can be demodulated.

4. The **original intelligence** (modulation) imposed upon the r-f carrier wave must be **recovered** in the form of an audio-frequency signal by a **detector** or **demodulator** circuit. For **frequency-modulated** waves, **limiting** of the carrier amplitude variations is usually also necessary. In the case of c-w (radiotelegraphy) reception, an additional **beat-frequency oscillator** is required to make the carrier-wave interruptions audible.

5. The audio signal from the detector must be strengthened by one or more stages of **audio-frequency amplification** to a level sufficient to operate a set of headphones or a loudspeaker.

6. The amplified audio-frequency signal must be **converted into corresponding sound waves by an electromechanical reproducer,** such as a loudspeaker or headphones.

AMPLITUDE-MODULATED (AM) RECEIVERS

Although the basic functions of any radio receiver are the same, certain differences in the receiving circuits make it convenient to consider receivers for different types of modulation separately. Let us take the standard AM receivers first. We can then point out significant differences in FM and CW reception.

ELEMENTARY CRYSTAL RECEIVER

Though it is more a toy than a well-functioning radio receiver, the **crystal receiver** (Fig. 147) illustrates the basic principles of radio reception and deserves brief consideration.

The signal **intercepted** by the antenna of the receiver (Fig. 147a) is coupled to the L-C tank circuit through a step-up radio-frequency transformer, which provides some **amplification. Signal selection** takes place in the L-C tank circuit, which is tuned to the frequency of the desired radio-frequency carrier wave (Fig. 147b). The signal is **rectified** by the crystal (Fig. 147c) which permits current to pass in one direction much more readily than in the other. The crystal is usually some mineral, such as galena, silicon, or carborundum. An elementary filter, consisting of a **bypass capacitor,** removes the remaining r-f current pulses and keeps them out of the headphones. The resulting audio-frequency variations (Fig. 147d) are translated by the **headphone reproducer** into sound waves corresponding to the original information at the transmitter. Since a crystal re-

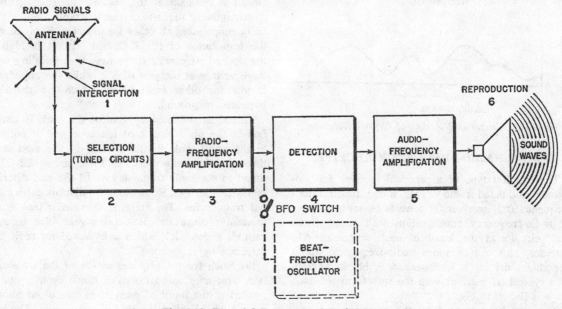

Fig. 146. *Essential Functions of Radio Reception*

ceiver is not sufficiently **selective** to discriminate well between adjacent r-f carrier signals and does not provide sufficient amplification to operate a loud-speaker, it is rarely used today.

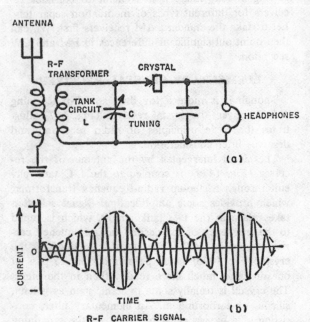

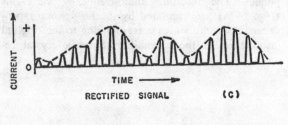

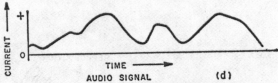

Fig. 147. *Circuit and Action of Crystal Receiver*

TUNED RADIO-FREQUENCY RECEIVER

The simplest type of a practical receiver for amplitude-modulated radio waves is the tuned radio-frequency (trf) receiver. It consists of several stages of radio-frequency amplification with tuned (L-C) tank circuits at the input of each stage, an AM detector, and one or more audio-frequency amplifier stages driving a loudspeaker. A block diagram of a typical trf receiver with the waveforms of each stage is shown in Fig. 148.

The signal intercepted by the receiving antenna, portrayed as an r-f carrier modulated by two cycles

of an audio tone, is selected and amplified by several stages of tuned radio-frequency amplification. The amplified r-f signal is demodulated in the detector stage, so that only audio-frequency variations of the carrier remain. The detector may be of the grid-leak, plate, or infinite-impedance type discussed in Chapter 13. (Diode detectors are not generally used in the low-gain trf receivers because of their insensitivity.) The a-f signal output from the detector is amplified by one or more stages of audio amplification until it is sufficiently strong to drive a loud-speaker. The loudspeaker (or headphones) converts the audio variations into sound **waves.**

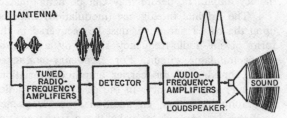

Fig. 148. *Block Diagram of TRF Receiver and Waveforms*

Selectivity of Tuned R-F Amplifiers. To refresh your memory of trf amplifiers (see Chapter 12), Fig. 149 illustrates a triode input stage of a typical trf receiver. A triode is shown for convenience, though the higher-gain pentodes are usually used. (See also Fig. 100.)

In Fig. 149, the signals intercepted by the antenna are coupled to the tuned grid tank circuit, L1-C1, through r-f transformer T1. Since the tank circuit is resonant at the desired carrier frequency, it strengthens the r-f carrier currents, while partially suppressing all other frequencies. Equivalently, the **impedance** of the L-C tank circuit is high at the desired r-f carrier frequency, thus building up a large voltage at the grid of V1, while the impedance is low for other frequencies, which are therefore bypassed to ground.

The signal selected by the L1-C1 tank is amplified in the plate circuit of the tube and is coupled to the input tank circuit, L2-C2, of the next stage through r-f transformer T2. Tank circuit L2-C2 is tuned to the same frequency as L1-C1 and discriminates further against unwanted, adjacent carrier signal frequencies. The stage is operated Class A for minimum distortion. It obtains grid bias through cathode resistor R1, which is bypassed for radio frequencies by capacitor C3.

So much for the physical action of the trf amplifier. You may wonder why a tuned tank circuit is needed at the input of *each* stage, since one should be enough to select the desired frequency. The reason for this is to obtain sufficient **selectivity;** that

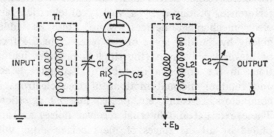

Fig. 149. *Tuned RF Input Stage of TRF Receiver*

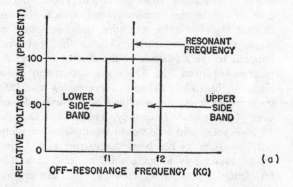

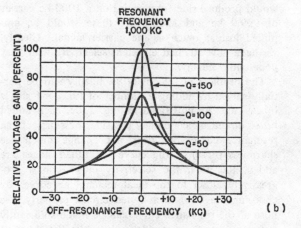

is, the ability to select **one frequency** and suppress all others. One tank circuit just won't do the job. You will recall from our discussion in Chapter 12 that it is the **sharpness of the resonance curve,** which makes a tuned circuit selective. The sharpness of resonance, in turn, is determined by the **Q-factor** of the tank circuit, which—you will remember—is the **ratio** of the coil reactance to the resistance of the tank circuit at the **resonant frequency.** The higher the Q of the tuned circuit, so sharper is its resonance curve and, hence, the more **peaked** is the response to the desired frequency. (See Fig. 150b). A sharply peaked response means, of course, that only the selected resonant frequency is amplified and all others are rejected. But there is a limit to the usable sharpness of the resonance curve and, hence, the selectivity, since not only the carrier frequency of the desired signal must be amplified, but a whole **band** of frequencies—the upper and lower sidebands—that contain the **audio modulation.** If the response of the tuned circuit is *too sharp,* a portion of the sidebands containing the audio modulation will be suppressed, and we shall gain selectivity at the price of audio quality (fidelity). Thus, a compromise is necessary.

Ideally, the resonance curve of the tank circuit in an r-f amplifier should be rectangular, flat at the top with vertically sloping sides, or **skirts,** as they are called. Such a rectangular **selectivity characteristic,** illustrated in Fig. 150a, would **pass the carrier** with its upper and lower **sidebands,** contained between frequencies f_1 and f_2, and *nothing else.* Unfortunately, actual tank circuits don't work that way. The response of a real tank circuit, tuned to a carrier frequency of 1000 kc, is illustrated in Fig. 150b for various Q-factors. An arbitrary voltage gain of 100 percent has been assumed for a Q of 150. Assume further that the desired carrier wave contains upper and lower sidebands of 10 kc each and, hence, extends from 990 to 1010 kc. This means that audio frequencies up to 10,000 cycles (10 kc) are contained in the transmitted signal, a situation typical for AM transmitters.

As you can see (in Fig. 150b) the selectivity (resonance) curve for $Q = 50$ is very **broad** and

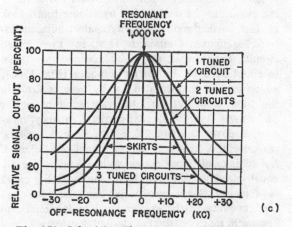

Fig. 150. *Selectivity Characteristics of TRF Receiver*

discriminates little against unwanted frequencies. The carrier frequency and its two sidebands are easily passed, but so are adjacent channels as much as 30 kc away. The curve for $Q = 100$ discriminates much more sharply against unwanted, adjacent carrier frequencies, but also cuts down the 10-kc sidebands considerably. Finally, the curve for $Q = 150$ has a very **sharp nose** and, hence, provides excellent selectivity against unwanted adjacent channels. The response to an adjacent carrier signal, say, 30 kc distant on either side of the resonant frequency (that is, either at 970 kc or 1030 kc) is only about 10 percent of the resonant response and, hence, the undesired signal will be amplified only *one-tenth* as much as the desired carrier. This is fine, but what happens to the sidebands of the *desired* carrier? As is apparent from Fig. 150b, the upper and lower limits of the 10-kc sidebands, representing a 10,000-cycle audio tone, are amplified only about 35 percent of the carrier frequency. This means that the 10,000-cycle audio note would be amplified, roughly, only *one-third* as much as the carrier and, hence, would sound only about a third as loud as a low-frequency (bass) tone, close to the carrier frequency. (A 100-cycle audio tone, for example, would produce side frequencies on a 1000-kc carrier of 999.9 kc and 1000.1 kc; these would be amplified just as much as the carrier signal.) Clearly, we have now attained **excellent selectivity** with very **poor audio fidelity.**

The problem of selectivity vs. fidelity can be partially solved by adding a number of tuned r-f stages in cascade. As illustrated in Fig. 150c, each added tuned circuit affects chiefly the sides or skirts of the resonance curve, but does not change much the sharpness (nose) of the curve. As tuned stages are added, therefore, the selectivity curve comes progressively closer to the ideal rectangular curve. By choosing the Q of each tuned circuit properly, the nose of the resonance curve can be made sufficiently broad to pass the two sidebands, while the skirts of the curve may be made sharp by adding tuned tank circuits, so that good selectivity against adjacent interfering signals is attained. Thus, for three tuned circuits, for example, the 10-kc sidebands are down to about 50% of the carrier response (Fig. 150c), which can be compensated for by **tone controls** in the audio amplifier. An interfering adjacent carrier 30 kc away from the desired carrier frequency, however, is down to about 2% of the desired resonant response and, hence, is almost completely rejected. By adding still more stages with a lower Q each, equal selectivity with even better audio fidelity could be achieved. Sometimes special **bandpass filters** are used in place of the conventional tank circuits, to approach more nearly the desired flat-topped, rectangular curve of Fig. 150a.

Sensitivity. The **sensitivity of any receiver is the amount of r-f input voltage needed to produce a specified amount of audio output power.** It would seem that any amount of sensitivity could easily be obtained by simply adding more r-f amplifier stages. But this is not so, because of the presence of **noise.** Noise is any undesired electrical disturbance within the desired frequency band. Noise is already picked up by the antenna in the form of atmospheric disturbances, called **static,** and **man-made** electrical interference, produced by a variety of electrical and electromechanical devices. More noise is added chiefly by the first stage of the receiver in the form of power-supply hum (the ripple), random fluctuations in the emission of tubes (called **shot effect)** and random thermal motions of free electrons in conductors, resistors, etc. (called **thermal agitation).** These external and internal receiver noises are amplified by all the stages of the receiver along with the desired signal and eventually tend to drown out and **mask** the signal.

It is evident, therefore, that the **usable sensitivity** of a receiver is not determined by the **absolute value** of the signal strength at the receiver input, nor by the number of stages of amplification, but rather by the **ratio** of the signal strength to all noise present at the input of the receiver. It is this **signal-to-noise ratio** at the input of a receiver, therefore, that **limits its maximum usable sensitivity.** Little can be done to improve the signal-to-noise ratio for a given receiver circuit or system of modulation. There is, obviously, no point in amplifying the signal *with the noise* beyond the point, where the noise itself becomes objectionable, since this will only make the signal (plus noise) louder, but not more intelligible. However, since noise is more or less **uniformly distributed** over the entire frequency spectrum, the noise pickup can be **reduced by limiting the bandwidth** and hence audio fidelity passed by the receiver. In communication-type receivers, where audio quality is of secondary importance, the bandwidth is generally cut down to the bare minimum for acceptable intelligibility. It now becomes evident, why frequency modulation—which does not respond to amplitude noises—is such a boon compared with AM reception. The signal-to-noise ratio and, hence, usuable sensitivity of FM receivers is far greater than that of AM receivers.

The number of stages that can be used in a trf receiver is also limited by possible **instability.** Since the entire amplification takes place at the *same* frequency, the slightest amount of coupling between input and output stages (through the power supply, for example) leads to large amounts of **regenerative feedback.** If sufficiently amplified, this regenerative (positive) feedback results in instability and oscillations, you will recall. In general, a trf receiver performs well only on a single low- or medium-frequency band, such as the broadcast band. It is not practical for use at high frequencies or as **multiband receiver,** since the **amplification and selectivity of r-f stages falls off rapidly at higher frequencies.** We shall see in the following paragraphs how the **superheterodyne receiver** overcomes some of these limitations of the trf circuit.

SUPERHETERODYNE RECEIVERS

The superheterodyne receiver, invented during the first World War by Major Armstrong, achieves its unique advantages over the trf receiver by **converting all incoming carrier frequencies** to a **fixed, lower value** (the intermediate frequency). At this fixed **intermediate frequency,** the amplifier circuits can operate with maximum stability, selectivity and sensitivity and are not subject to the variable amplification and instability of the trf circuit. The conversion of the received signal frequency to the lower (i-f) frequency is achieved by **heterodyning** or **beating** the carrier frequency against a locally generated frequency.

A block diagram of the basic superheterodyne circuit is shown in Fig. 151. As in the trf receiver, the modulated r-f carrier signals intercepted by the antenna are coupled to a tuned r-f stage, where the initial signal selection and amplification takes place. Because of the inherently high gain (sensitivity) and selectivity of the superheterodyne circuit, the input r-f stage is sometimes **untuned** (i.e. without an L-C tank circuit) and, occasionally, it is omitted altogether.

The amplified r-f signal is coupled to the input of a **mixer** stage, where it is combined with the output of a **local oscillator.** The two frequencies **beat** together in the mixer, as we shall see presently, and generate an intermediate (i-f) frequency equal to the **difference** between the r-f signal and local oscillator frequency. The frequency of the local oscillator may be either above or below the r-f signal frequency by the amount of the desired intermediate frequency. (In practice, it is usually *above* the signal frequency.) Note that the i-f signal at the output of the mixer, though lower in frequency, contains the **same modulation** as the original r-f carrier signal. (See waveform in Fig. 151.)

The i-f signal from the mixer is then amplified by several stages of fixed-tuned i-f amplification (see Fig. 101) and is coupled to the input of a detector stage, where it is demodulated. Since the circuit has plenty of gain, a **diode detector** is usually used because of its low distortion and excellent audio fidelity. The audio signal from the detector is amplified by one or more stages of audio amplification until it is sufficiently strong to drive a loudspeaker. The loudspeaker converts the audio signal into sound waves corresponding to the original sound at the transmitter. (We shall have more to say about loudspeakers in the next chapter.)

Frequency Conversion (Heterodyning). If you have ever listened to the throbbing, pulsating sound emanating from two tuning forks (or organ pipes) differing slightly in frequency, you know what **beats** or **heterodyning** sounds like. The pulsations or beats, in this case, are caused by alternate destructive interference and reinforcement between the two tones. Since the notes differ only slightly in frequency, their waves will be alternately **in phase** and **out of phase** with each other. Whenever they are *in* phase, the notes **reinforce** each other and the sound is strengthened; whenever they are *out* of phase, the sound

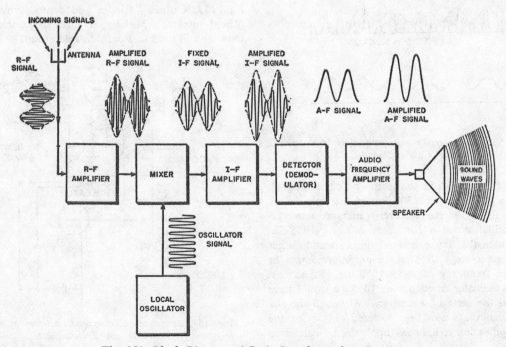

Fig. 151. *Block Diagram of Basic Superheterodyne Receiver*

waves tend to cancel each other, and the sound is weakened. The in-phase and out-of-phase conditions take place at the difference between the frequencies of the two tones. Thus, a beat frequency tone is heard equal to the difference between the two frequencies.

The same principle of beats is utilized to convert a high-frequency wave into a lower-frequency one. It does not matter whether or not the high-frequency signal carries amplitude or frequency modulation. The process of receiving a modulated r-f signal by the heterodyne (beat) principle is illustrated in Fig. 152.

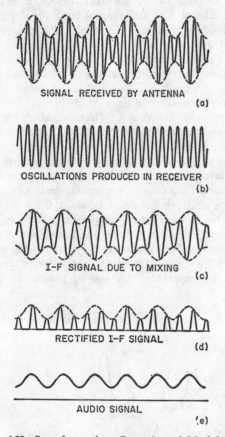

SIGNAL RECEIVED BY ANTENNA

(a)

OSCILLATIONS PRODUCED IN RECEIVER

(b)

I-F SIGNAL DUE TO MIXING

(c)

RECTIFIED I-F SIGNAL

(d)

AUDIO SIGNAL

(e)

Fig. 152. *Superheterodyne Reception of Modulated R-F Carrier Signal*

Part (a) of Fig. 152 shows the r-f carrier signal, say, at a frequency of 1500 kc, at the antenna. A local oscillator in the receiver generates **unmodulated** oscillations at a frequency of, say, 1955 kc, as shown in (b). These heterodyne or beat together in the mixer stage to produce an intermediate or difference frequency of 1955–1550, or 455 kc (c). (A local oscillator frequency of 1045 kc would have produced the same i-f frequency.) Although the signal frequency has now been lowered to 455 kc, the **same modulation** is imposed upon the i-f carrier as

on the original r-f signal, as shown by the peak amplitudes of the i-f signal in Fig. 152. The i-f signal is then amplified by several stages, rectified (d) and filtered in the detector, and the resulting audio signal (e) is amplified to drive a speaker. It is of interest to note that the process of heterodyning is essentially the same, except for the different waveforms involved, for the reception of amplitude-modulated, frequency-modulated, or continuous waves (cw).

Frequency Converter Circuits. As we have seen in the case of non-linear (harmonic) distortion and amplitude modulation, a **non-linear** characteristic is required to produce **new** frequencies. Similarly, beat or difference frequencies between the r-f and local oscillator signal can be produced only by mixing the two signals in a non-linear device; that is, one that does not follow Ohm's law. Again electron tubes come to the rescue by providing the required non-linearity in the curved (bottom) portion of their characteristic. To obtain beat frequencies, therefore, we simply apply the r-f and local oscillator signals to a mixer tube, operated non-linearly, so that its plate current varies at the desired intermediate frequency. As is so often the case, however, not only the desired *difference* (beat) frequency is produced by the non-linear mixer, but also the original frequencies at the input, as well as a frequency equal to the *sum* of the originally applied frequencies. These undesired frequencies must be suppressed by **tuning** the output of the mixer to the intermediate (difference) frequency only.

A typical frequency converter circuit using a triode mixer and a triode local oscillator is shown in Fig. 153. Ordinary triodes and pentodes make excellent mixers, especially at high frequencies, and they are still used to a considerable extent.

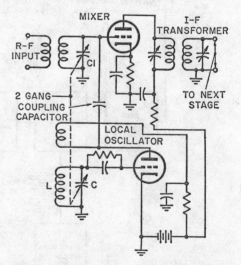

Fig. 153. *Frequency Converter Consisting of Triode Oscillator and Triode Mixer*

The mixer tube functions as an ordinary plate **detector** biased approximately to cutoff; i.e., in the **non-linear** portion of the tube's characteristic. (The use of this simple detector circuit in the early days led to the name of **first detector** to distinguish it from the later demodulator stage, often called **second detector.)** The input tank circuit of the mixer is tuned to the frequency of the incoming signal, while the oscillator grid tank circuit is tuned above (or below) the r-f signal frequency by an amount equal to the desired beat or intermediate frequency. In AM superheterodynes, this is usually 455 kc. The i-f transformer in the plate circuit of the mixer is then tuned to the desired intermediate (difference) frequency. The oscillator and mixer tuning capacitors are "ganged together," as shown, to permit tuning by a single knob.

A conventional tickler feedback oscillator is shown in Fig. 153. Any of the other types of oscillators discussed in Chapter 11 could be used equally well as local oscillator, since little power is required. Oscillations from the plate circuit of the local oscillator are coupled to the **grid** of the mixer, in this case, by means of a coupling capacitor, a method known as **grid injection.** Any convenient means can be used, however, to couple the local oscillator voltage to the mixer tube. The oscillator voltage may be **inductively** coupled (by means of a coil) to the cathode circuit of the mixer, a popular method known as **cathode injection,** or the oscillator grid tank can be inductively coupled to the mixer input tank circuit.

Separate triode or pentode mixers and oscillator perform very well and have an excellent **signal-to-noise ratio,** which makes them suitable for use at high frequencies. Triode and pentode converters have the disadvantage, however, of producing undesirable **coupling** between the mixer and oscillator portions, which may result in the oscillator **locking in** to some strong interfering signal. Special tubes with five grids, called **pentagrid** tubes, have been developed to provide the necessary isolation between the oscillator and mixer portions of the converter. In the **pentagrid mixer** isolation is achieved by means of two independent control grids, one for the r-f signal and one for the local oscillator signal. A separate triode or pentode is used in the local oscillator circuit. A more elegant solution is the **pentagrid converter,** which combines the functions of the local oscillator and the mixer within a single pentagrid tube, coupling between the two circuits being accomplished by means of the common electron stream. (This is similar to the electron-coupled oscillator discussed in Chapter 11.) A block diagram and actual circuit of a pentagrid converter is shown in Fig. 154.

The cathode, grid 1 and grid 2 of the pentagrid

converter are connected to an ordinary triode tickler feedback oscillator. Grid 1 functions as oscillator control grid and grid 2 as oscillator plate. The local oscillator frequency is controlled by the variable capacitor in the oscillator tank circuit. This capacitor is ganged with the input signal tuning capacitor in such a way that it maintains a constant frequency difference equal to the desired intermediate frequency.

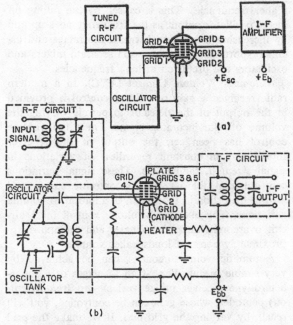

Fig. 154. *Block Diagram (a) and Circuit (b) of Pentagrid Converter*

The incoming r-f signal is selected by the tuned input circuit and is applied to grid 4 of the pentagrid tube, so that it modulates the electron stream. The action is approximately as follows. Most of the electrons emitted by the cathode pass through the oscillator plate (grid 2) and go on to the screen grid 3, which accelerates the electron stream and serves as electrostatic shield. Again, most of the electrons pass through the screen and travel toward the r-f signal grid (grid 4), which is biased **negatively** through the self-bias cathode resistor. Because of this negative bias, the electrons are retarded and form a space charge between grid 3 and 4 that constitutes a **virtual cathode** for the mixer section. The number of electrons available from this virtual cathode depends on the oscillator pulses; thus, the electron stream available to the mixer **varies at the oscillator frequency.** Since the electron flow between the r-f signal grid 4 and the plate of the tube depends also on the r-f signal voltage, the electron current actually arriving at the plate is modulated by *both* the oscillator and r-f signal voltages. The two voltages are mixed in the output of the tube and, because of the non-linear characteristic, sum and difference frequencies appear in the plate current of the tube. The i-f transformer in the plate circuit is tuned to the correct difference-frequency (i-f) component of the plate current and this component is coupled to the succeeding i-f amplifier.

Although the pentagrid converter performs well in the medium-frequency range, and is the standard circuit in all broadcast superheterodyne receivers, its performance becomes increasingly poor at the higher frequencies. This is caused by the falling off in the oscillator output as the frequency goes up and by increasing interaction between the mixer and oscillator portions. For this reason separate mixers and oscillators are preferred at high frequencies.

Automatic Volume Control (AVC). In a modern radio receiver, a **gain** or **volume control** is provided in the output of the detector to permit varying the volume of the incoming signal. Once the volume control has been set, the output of the receiver should remain constant regardless of variations in signal strength, fading, etc. These signal variations can overload the r-f, i-f, or detector stages and lead to distortion in the signal, in addition to volume variations. The **automatic volume control** (avc) circuit overcomes these deficiencies and maintains approximately constant loudspeaker volume.

Automatic volume control can be achieved by very simple means. The r-f an i-f stages of a superheterodyne receiver utilize **variable-mu (remote-cut-off)** pentodes, whose gain can be controlled, you will recall, by varying the grid bias. If we make the grid bias of these r-f and i-f stages **more negative** when the signal strength **increases,** thus cutting down the gain, and **less negative** when the signal strength is fading, we will attain more or less constant output volume, regardless of signal strength. All that is needed, therefore, is a **negative bias voltage whose value is controlled by the signal strength.** As we have seen in Chapter 13, the **load resistor** of a diode detector is an excellent source of this variable voltage, since the rectified signal voltage will increase and decrease with variations in the signal strength. A typical diode detector with an automatic volume control circuit is shown in Fig. 155. **Manual** volume control is also provided.

As in the conventional diode detector, the signal is rectified by the diode and the rectified current flows through the load resistor, R, causing a voltage drop with the polarities indicated. The load resistor is in the form of a **potentiometer,** so that a portion of the rectified voltage can be tapped off and fed to the first audio amplifier stage through capacitor C_o. This provides manual volume control.

The output from the negative end of the load resistor is fed through filter R1-C1 to the grids of

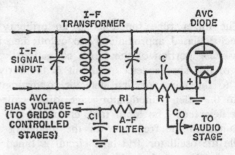

Fig. 155. *Automatic Volume Control (AVC) Circuit*

the controlled r-f and i-f stages and provides variable bias in accordance with the signal strength. The filter R1-C1 removes the audio-frequency component of the signal, so that the **avc** voltage cannot follow the audio modulation of the signal, which would result in distortion. The avc thus follows only the slower variations in signal strength due to fading and other causes.

The avc diode is usually combined with the regular diode detector in the envelope of a single **duo diode,** and sometimes the first audio amplifier is thrown in also by using a **duo diode-triode** tube Since some avc bias is developed even for weak input signals where maximum amplification is desired, a special **delayed avc** circuit is occasionally employed, which prevents the application of the **avc** bias until the signal strength exceeds a certain *predetermined* value. This necessitates another diode.

Complete Superheterodyne Receiver. Let's look at a complete superheterodyne receiver, typical of those found in almost every home. Fig. 156 illustrates a high-quality receiver. We say "high quality" because of the presence of push-pull audio amplification and an initial r-f amplifier; both features are frequently omitted in cheap sets.

Although the circuit diagram on first glance looks rather forbidding, a closer inspection reveals only familiar circuits. The receiver is seen to consist of a stage of tuned r-f amplification, a pentagrid mixer and local Hartley oscillator, one stage of i-f amplification, a combined detector, avc, and first audio amplifier (V4), a push-pull audio power amplifier, and a loudspeaker. The set is operated from the 115-volt a-c line through a full-wave rectifier power supply with a two-stage capacitor-input filter circuit. The power supply provides the high-voltage d.c. for the plates and screen grids of the tubes as well as the low-voltage a.c. for the heaters of the tubes.

There are no unusual features in the receiver, but some points are worth noting. The tuning capacitors C1, C2, and C3, of the r-f amplifier, mixer, and local oscillator, respectively, are all ganged together to provide single control tuning. The small variable capacitor C24—called a trimmer—in parallel with

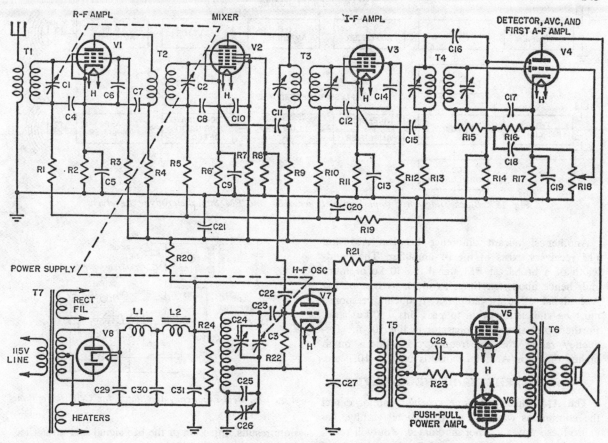

Fig. 156. *Typical High-Quality Superheterodyne Circuit*

oscillator tuning capacitor C3 is factory-adjusted or **aligned** to assure that the oscillator tuning capacitor follows or **tracks** the r-f signal tuning capacitors (C1 and C2) at exactly the right i-f difference frequency, *above* the r-f signal frequency. Since this trimmer capacitor affects the tracking chiefly at high frequencies, another preadjusted capacitor, called a **series padder,** is needed to insured correct **tracking at low frequencies.** Padder capacitor C26, in parallel with fixed capacitor C25, fulfills this purpose. Note that the oscillator output signal is injected through capacitor C22 into grid 3 of the pentagrid mixer tube, V2.

Duo diode-triode V4 performs the functions of a diode detector, **avc** tube, and first triode audio amplifier. The rectified signal voltage from the diode detector is developed across load R16 and C18, and is coupled through manual volume control R18 to the grid of triode V4. Here it is amplified and then applied to the push-pull power amplifier, which drives the speaker. A **separate diode** in V4 provides the **avc** bias voltage across resistor R14. This voltage is applied, through filter R19-C20, to the grids of the r-f amplifier (V1), mixer (V2), and i-f amplifier (V3).

FREQUENCY-MODULATED (FM) RECEIVERS

Receivers for frequency-modulated signals *always* use the superheterodyne circuit because it is the only circuit that provides sufficient gain to boost the amplitudes of weak signals up to the point where they can be maintained *constant* by the limiter. You will recall that the noise-free feature of FM is only attained, when the carrier amplitudes are limited to a fixed value. Fig. 157 shows a comparison between an AM and FM superheterodyne receiver.

It is evident from Fig. 157 that the AM and FM superheterodyne circuits are very similar, except for the presence of the limiter and discriminator in the FM receiver in place of the conventional detector in the AM receiver. As was pointed out in Chapter 13, the discriminator is sometimes replaced by the similar ratio detector circuit, and the limiter stage may then be omitted. In either case, the functions of the limiter-discriminator or the ratio detector are to maintain the amplitude of the carrier constant, while at the same time converting the carrier frequency variations into equivalent amplitude (audio) variations.

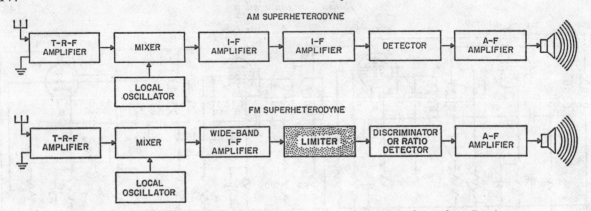

Fig. 157. *Block Diagram Comparison Between AM and FM Superheterodyne Receivers*

Another significant difference between AM and FM receivers exists in the i-f amplifier. The bandwidth of a broadcast FM signal is 150 kc or more and, hence about ten times as much as that of the AM signal. The i-f amplifier frequency response must be sufficiently wide to pass this 150-kc band. Furthermore, since FM operates in the 100-mc frequency range, the i-f frequency is chosen much higher than for AM; its value is usually 10.7 mc.

C-W SUPERHETERODYNE RECEIVER

The AM superheterodyne receiver cannot detect the continuous waves (c.w.) of radiotelegraphy, because these waves are **not modulated.** You will recall that c.w. consists of an **unmodulated** carrier signal which is interrupted in accordance with a telegraphic code. To make these continuous waves audible, they must be **modulated locally** in the receiver with a convenient audio tone. The modulated waves may then be detected by a conventional detector circuit and the interrupted audio tone may be listened to in a headset. As already indicated in Fig. 146, the addition of a **beat-frequency oscillator (bfo)** is all that is required to convert an AM superheterodyne receiver for the reception of c-w signals. When the oscillator is switched on with the bfo switch, its output heterodynes with any incoming c-w signal to produce an audio beat tone. Any standard oscillator circuit may be used as beat-frequency oscillator. Fig. 158 illustrates a typical Hartley pentode BFO.

In operation, the main tuning capacitor, C4, of the oscillator is **preset** to the intermediate frequency of the receiver. A parallel trimmer capacitor, C5, then permits **detuning** the oscillator frequency over a band ranging from a few hundred to about 1000 cps. The output of the beat-frequency oscillator is injected, in this case, through C6 and the bfo switch into the secondary of the last i-f transformer (the plate of the detector). Here the off-tune oscillations are mixed with the i-f current, and the audio beat

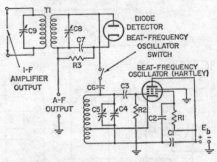

Fig. 158. *Beat-Frequency Oscillator for C-W Reception*

note results. Injection of the bfo signal into an earlier i-f stage is equally suitable.

SUMMARY

The basic elements of a radio communication system are 1. a radio transmitter to **generate** radio-frequency waves; 2. a telegraph key or microphone to **control** the waves in accordance with the information to be conveyed; 3. a transmitting antenna to **radiate** the waves; 4. a receiving antenna to **intercept** the waves; 5. a radio receiver to **select** and **amplify** the desired signal, and recover **(demodulate)** the information contained in it; 6. a loudspeaker or headphones to **convert** the demodulated signal into sound, **reproducing** the original information.

In **c-w (radiotelegraph) transmission** the r-f carrier wave is periodically interrupted by a key in accordance with a telegraphic (Morse) code. The c-w signal may be made audible at a distant receiver by converting it to a low-frequency (audio) tone.

Neutralization is used to balance out the feedback produced by the grid-to-plate interelectrode capacitance of a triode.

In amplitude- or frequency-modulated transmitters the required modulation of the carrier wave is obtained by **microphones** that change the sound waves into corresponding electrical variations. Three types of microphones, in ascending order of quality, are the **carbon, crystal,** and **dynamic or moving coil** microphones.

Accelerated electric charges (electrons) in **antennas radiate energy in the form of electromagnetic waves.**

A Hertz antenna, or **dipole,** is a half-wavelength long and is **resonant** to the r-f oscillations. The **electric field** of a half-wave dipole is **parallel** to the plane of the antenna; the orientation of the electric field with respect to earth is called the **polarization** of the radio wave.

The **radiation pattern** of an antenna is a measure of the spread of the **magnetic field.** The radiation pattern of a half-wave antenna in the antenna plane is a **figure-8,** with **maximum radiation** occurring broadside to the plane of the antenna, and **minimum radiation** taking place along the **length** of the antenna (from tip to tip). The complete **free-space radiation pattern** of a half-wave dipole is doughnut- or pincushion-shaped.

The **Marconi** or **grounded antenna** is a quarter-wave long and is grounded at one end. The earth provides a quarter-wave **image.**

Electromagnetic (radio) waves may travel from transmitter to receiver either as **ground or surface waves, ground-reflected waves, direct or line-of-sight waves,** or as **sky waves.** The ground (surface) wave is rapidly absorbed by the ground over which it travels and is useful only at low and medium frequencies up to a few megacycles. The ground-reflected wave is useful also at high frequencies, but may interfere with the direct wave, causing **selective fading.** Direct, line-of-sight transmission over the **radio horizon** starts at about 30 mc and is the main path for VHF and UHF transmissions. **Sky-wave** transmission is used for **long-distance** (short wave) communication in the frequency range between 2 and 30 mc. It is subject to erratic daily and seasonal changes due to variations in the number, density, and height of the ionized layers in the upper atmosphere (the **ionosphere).**

A radio receiver must have 1. a **receiving antenna** to **intercept** the radio waves; 2. **tuned L-C circuits** to **select** the desired signal; 3. **tuned r-f amplifiers** to **amplify** the weak r-f signals; 4. a **detector** for **demodulating** the r-f carrier and recovering the audio signal; 5. an **audio-frequency amplifier;** and 6. a **reproducer.**

A **tuned radio-frequency (trf) receiver** consists of several stages of tuned r-f amplification, an AM detector, and one or more audio-frequency amplifier stages driving a loudspeaker (or headphones).

Selectivity and **audio fidelity** are interdependent. The sharper the resonance curve (Q) of the tuned circuit, the greater is its selectivity (to discriminate against unwanted signals), but also the poorer is its sideband (audio) response.

Sensitivity is a measure of the r-f signal voltage needed to produce a specified amount of audio power. **Usable sensitivity** depends on the **signal-to-noise ratio** at the input of the receiver.

The **superheterodyne receiver** converts all incoming carrier frequencies to a fixed **intermediate frequency** and thus attains maximum sensitivity, stability, and selectivity. It comprises a tuned r-f amplifier (usually), a **frequency converter**—consisting of **mixer** and **local oscillator**—an i-f amplifier, a detector stage, an audio amplifier, and a loudspeaker.

Frequency conversion in a superheterodyne is achieved by beating or **heterodyning** the output of a local oscillator, tuned above or below the incoming r-f signal by the difference or i-f frequency, with the r-f carrier signal in a **non-linear** mixer stage. The output of the mixer is tuned to the resulting difference (i-f) frequency.

A **pentagrid converter** combines the functions of a local oscillator and mixer tube within a single five-grid tube.

Automatic volume control (avc) to compensate for varying signal strength is obtained by rectifying the i-f signal in a diode detector and using the variable bias voltage to control the gain of the r-f, i-f, and mixer stages.

A **beat-frequency oscillator (bfo)** makes possible superheterodyne reception of c-w signals by providing an audible beat note.

Chapter Sixteen

HIGH FIDELITY AND STEREO

High fidelity is the attempt to reproduce by electronic and electromechanical means and without distortion the full range of audible sound, from a low of about 20 cps (cycles per second) to a high of about 15,000 cps. A reproducing system is considered "high fidelity" if it fulfills these conditions and thus comes as close as possible to sounding like the original. We might as well say right now that *no* "Hi-Fi" reproducing system, regardless of price, will sound in your 12′ × 20′ living room *exactly* like a 110-piece orchestra in Carnegie Hall. *Absolute* high fidelity is as yet not possible because our definition above embraces too many variable and uncontrollable factors which are *unlike* the original, such as the **size** and **acoustics** of your living room, the **level** of reproduction (which cannot be as high as the original), the fact of **single-channel (monaural)** reproduction versus listening with *two* ears in a concert hall (**binaural** hearing). *Practical* high fidelity is always some sort of compromise, though it may sound remarkably good at times.

Despite the faddist proportions Hi-Fi has reached of late, high-quality reproduction of music is not new. The principles of designing high-quality amplifiers and radio receivers have been known for many years, and we have discussed them fully in earlier chapters. However, there has been a remarkable improvement in recent years in the weakest links of high-fidelity, which are the **electromechanical components,** such as loudspeakers, phono pickups and records, and tape recorders. These improvements have given new inpetus to the quest for Hi-Fi and it might therefore be of interest to explore the subject further.

FUNDAMENTALS OF SOUND

We have repeatedly stated that the range of human hearing extends from about 20 to 15,000 cps. This is an average figure, since many of us can't hear sounds much below 50 cps or above 12,000 cps. As we get older, our range of hearing is likely to become more restricted, while many children, in contrast, can hear musical sounds up to 18,000 cps

or higher. It is no accident, of course, that the frequency range of musical sounds fits comfortably within the range of human hearing, since musical instruments have been adapted to our ears. Fig. 159 illustrates the frequency range of some musical instruments compared with the range of human hearing.

As you can see, the range of the pipe organ, for example, encompasses the entire range of human hearing, while the piano keyboard has a range of **fundamental** notes from 28 to 4186 cps. Note the stress on "fundamental." When you strike the key to sound an A-note, for example, the string will vibrate at a **fundamental** frequency of 440 cycles, which is the characteristic pitch to which all instruments of an orchestra are tuned. In addition to the fundamental, however, the string will also be excited to "sympathetic" vibrations or **overtones** at **multiples** of the fundamental; that is, an overtone of 880 cps (called the **second harmonic),** one of 1320 cps (the **third harmonic),** one of 1760 cps (the **fourth harmonic),** and so on. The fourth harmonic of the 4186-cps piano note thus reaches up to 16,-744 cps—the upper limit of human hearing. Although the intensity (loudness) of these harmonics is much less than that of the fundamentals, and varies from instrument to instrument, their presence determines the characteristic quality and timbre of different instruments. A high-fidelity system, therefore, must reproduce the full range of fundamentals *and* harmonies of all musical instruments.

Loudness. How loud a sound *appears to us* depends on the **intensity** of the sound, the sensitivity

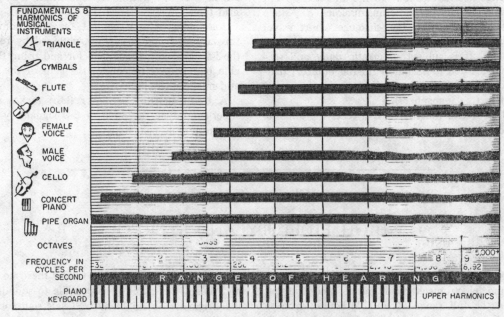

Fig. 159. *Frequency Range of Musical Instruments*

of our ears, our state of sobriety, interest, attention, and other psychological factors. In other words, the **loudness** of a sound is a **subjective,** psychological sensation, depending to some extent on the listener, while the **intensity** of a sound is an **objective** indication of its power, measurable by scientific instruments. The distinction is important, since loudness and intensity do *not* go hand in hand, as it might appear. The human ear is rather slow to respond to increased stimulation. Thus, a sound that is increased in intensity to *twice* its original value (twice the power) is *not* heard twice as loud, but seems to have hardly changed in loudness. To make it *sound* twice as loud, the intensity must be increased *ten times*. Moreover, to make the sound appear *three* times as loud as before, the intensity must be increased a *hundred-fold* over the original value, and to make it appear *four* times as loud, the sound intensity must be increased a *thousand times* over the original value. This is a little like lighting candles in the darkness: the first candle brings a flood of light into the dark room; the second candle has much less effect, and you may need *ten* candles to get the same sensation of increased brightness as was produced by the first.

A relationship between two quantities, where one goes up by jumps of 1, 10, 100, 1000, etc., while the other (dependent) quantity increases by 1, 2, 3, 4, and so on, is called **logarithmic.** The response of the ear to increased stimulation is thus **logarithmic** in nature: To obtain equal increases in loudness of a sound, its intensity must go up by ratios of *ten* each time. It is convenient, for this reason, to measure the sound intensity in **logarithmic units** which go up in equal steps for ten-fold increases in intensity. Such a unit is the **decibel,** which is used to express **ratios** of intensity and increases in loudness (as well as many other logarithmic quantities). Thus, when the ratio between two sound intensities is ten, they are said to differ by 10 decibels (db), when it is a hundred, they differ by 20 db, a ratio of 1000 equals 30 db, and so on. Moreover, each 10-db intensity increase boosts the loudness by a factor of 1.

Fig. 160 illustrates the relative response of the ear at different frequencies and sound intensity levels. The two curves in Fig. 160 tell a big story. The lower curve, entitled "threshold of hearing," is a measure of the minimum sound intensity required to barely move the eardrum and thus become audible. The upper curve, entitled "threshold of pain," refers to a sound intensity so great that it causes a tingling or painful sensation in the ear. All sounds lie between these two loudness or intensity limits. The range encompassed by these limits is truly tremendous. You will note that the point where the lower (threshold) curve crosses a frequency of 1000 cps has been arbitrarily marked **zero decibels.** The

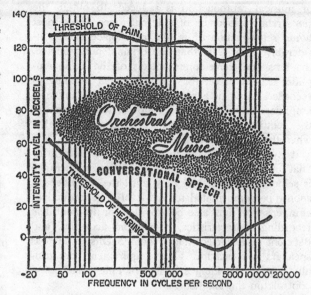

Fig. 160. *Relative Response of Human Ear*

intensity at which a 1000-cycle tone becomes barely audible, thus, is our **reference** or **zero-db level.** Since we are interested only in **ratios** of intensity or loudness, and not in absolute values, this is a logical point for the zero reference level. Note further that to boost the same 1000-cycle note to the threshold of pain, the intensity of the sound must go up by about 120 decibels. An increase of 120 db expresses an intensity ratio of a *thousand billion to one* between the loudest and softest sound, a fantastic figure. However, since the intensity has gone up by only 13 logarithmic (10-db) steps, from 0 to 120 db, the loudness of the loudest possible sound appears only about 13 times as great as that of the softest, barely audible sound.

The two curves of Fig. 160 are also known as **loudness contours,** because anywhere along a curve a sound will be heard at the *same loudness* as the corresponding 1000-cps note. (There are many more loudness contours than we have shown.) It is evident from the distorted shape of the contours that the ear does not respond in the same way to sounds of differing frequencies and, hence, the **apparent loudness of a sound depends on its frequency as well as on its intensity.** For example, to make a 40-cps tone just barely audible requires a 60 db increase in sound intensity over the 1000-cycle tone (at 0 db). This is an intensity ratio of a million to one. A 10,000-cps tone, in contrast, will require only a 10-db boost over the 1000-cycle note to make it barely audible. Note that at the other extreme, the threshold of pain, the loudness variations with frequency are not nearly as great as for the lower curve. This means that both the **low and high frequencies must be boosted** in intensity, when the level of musical reproduction is reduced, to obtain the *same* relative loudness of different tones. Since most home reproduction of music takes place at a considerably lower level than the original performance, the bass and treble tone controls of the audio amplifier should be set for boosting, when listening to recorded music or a radio reproduction. Many hi-fi amplifiers have "loudness" controls, which perform the required bass and treble boost automatically, when the volume level is reduced.

It is evident from the foregoing that a uniform or "flat" amplifier response over the entire audio range is seldom desirable. Not only must the amplifier response be compensated for the level of reproduction, compensation must also be provided for different disk recording characteristics, poor room acoustics, and deficient electromechanical components, such as speakers and pickups. These requirements are usually met by providing a variety of **tone equalization** controls on the amplifier.

BASIC COMPONENTS

A complete high-fidelity system usually consists of a number of distinct, separately-bought components, which makes for flexibility in fitting various needs and pocketbooks. A *minimum* system must consist of at least one **input** source, such as a record player or radio tuner; an **audio amplifier;** and a **loudspeaker** in an enclosure (sometimes called **baffle).** A more elaborate system, illustrated in Fig. 161, may contain many input sources, separate pre- (voltage) amplifier and main (power) amplifier, and a multiple speaker system at the output. You may be aware of the fact that many radio-phono-tv combinations contain the same or similar components in a single cabinet; this might appear to be a blessing, considering the mess of wires and components strewn about in the average hi-fi system. The reason why people bother with the mess and considerably higher cost of separate components, is the presumably much better quality of each individual component, especially the amplifiers and loudspeakers, and also because of the absolute need for the speakers to be housed separately for high-quality sound reproduction.

Note that the hi-fi system of Fig. 161 not only contains two types of record players, an AM-FM radio tuner and a television set, but also a **tape recorder.** The tape recorder, which records sound on magnetized tape, is *both* an input and output device. When a recording is to be made, the audio signal from a microphone or other source (such as a radio tuner) is fed through the preamplifier and its "Record" output jack to the input of the tape recorder, where the signal is impressed on magnetic tape. When a tape recording is to be played back, the output of the tape recorder feeds to the input of the preamplifier, and from there to the main am-

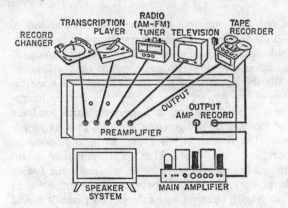

Fig. 161. *Components of a High-Fidelity System*

plifier and speaker system, where the recording is reproduced.

Let us now consider separately the various components of a high-fidelity system, with the exception of the television receiver, which will be explained in detail in the next chapter.

THE RADIO TUNER

A radio tuner is a radio receiver *minus* audio amplifier. It therefore performs all the functions of the radio receiver in selecting (tuning), amplifying, and demodulating a signal, except that it does not amplify the audio signal from the detector to a point where it is capable of driving a loudspeaker. Since a hi-fi system has its own audio amplifiers, this latter function, obviously, is not required. If you want to listen to both AM and FM broadcasts you must buy a tuner that combines the functions of an AM and an FM receiver. However, FM provides such superior quality over AM, that many audiophiles make do with an FM tuner alone. After checking the cost of a good AM-FM tuner, you'll understand why.

To qualify as a hi-fi component a radio tuner (AM or FM) must incorporate special high-quality design features. Most important the r-f, i-f, and detector circuits of the tuner must pass the entire audio modulation band impressed on the carrier. Despite this large bandwidth, the tuner must be sufficiently selective to separate two closely spaced stations, to prevent any interference between them. The sensitivity of a tuner should be in the order of a few microvolts to pull in weak signals. Moreover, a tuner should incorporate **automatic frequency control** (AFC), to prevent the tuner frequency from drifting away from the selected station. The output circuit of the tuner should be a **cathode follower** (see Chapter 10) to provide isolation and low loading of the detector stage, when a long output cable is connected. Finally, a hi-fi radio tuner should have complete band coverage and exceptionally low distortion, hum, and noise level.

Fig. 162 illustrates two general types of commercially available FM or AM-FM tuners. The basic tuner, shown in (a), consists of the setup just dis-

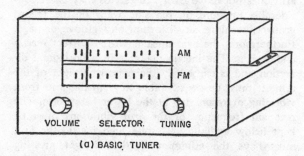

(a) BASIC TUNER

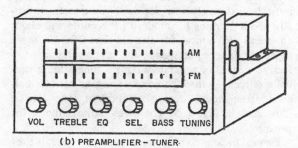

(b) PREAMPLIFIER – TUNER

Fig. 162. *Radio Tuners; (a) Basic Tuner, (b) Preamplifier-Tuner*

cussed; it provides an audio signal from the detector, which is to be fed to the preamplifier of the hi-fi system. A basic tuner is usually equipped with a volume control, a tuning control, and a control for selecting AM or FM stations. The type of tuner shown in (b), incorporates in addition an **audio preamplifier,** in case you do not have a separate one. As we shall see, such a preamplifier strengthens and **equalizes** the output signal from your phono cartridge, so that it can be applied to a main amplifier. It also acts as a general **control center** of your hi-fi system, permitting front-panel control of the main amplifier and various other inputs, plugged into the back of the tuner. The preamplifier-tuner is therefore far more expensive and has many more controls than the basic type.

PHONO PICKUPS AND RECORD PLAYERS

Any record player, whether it be a single record turntable (**transcription** player) or a changer type, must have a **motor** and **turntable** for rotating the records at the correct, uniform speed; a **tone arm** that moves over the record surface; and a **pickup** consisting of phono **cartridge** and **stylus** to pick off the sound vibrations from the record grooves and translate them into a corresponding audio signal. A **record changer** has, in addition, a more or less complicated mechanism for playing a number of rec-

ords automatically and shutting itself off after the last one. Let us say at the outset that a true audiophile scorns a record changer, since it provides inferior quality of reproduction, is too rough on the records, and has too many things that can go wrong. He much prefers a single turntable or transcription-type of player, and this is the only type we shall talk about.

Motor and Turntable. The function of the motor is to rotate the turntable and disk at the *same* speed at which the music was recorded. This sounds simple, but you will be surprised to learn how many motors cannot do this job. Motors and turntables are subject to a number of mechanical distortions and speed variations, which are quite noticeable to the discerning ear. If the speed of the motor is not absolutely uniform, the turntable is uneven, the record has an off-center hole, or is warped, you will hear unpleasant variations in the pitch of the sound. Pitch variations that take place at a relatively low rate (less than ten times per second) are described as low-frequency **wow,** while the term **flutter** is used for higher-frequency pitch variations. Furthermore, if the motor and drive mechanism are not perfectly machined and aligned, and the motor and turntable are not properly **shock-mounted,** a low-pitched growling noise, called **rumble,** is transmitted to the turntable and superimposed on the music. You'll know when you have it.

A really well-made motor drive mechanism and single turntable can get around all these deficiencies. The motor is usually of the **induction** type, used for a.c. only. Cheaper units employ a **two-pole** induction motor, while more expensive units use a **four-pole** motor, which provides more power and uniform speed. The ultimate in the hi-fi field is the **hysteresis-synchronous** motor, which provides a speed as constant as that of an electric clock, because it is geared to the frequency of the a-c line.

The **drive mechanism** of a transcription (single) turntable generally consists of a **motor pulley** driven by the motor shaft and a number of differently sized **idler wheels** that drive the turntable, usually at the rim. The idler wheel functions as a speed reducer to convert the rapid motor rotation to the correct turntable and record speed. A **speed shift** knob at the top of the turntable brings idlers of varying diameters into contact with the motor pulley, according to the desired record speed. Up to *four* speeds may be provided, for the old 78 rpm shellac records, the 33 rpm and 45 rpm long-play (LP) records, and the 16-⅓ rpm transcription-type records. A typical motor and drive mechanism is illustrated in Fig. 163b.

A heavy, cast-aluminum turntable (Fig. 163a), well balanced and machined on a lathe, will give the stability and smoothness required for high fidel-

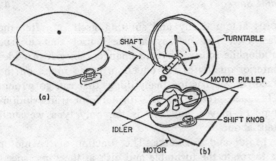

Fig. 163. *Typical Turntable (a) and Drive Mechanism (b)*

ity. The rim of the casting is made very heavy to give a flywheel action to the turntable, which helps to smooth out minor speed variations in the motor and drive mechanism. The turntable must be **concentrically** mounted on the drive shaft and must run absolutely true. You can check this by observing the lower rim of the turntable as it revolves. There should be no up and down or sideways motion. One reason why aluminum is preferred to any other turntable material is that aluminum is **non-magnetic** and hence does not "pull" a magnetic phono cartridge.

Tone Arm. The tone arm must move the stylus of the pickup across the record surface, following as closely as possible a **straight line along a radius** of the turntable circle. (See Fig. 164a) This is necessary so that the tone arm always rides tangent to the record grooves at the point of contact, regardless of its position along the record surface. If the stylus does not "track" well in the grooves, it may exert a side pressure on the grooves which results in distortion and eventually destroys the record. Of course, any arm **mounted in a central hole cannot track perfectly** along a straight line, as in (a), since the end of the arm must describe a **circle.** A centrally mounted tone arm, therefore, is tangent to the record grooves at only *one point,* usually at the center of the record surface, and always has some "tracking error" at the extremes of its motion. However, a short tone arm mounted close to the turntable, as is done in record changers, has a much larger tracking error at the start and end of the record (Fig. 164b) than a long, 9 to 11 inch arm mounted farther away from the turntable (Fig. 164c).

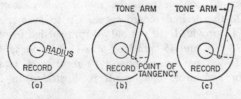

Fig. 164. *Tracking Error; (a) Ideal—No Error, (b) Large Error of Short Arm, and (c) Small Error of Long Tone Arm*

Besides it length, a tone arm must have other desirable characteristics for hi-fi use. The arm should have high-precision bearings with a minimum of **friction** in horizontal and vertical directions. If the arm does not move freely, the stylus may exert considerable pressure on the side walls of the record grooves, resulting in distortion and groove damage. Furthermore, the arm must have the correct **inertia** and **weight.** The inertia operates in a horizontal direction and is caused by the mass or weight of the arm. It must be considerable to keep the arm from **vibrating** or **resonating** at the lower bass notes. The **resonant** frequency of the arm vibrations must be kept below audibility. Despite high inertia the arm must have the **minimum vertical weight** pressing down on the record. The pressure exerted on the stylus by the weight of the tone arm and pickup must be extremely low for hi-fi reproduction, usually about 5 grams for modern LP records and about 15 grams (½ ounce) for the old-time 78s. To serve these varying requirements, the stylus pressure (effective weight) of separately bought tone arms is usually adjustable through some spring or counter balance arrangement. A typical modern tone arm is illustrated in Fig. 165. The screw adjusts the stylus pressure.

Phono Pickup. The phono pickup, consisting of **cartridge** and **stylus** mounted at one end of the tone arm, constitutes the electronic starting point of the hi-fi system. It must faithfully translate the mechanical (sound) modulation contained in the record

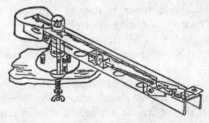

Fig. 165. *Tone Arm Showing Interior Construction*

grooves into corresponding audio variations. Not so long ago this job was done more or less satisfactorily by a **Rochelle crystal** cartridge to which an ordinary steel or osmium needle was attached. You will recall that the action of crystals is based on the **piezoelectric** principle. When the needle follows the lateral motion of the record groove modulation, the crystal faces become slightly deformed and generate a voltage proportional to the deformation and, hence, needle excursion. Crystal pickups have a high voltage output (1 to 2 volts) and compensate, to some extent, automatically for the recording frequency characteristics. They therefore do not need a preamplifier, which makes them popular in ordinary phonos. However, the non-uniform frequency response of crystal pickups (they have a large bass boost) cannot compensate for all types of recording characteristics used by different record manufacturers. This compensation, hence, is best left to the specially designed **equalizer circuits** in the preamplifier, and a pickup with a "flat" frequency response over the audio band is preferred. Crystal pickups, moreover, function poorly in warm, damp climates and the crystal may be destroyed by high temperatures.

The most popular pickups for hi-fi use consist of a **magnetic cartridge** to which a sapphire or diamond stylus is permanently attached. Magnetic pickups depend on variations in a magnetic field to generate an electrical signal proportional to the stylus motion. They are, in general, characterized by a low output voltage and impedance, and excellent, uniform frequency response over the entire audio range. Because of their low output voltage and their lack of compensation for the recording characteristic, magnetic pickups always need a preamplifier that will boost the voltage and equalize the frequency response.

Two types of magnetic cartridges are commercially available. One is the **variable-reluctance** type, the other the **moving-coil** or **dynamic pickup.** The moving-coil (dynamic) pickup works like the dynamic microphone discussed in the last chapter and is similar to it in quality. An **armature,** consisting of a stylus and a small coil wound on a thin steel sleeve, moves in the airgap between the poles of a strong permanent magnet. The lateral motion of the stylus in the record groove moves the coil across the lines of force of the magnetic field, thus generating a voltage in the coil that is proportional to the **velocity** of the motion. Since the moving system is very light, substantially flat frequency response from about 10 to 30,000 cps can be obtained. The quality of the dynamic pickup is high, but so is the price!

Almost equal quality at less expense is provided by the variable-reluctance cartridge, which operates on a slightly different principle. Here the stylus is mounted on a magnetic armature and moves freely in the airgap between the pole pieces of a strong permanent magnet. **Two coils** are fixed-mounted on the pole pieces, one on each side of the stylus and armature, as illustrated in Fig. 166. When the magnetic armature is deflected by the stylus motion toward either side, the resistance, or **reluctance,** of the magnetic circuit on that side is *decreased* and, hence, the magnetic field **(flux)** is *increased*. At the same time, the reluctance on the other side—away from the stylus—is increased and the magnetic flux on that side is decreased. As a result, the magnetic flux increases through one of the coils and decreases through the other, and

voltages proportional to the *difference* in magnetic flux are induced in the coils. Since the coil windings are in series, a **push-pull** action results, with the total coil voltage being proportional to the **lateral** stylus motion. No voltage is generated by vertical (up and down) motions of the stylus, which considerably reduces **record scratch.** The output voltage of a variable-reluctance cartridge is about 22 *milli*volts and the frequency response is smooth from about 20 to 20,000 cps.

Fig. 166 shows a G.E. variable-reluctance cartridge. The single-stylus cartridge will hold a single, fixed stylus for one type of record only (either 78 rpm or LP record). A triple-play cartridge permits either one of two stylus tips to be brought into position by depressing and rotating the stylus knob on top of the cartridge. One stylus has a tip thickness of .0025 to .003 inch (3 mils) and will fit the grooves of 78 rpm records; the other stylus has a .001-inch (1 mil) point for playing either 45 rpm or 33-1/3 rpm LP records. Either

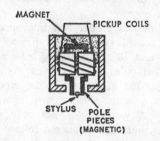

Fig. 166. *Variable-Reluctance Cartridge Construction*

stylus can be replaced easily be sliding it out from the end of the rotatable sleeve.

While we're on the subject, let us make a few observations on styli, which replace the old-fashioned, sharply pointed needles. In general, the cheaper the stylus, the more expensive it will be in the long run. As pointed out above, the stylus must have the proper tip radius to fit the groove of the record for which it is designed. The cheaper styli may fit the groove when first used, but they are rapidly worn down by the constant abrasion, and end up scratching your records badly. By the time the distortion and scratch noise are loud enough to be observed, the damage is usually done. The cheapest stylus is made of osmium and it is good for about 10 to 15 hours playing time. If you must, buy it, but throw it away after that time; don't wait for distortion and noise to show up. The **sapphire** stylus is moderately expensive and will last for about 25 to 30 hours of playing time. It is popular for 78-rpm records (requiring a 3-mil stylus), which are rarely played by the hi-fi fan. Thus, when used in combination with a 1-mil **diamond** stylus for LP records in a triple-play cartridge, both

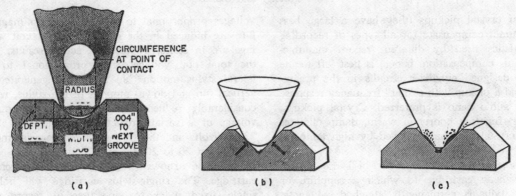

Fig. 167. *The Stylus in the Groove.* (*a*) *Dimensions of 78-rpm Record Groove and Stylus;* (*b*) *Proper Fit;* (*c*) *Worn Stylus*

needles may wear out at about the same time. Being the hardest, the diamond stylus is, of course, the best and most expensive you can buy. It is good for at least 800 to 1000 hours playing time and may last much longer. It is a good idea to have a diamond stylus occasionally inspected for wear under a microscope.

Fig. 167 illustrates proper and improper fit of a stylus in the record groove. The dimensions of a 2.5 mil stylus and 78-rpm record groove are shown in (a). The "ball-point fit" of a new stylus in the groove is illustrated in (b), while the groove damage caused by a worn, chisel-shaped stylus is portrayed in (c). A worn stylus not only ruins the record grooves, but since it cannot follow the rapid vibrations of the sidewall modulation, it cuts off the higher frequencies and generally distorts the sound. Of course, the scratch level of a worn needle also goes up noticeably.

PREAMPLIFIERS

Originally designed to provide voltage amplification and tone equalization for a magnetic phono pickup, the modern preamplifier has evolved into a **control center** for the entire hi-fi system. The preamp still provides voltage preamplification for microphones and pickups, as well as tone equalization, but it now has the added functions of tying together all the components of the hi-fi setup and controlling them from a convenient location. The flexibility of location of the preamplifier derives from the fact that its cathode follower output may be connected with a relatively long cable to the main (power) amplifier and speakers.

A typical preamplifier is illustrated in Fig. 168. Since most units have somewhat similar controls, let us study their functions a little more closely. The **Volume** control usually combines the function of an **ON-OFF** power switch with a volume (audio gain) control. The simplest type of volume control

sets the voltage gain of the preamplifier equally for all frequencies, leaving all tone compensation to the individual tone controls. Somewhat more sophisticated controls provide for "bass compensation," that is, they boost the low-frequency response as the volume level goes down to compensate for the falling off in the ear's response. (See Fig. 160, threshold of hearing curve.) The cleverest kind of control is the **loudness control,** which compensates for the ear's deficiencies at both low and high frequencies by inverting the ear's loudness contours at a particular volume level. (See Fig. 160, upper and lower loudness contours.)

As illustrated in Fig. 161, a number of audio input devices may be plugged into the rear of a preamplifier. The **Selector** switch on the preamp front panel selects the desired input device, whose output is to be amplified. The Selector switch shown has five positions, which may be used for a microphone, phonograph, radio tuner, TV set, and a tape recorder.

The tone controls of a preamplifier are generally of two types: **bass** and **treble** controls, and tone **equalization** controls to compensate for different recording characteristics. The bass and treble controls permit **boosting** or **attenuating** (cutting down) the low- and high-frequency response, respectively, in accordance with your desires. Besides satisfying your personal taste, the proper use of the tone controls permits achieving the proper **tonal balance**

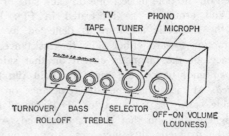

Fig. 168. *Typical Hi-Fi Preamplifier*

of the sound, taking into account the deficiencies of your system, loudspeakers, room acoustics, etc. A typical set of frequency response characteristics for various settings of the bass and treble controls is illustrated in Fig. 169. The center setting of each control usually provides "flat" (uniform) frequency response, while the extreme settings provide maximum boost or attenuation. The particular curves shown provide up to 20 db maximum boost or attenuation of the bass and treble response. This means that the output voltage of the preamplifier is up to *ten times* as great at the extreme bass or treble frequency compared with the mid-frequency (1000 cps) response for a *boost*, or only *one-tenth* of the mid-frequency response for maximum *attenuation*. While this range may appear somewhat excessive, a control range of 12 to 15 db minimum is considered desirable.

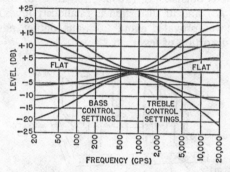

Fig. 169. *Preamplifier Frequency Response for Various Settings of the Bass and Treble Tone Controls*

We have previously referred to the **recording characteristics** of various record manufacturers. Fig. 170 shows *generalized* recording and playback curves of modern records, not taking into account individual differences of various record manufacturers. As you can see, the bass frequencies are progressively cut during recording, while the treble frequencies are boosted. The bass frequencies must be reduced during the cutting of the original **master record** to prevent the cutting stylus from making too wide excursions, which might result in cutting over from one groove to the next. Any disk recorder has the inherent characteristic that the amplitude of the stylus motion increases as the frequency goes *down*. Hence, by limiting the amplitude of the stylus excursion to the maximum permissible value, the bass response is correspondingly reduced. Furthermore, the high-frequency response (beyond about 1600 cps) is deliberately **boosted** during recording to reduce the needle scratch (consisting mostly of high frequencies) and, thus, obtain a better **signal-to-noise ratio** during playback. This is called high-frequency (treble) **pre-emphasis**. To obtain a uni-

formly flat frequency response during playback, therefore, the preamplifier must provide an **equalization characteristic**—that is the exact inverse of the recording characteristic, as shown by the dotted amplifier curves in Fig. 170.

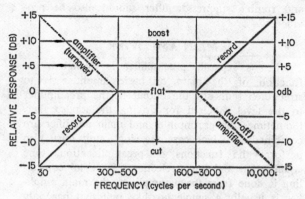

Fig. 170. *Recording and Playback Characteristics of Modern Records*

The recording characteristics of different records, especially those made before 1953, differ somewhat from the generalized curves of Fig. 170. To compensate for different recording characteristics, a minimum of 16 equalization curves must be provided by the preamplifier. This is most conveniently done by means of four low-frequency or **turnover** positions and four high-frequency or **rolloff** positions. Separate turnover and rolloff controls may be provided (as in Fig. 168), or both functions may be combined in a single control with the particular recording characteristic marked next to each position. Thus, you may have equalization positions labeled "early 78," "LP" (Columbia), "AES" (Audio Eng. Society), and "NAB" (National Association of Broadcasters). Recordings made after 1953 are practically all recorded according to a new standard curve, called the "RIAA" (Recording Industry Association of America). Don't worry, however, if your preamplifier doesn't have all these positions, as long as it has the new RIAA curve and a couple for the older records. The differences between the older curves are so minute, that you probably won't be able to discern them.

The quality of the preamplifier is important, since it controls what goes *into* your main power amplifier. There is no point in buying an expensive main amplifier costing several hundred dollars and then skimping on the preamp. As a matter of fact, one consumer testing organization discovered that the preamplifier almost exclusively controlled the *listening* quality of the system, regardless of the type and expense of the main amplifier. A good preamplifier should have a uniform frequency response (without tone controls) of at least ± 1 db from 20 to 20,000 cps (many do better than that), harmonic and intermodulation distortion of no more than 1 to 2%, and a combined hum and noise level of at least 50–85 db below the maximum output level of the amplifier. If possible, a phono noise and rumble suppressor filter should also be provided.

MAIN AMPLIFIER

When a separate preamplifier is available, the function of the **main** or **basic** amplifier is to strengthen faithfully the output of the preamplifier to the level required for driving the loudspeaker. (Sometimes the preamplifier and main amplifier are combined, in which case the single unit must also perform the functions of preamplification, tone equalization and input selection.) Since all controlling is done from the preamplifier, the main amplifier is usually a simple box-like unit that has only audio input and output connections, a 117-volt power cable, and possibly one or two adjustment controls for "balancing" the output tubes. The output of most preamplifiers is about 1 to 2 volts, while the input voltage required to drive the audio power amplifier stage is anywhere from 15 to 60 volts. The basic amplifier must provide, therefore, for several stages of **voltage** amplification as well as sufficient power amplification to drive the speaker. The question is, what *is* sufficient power amplification? Speaker systems will work with audio output powers from a fraction of a watt to 30 watts or more. Commercial amplifiers are available with power output ratings from 5 to 100 watts, with the cost rising steeply as the power goes up. Ironically, the level of speech and music performed in the home is usually no more than about one-half watt, and often only a few *milli*watts. Wildly conflicting claims are issued in audio circles about how much power is required for proper reproduction, with the high-power, expensive amplifiers being favored. It's very confusing.

First of all, a power rating without specifying the total **distortion** at maximum power, is meaningless. Even cheap amplifiers can be driven to substantially greater outputs than their ratings, if you're willing to accept about 10% distortion in the bargain. The maximum distortion for which the amplifier power output is specified should never be more than 1 to 2 percent. (This includes **harmonic** as well as **intermodulation** distortion.) Secondly, an amplifier must have considerable **reserve power** to accommodate the tremendous **dynamic range** between the loudest crescendos of a symphony orchestra and the softest whisper of a human voice. (See Fig. 160.) Although this range is about 120 db, you could never reproduce such loudness variations within the confines of your living room and, hence,

there is no point in providing it. Moreover, the power output of an amplifier is to some extent a question of individual preference and depends on the average level or reproduction, as well as its peaks. It has been the author's experience that an audio output power of about 20 watts (with no more than 1% distortion) provides ample reserve power for the maximum dynamic range you would care to listen to. An amplifier with 40 watts output is more than ample and anything above that is needless luxury. The high-power units have their uses in theaters and halls, but don't buy a 300-horsepower Cadillac to drive at 30 miles per hour.

A basic amplifier employing the popular Williamson circuit is shown in Fig. 171. The Williamson circuit uses triodes in push-pull throughout the amplifier, has a large amount of inverse feedback and oversized transformers. As a result its frequency response is outstanding (about ± 1 db from *10 cps to 100 kc*) and its distortion is fabulously low. The unit shown provides about 20 watts output power with a distortion of less than 2%.

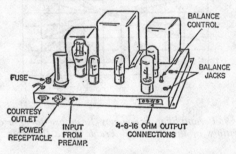

Fig. 171. *Basic Williamson Amplifier*

Note the simplicity of the interconnections in this unit. A power receptacle permits plugging in the power cable from the preamplifier to supply the tubes. A courtesy outlet is provided to control the power to a phonograph or other unit from the **preamplifier** control panel. The output transformer is equipped with 4-, 8-, and 16-ohm terminals to match the impedance of the speaker to the output stage. Two jacks are provided for inserting a milliammeter into the plate circuit of either output tube. The push-pull output stage is balanced for minimum distortion by adjusting the "balance control" until the milliammeter reads the same plate current for both output tubes.

LOUDSPEAKERS AND ENCLOSURES

A loudspeaker is an electromechanical device for converting a varying audio voltage into corresponding sound waves. Being electromechanical, loudspeakers are the weak sisters of any hi-fi system. Despite all the care taken in preserving the fidelity of an audio signal from pickup to the output of the amplifier, loudspeakers invariably add some form of distortion to the signal. You can buy loudspeakers anywhere from simple $5 units to $1000 multiple-speaker systems, but chances are that even the most expensive system will not have an audio frequency response as "flat" as that of an amplifier. Loudspeaker enclosures, or **baffles,** as they are often called, assist the speaker in doing its job properly and, thus, are almost equally important.

Basic Dynamic Loudspeaker. Practically all loudspeakers at the present time are of the **dynamic** type, in which a **voice coil** carrying the audio signal moves in and out of a strong magnetic field produced by a permanent magnet. Fig. 172 illustrates the construction of the dynamic speaker.

As is apparent from the figure, a large permanent (alnico) magnet with a central pole piece is used to provide the magnetic field. In the narrow airgap between the central pole piece and the outer poles, where the field is most concentrated, a small **voice coil** is suspended, consisting of a few turns of wire wound around a paper, plastic, or aluminum form. This voice coil carries the audio currents from the output of the audio amplifier. At one end of the

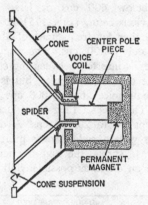

Fig. 172. *Simplified Construction of Dynamic Loudspeaker*

voice coil is mounted a fairly large paper or cloth cone, which radiates the sound. The cone is held at its edges by a flexible suspension ring that is fastened to the metal frame of the speaker. Another suspension at the other end, mounted at the junction of the cone and voice coil, keeps the voice coil **centered** with respect to the pole piece. This suspension, known as the **spider,** is made of flexible material and permits forward or backward (longitudinal) motion of the voice coil, but no sideways (lateral) motion. The entire cone and voice-coil assembly, thus, can move freely **in or out** of the central pole piece, but not sideways.

When audio currents flow through the voice coil, a magnetic field is produced around the voice coil that is at **right angles** to the field of the permanent

magnet and proportional to the strength of the audio currents. As is explained in basic magnetism, the two fields **attract or repel** each other, depending on the instantaneous polarity of the audio currents. Since the position of the permanent magnet is fixed, the voice coil and cone assembly moves inward or outward from its central position in proportion to the attraction or repulsion of the two magnetic fields—that is, in proportion to the momentary strength and polarity of the voice coil currents. The resulting cone vibrations produce air-pressure variations (sound) in correspondence with the audio signal.

The frequency response of such a basic speaker is generally quite irregular, with a number or **resonant** peaks and valleys, and rarely extends beyond a range of about 60 to 8000 cps. By using a fairly large (12 to 15 inch diameter), heavy cone, the low-frequency response of a single speaker can be extended downward to 45 or even 30 cps, but then the high-frequency response suffers. In contrast, by using a small (3 to 5 inch diameter), light speaker cone, the high-frequency response of a single speaker can be pushed to the limits of audibility, but then the bass response will cut off around 100 to 150 cps because of the insufficient mass of the speaker cone. It is almost impossible to make a single speaker perform well over the entire audio range. Nevertheless, various tricks and ingenuous design have made available in recent years single-unit **wide-range speakers** that will perform well over an audio-frequency range from about 50 to 12,000 cps.

Coaxial and Triaxial Speakers. The difficulty in designing a single speaker to cover the whole audio range has resulted in two alternate approaches. One is to use separate speakers, each designed for a specific audio band, and connect them together to cover the entire audio range. The other approach is to combine the large speaker required for low-frequency reproduction and the small one needed for the high frequencies into a single unit, mounted in line, or **coaxially. A coaxial speaker** may consist of two completely separate speakers mounted on the same axis, as shown in Fig. 173(a). Alternatively, a single electromechanical **driver** may be employed to operate two differently sized cones or **diaphragms,** one for low and the other for high frequencies, as illustrated in Fig. 173(b). A mechanical or electrical **crossover network** is associated with the speaker to divide the audio input signal into the proper low- and high-frequency bands, which are fed separately to each speaker unit.

The coaxial speaker principle may be extended to three speakers by further subdividing the audio range. Either three separate speakers or a single driver with three diaphragms may be mounted

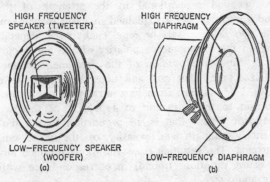

Fig. 173. *Two Coaxial Speakers; (a) Separate Drivers, (b) Single Driver with Dual Diaphragms*

along the same axis in the **triaxial speaker.** Fig. 174 illustrates a triaxial unit that has a single driver for a low- and medium-frequency diaphragm and, in addition, a separate high-frequency "tweeter." Each

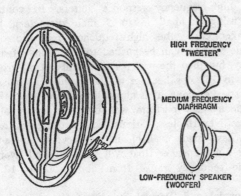

Fig. 174. *Triaxial Speaker with Low- and Medium-Frequency Diaphragms and Separate High-Frequency "Tweeter"*

unit performs at maximum efficiency in a particular frequency range.

Multiple Speaker Systems. The best way to overcome the limited frequency response of a single speaker is a multiple speaker system made up of completely separate speakers, each designed for maximum efficiency in a specific frequency band. The simplest type of multiple speaker system is a two-way system, consisting of a large, low-frequency speaker, or **woofer,** a small (horn-type) high-frequency speaker, or **tweeter,** and a frequency-dividing **crossover network.** The speakers must be **matched** so that the tweeter supplements the frequency range of the woofer and the whole audio range is covered smoothly. The crossover network is designed to split the audio range smoothly at the **crossover point,** so that each speaker works only with frequencies for which it is designed and is not overloaded. Depending on the system, the woofer may be designed to reproduce frequencies from about 30 cps to a crossover point anywhere between 400 and 3500 cps. The

crossover network, which may be a simple **high-pass filter** for a two-way system, will then feed the balance of the audio frequencies up to 15,000 cps or higher to the tweeter. A typical two-way setup is illustrated in Fig. 175.

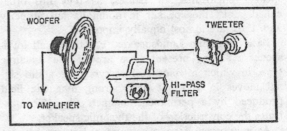

Fig. 175. *Two-Way Speaker System Consisting of Woofer, Tweeter, and High-Pass Filter*

A somewhat more ambitious approach to hi-fi reproduction is the **three-way system,** illustrated in Fig. 176. This particular system has been derived from the two-way setup of Fig. 175 by adding a dual-horn high-frequency tweeter and an L-C crossover network, which limits the woofer to audio frequencies below 600 cps or less. The tweeter from the previous setup (Fig. 175) is now used as mid-range speaker for frequencies of about 600 cps to approximately 4000 cps. The previously used high-pass filter is inserted between the crossover network and the dual-horn tweeter, but is adjusted to pass only frequencies above 4000 cps (approx.) to the tweeter. The tweeter then reproduces the range from 4000 cps up to 20,000 cps. A dual horn or other speaker with a **wide flare** is preferred for the tweeter so as to spread the high-frequency "beams" over as wide an angle as possible. Still more elabo-

Fig. 176. *Three-Way Speaker System Consisting of Woofer, Tweeter, Mid-Range Speaker and Crossover Networks*

rate and expensive multiple-speaker setups are used sometimes to achieve uniform response.

Loudspeaker Enclosures (Baffles). The sound you hear from a speaker is produced by the **combination** of the loudspeaker and its enclosure or baffle. A loudspeaker cannot perform its job properly without a suitable baffle to **rout the sound energy.** You can discover this for yourself by listening to the tinny sound emerging from even the best hi-fi speaker, when it is simply held in the hand. The bass response, which gives "body" to the sound, is almost completely absent in a hand-held speaker.

The reason for this phenomenon is simple. Any loudspeaker radiates sound from both the front and the rear of the speaker cone. When the speaker cone is pushing the air in front of it, thus **compressing it,** the air in the rear of the cone is simultaneously thinned out or **rarefied,** and vice versa. Since sound compressions and rarefactions are 180 degrees out of phase with each other, the sound from the front of the speaker is **out of phase** with that radiated from the rear. When these two out-of-phase sound waves meet, they effectively produce a short circuit, with the rear rarefaction **sucking in** the front compression, and they cancel each other out. This cancellation affects primarily the low-frequency sound (long wavelengths), since the high-frequency sound is radiated from the front and rear in the form of beams, which do not meet.

The obvious answer to this situation is to put some sort of obstacle around the speaker to prevent the front and rear sound waves from reaching each other. An obstacle sufficiently large to do this completely is rarely practical nor required. It is only necessary to **elongate** the sound path from front to the rear of the speaker sufficiently to build up a sound wave in front of the speaker, before it has a chance to be pulled back to the rear. The **longer the path length** from front to rear **compared to the wavelength** of the sound to be radiated, the **less cancellation** takes place. When the wavelength of the sound becomes comparable to the front-to-rear path length of the baffle, cancellation occurs, and wavelengths longer than this (lower frequencies) will not be radiated to any extent.

This is the sort of reasoning that led to mounting the speaker on a simple flat wooden board, or "baffle," shown in Fig. 177a. The term "baffle" originated from this open structure; complete enclosures were not used in the early days of hi-fi. The flat, open baffle works well, but has the drawback of requiring a very large size to prevent cancellation at low frequencies. To radiate frequencies in the 30–50 cps range, for example, the baffle board would have to be from 20 to 40 feet in length and width. That is obviously not practical.

The attempt to reduce the physical size of the baffle while retaining the required path length from front to rear has led to the design of a variety of structures. The simplest is the open back enclosure (Fig. 177b), which is essentially a flat baffle with its edges folded back. This open box or cabinet used to house practically all commercial radio-phonographs, has the serious defect of **cabinet resonance,** which makes it unsuitable for hi-fi reproduc-

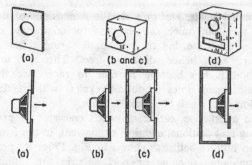

Fig. 177. *Basic Types of Baffles:* (*a*) *Flat Open Baffle;* (*b*) *Open Back Enclosure;* (*c*) *Infinite Baffle;* (*d*) *Bass-Reflex Enclosure*

tion. The entire box resonates for low-frequency sounds reproduced by the speaker, generally in the range from 100 to 200 cps. The resonant peaking of these frequencies produces a characteristic boominess, which **masks** the balance of the bass response.

By closing the rear opening of the box, the rear sound wave is entirely suppressed, and an "infinite" front-to-rear path length results. This **infinite baffle** (c) suppresses cabinet resonance and would at first appear to be ideal, except for some other deficiencies. Apart from wasting the entire sound output from the rear of the speaker cone, the infinite baffle has some undesirable damping effects on the speaker. The compressed air in the baffle box tends to raise the fundamental **resonant frequency** of the speaker assembly in the low-frequency range. Since a speaker reproduces very little below its resonant frequency (generally between 30 and 50 cps in a hi-fi woofer), the raising of this frequency by the infinite baffle results in the elimination of a portion of the bass response. The response to this lower limit is very smooth, however. A very large box avoids this defect, but the size required again makes it impractical. The best approach to an infinite baffle, consequently, is to mount the speaker either in a large unused closet or in the wall between two rooms.

More recently, the inherent defects of a small infinite baffle have been made a virtue of by a revolutionary principle of speaker and baffle design known as **acoustic suspension.** The elastic suspensions of the conventional loudspeaker cannot reproduce the large excursions demanded by low bass frequencies without considerable distortion. Acoustic suspension gets around this difficulty by vastly increasing the "compliance" (flexibility) of the mechanical spring suspensions and then making up for the lost spring action through a pneumatic spring consisting of the compressed air in the sealed box. The speaker is acoustically suspended inside a small infinite baffle, which is calculated to provide the proper amount of pneumatic spring action through its sealed-in air. Not only do the dimensions of the box turn out much smaller than for an infinite or any other baffle, but the pneumatic spring action is linear, and hence distortion-free. Through this method it has become possible to reproduce the lowest organ notes (about 30 cycles) without distortion in a small "bookcase" baffle.

An alternative, earlier approach consisted of providing an additional opening, or air vent, in the front of an infinite baffle, as shown in Fig. 177c. By providing some opening or **port** hole in the infinite baffle, the sound energy from the **rear** of the speaker can be used and the size of the enclosure can be drastically reduced. The resulting **bass-reflex** enclosure has other advantages which make it one of the most popular hi-fi baffles. The bass-reflex cabinet is essentially a **phase inverter for sound waves.** By making the path length from the rear of the speaker to the port just right, the rear wave can be delayed sufficiently (by one half wavelength) so that it emerges from the port **in phase** with the front wave and thus reinforces it. As a result of this reinforcing action, the low-frequency output is *twice* that of an infinite baffle. Moreover, the bass-reflex enclosure makes use of its **own resonant frequency** to extend and smooth out the low-frequency response of the speaker. By designing the box so that its own cabinet resonance frequency is somewhat below the fundamental resonance of the speaker that it houses, the resonant peak of the speaker is damped out and the bass response is **broadened.**

You may have gathered from this description that the bass-reflex cabinet and its speaker must be carefully **matched** to secure the proper response. A definite relationship must be maintained between the **size of the enclosure,** the **diameter of the port,** and the **resonant frequency of the speaker.** This is the reason why many speaker manufacturers specify a particular type of bass-reflex or other enclosure when stating the speaker response. There is a way of matching a bass-reflex enclosure to a particular speaker by "tuning" the port for the proper resonant response, but this goes beyond the scope of our discussion. You can find out about tuning a port by consulting one of the many excellent books on high-fidelity, which will also acquaint you with a variety of additional loudspeaker enclosures.

TAPE RECORDERS

Tape recorders have become increasingly popular in recent years, as the technique of high-fidelity recording on magnetized tape has been mastered. Furthermore, tape machines are at present the only practical means of achieving **stereophonic (binaural)** sound reproduction in the home.

Fig. 178 illustrates the essentials upon which modern tape recording is based. The audio signal that is to be recorded is fed into the winding of a ring-shaped electromagnet, called the **recording head.** Between the two ends, or **poles,** of the magnetic recording head is a small airgap, from which a **fringing flux** is emitted with a magnitude and polarity proportional to the electrical signal. The tape, which is drawn past the poles of the ring magnet, has a thin magnetic coating of iron oxide consisting of myriads of small magnetic particles, each acting as a tiny magnet. In the unmagnetized tape these magnetic particles have a random distribution and, hence, their combined magnetic field cancels out. When the tape is moved past the poles of the recording head, however, the individual particles are subjected to a magnetizing force *along* the direction of tape travel and they are turned or aligned in that direction. The constantly varying audio signal in the recording head thus produces corresponding degrees of magnetization *along* the moving tape. This is known as **longitudinal magnetization.**

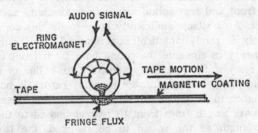

Fig. 178. *Longitudinal Recording on Magnetic Tape*

The magnetized tape emits itself a small magnetic field proportional in polarity and magnitude to the original signal. Consequently, when the tape is drawn past the airgap of a similar **reproducing head,** the magnetic field variations induce changing voltages in the winding of the ring magnet, corresponding to the originally recorded signal. The audio signal may then be amplified and reproduced by a loudspeaker.

Practical Tape Recorder. A practical tape recorder must perform *three* basic functions, as illustrated in Fig. 179. These are: **recording, reproduction, and erasure.** Erasure is required before a new recording can be made on a previously magnetized tape.

In tape recording, a low-level audio signal from a microphone, radio tuner, or other source, is first strengthened by the recording amplifier to the proper recording level and then is fed to the recording head. Also applied to the recording head is a **bias signal** from a high-frequency oscillator. The bias places the recording signal into the linear portion of the tape's **magnetization curve,** which brings about improved linearity, lower distortion,

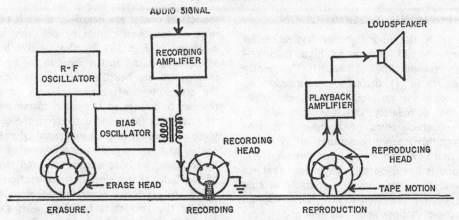

Fig. 179. *Block Diagram of Tape Recorder Functions*

and a **reduction of tape noise.** It has been found that an a-c signal does this job better than steady d-c bias. The frequency of the bias signal must be above audibility (usually between 30 and 150 kc) to avoid interference with the audio signal.

The **reproduction** process is the inverse of recording. The weak signal induced in the reproducing head by the traveling magnetic tape must first be amplified by the playback amplifier before it can be reproduced by a loudspeaker. Since the requirements for optimum performance differ for recording and playback, completely separate amplifiers and magnetic heads are preferred. However, in the cheaper machines the two amplifiers are always combined in a single unit, with proper equalization being switched in for recording and playback. Often the two heads are also combined into a single compromise recording-playback head, a compromise which any hi-fi fan should shun. You can get a **tape deck** without any amplifier, but with all other necessaries including three heads, and then use your hi-fi amplifier with proper equalization for both recording and playback. This is often the best solution.

Finally, the tape recorder must have provisions for **erasing** previous recordings. For this a strong magnetic field from a third magnetic head, the **erasure head,** is required. Again, an r-f field is preferred to d.c. for uniform, low-noise erasure. The field may be produced by a separate r-f oscillator with **added amplification,** or the bias oscillator may be used with an extra stage of amplification switched in. In either case, the r-f frequency for erasure must be **ultrasonic**—i.e., above audibility.

Tape Mechanism. Some sort of mechanical drive must be provided to move the tape at an even speed past the three heads. The speed with which this is done affects the **high-frequency response** of the recorder, higher speeds permitting an extended treble response. Hi-fi tape machines have tape speeds

of 7½ inches per second (ips), or even 15 ips, while those used primarily for voice (dictation) employ speeds as low as 1⅞ ips. Tape mechanisms are apt to be complicated affairs. The bare essentials of a typical machine are shown in Fig. 180.

The tape to be recorded is drawn off a **supply reel,** threaded around and **idler pulley** and then is passed over the three magnetic heads. The tape must be pulled at **constant speed,** which is achieved by the combination of **capstan** and **pressure roller.** The capstan is a rubberized rotor, which is driven by an electric motor with sufficient torque to draw the tape at constant speed. The pressure roller is adjusted to press the tape evenly against the cap-

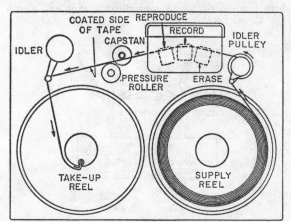

Fig. 180. *Mechanism of Typical Tape Recorder*

stan surface to prevent tape slippage. The tape then travels past a second idler, which keeps it under constant tension, and is taken up by the **take-up reel.** The functions of the reels are reversed, of course, during playback. While this explanation makes the mechanical portion of the tape recorder appear relatively simple, its job is actually very difficult, and good tape mechanisms are quite expensive.

STEREOPHONIC REPRODUCTION

In recent years the hi-fi industry has achieved the final step to add realism to high fidelity—**stereophonic reproduction** of sound (frequently called "stereo" for short). Stereo sound reproduction requires recording (or broadcasting) and reproduction over **two** completely separate channels, to simulate the characteristics of hearing with two ears. (It is, for this reason, sometimes called **binaural** reproduction.) Our ability to hear direction and depth of sound is based on the simple fact that we have two ears. You can convince yourself of that by plugging up one ear for a few hours. Sound will soon become shallow and confused and appear to lack realism. Stereo satisfies our instinctive need for two-eared, three-dimensional hearing. The older, conventional method of **monophonic** or **monaural** reproduction over a single channel cannot preserve the spatial relationships or depth of the sound, which we normally observe through sound waves striking our two ears at slightly different times. Mono reproduction makes no distinction between left and right, up and down, or front and back of a sound source. These spatial characteristics of direction and depth can be restored only by using two or more separate sound channels for recording or broadcasting and reproduction. This is what stereo is designed to do. Stereo, thus, is a completely different technique of sound reproduction. It is not necessarily high fidelity, but if it is, stereo hi-fi is the best method of sound reproduction that can possibly be devised.

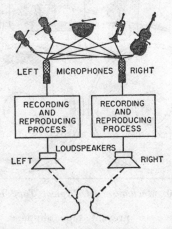

Fig. 181. *The Essentials of Stereophonic Reproduction*

To understand the essentials of the stereo technique, consider a "live" concert, as symbolically indicated in Fig. 181. If you were sitting in the center of the hall in front of the orchestra, the sound reaching your two ears would—in spite of the center position—**not be identical.** Instruments playing to the left of center are heard by the left ear a fraction of a second **sooner** than by the right ear. (This is indicated in Fig. 181 by the relative length of the sound-path lines.) Instruments to the right of center, in contrast, are heard sooner by the right ear. Only the sound of instruments playing in dead center reaches both ears at the same instant. (That's how we know that the sound came from the center.) Thus, these differences in time of arrival of various sounds enable us to distinguish the direction of each source. In addition, we hear sounds reflected from the walls of the hall, giving rise to reverberation or an echo effect. Again these reverberated sounds are sorted out by the two ears, blending with the earlier-received direct sounds, and these directional echo effects help us judge the size of the concert hall and contribute to the depth of the sound. The actual process is very complex, but the important idea is that the **two** ears are capable of judging **differences in arrival time** of sounds emanating from various sources, either directly from the instruments or indirectly through reflection and echo effects.

By placing two microphones in front of the orchestra, one taking the place of the left ear, the other that of the right ear, the effects of binaural hearing are simulated, as shown by the differences in length of the sound-path lines in Fig. 181. (In practice, several microphones may be used in place of each ear.) The sound pattern picked up by the *left* microphone is fed to its own recording and reproducing channel, which may be a tape recorder or a radio transmitter and receiver. This pattern is reproduced by the left loudspeaker. Similarly, the sound pattern picked up by the **right** microphone is fed to a separate recording and reproducing channel and is heard on the right loudspeaker. The listener sitting in the center between the two stereo loudspeakers then hears the differences between the two sound patterns, just as if he were listening in the concert hall. This is stereo.

How Stereo is Recorded. Until 1957, stereo sound could be recorded only on tape. Two or more microphones were set up in front of the sound source in the best acoustical positions. The sound picked up by each microphone was then recorded on two **separate** magnetic tracks of a single tape. (See Fig. 182.) Since most tape recorders had provisions to record two separate, parallel tracks (one forward, one reverse), it was no great shakes to add a separate electronic channel for **simultaneous** stereo recording on the two magnetic sound tracks. Both tracks start and end together on the same tape reel, of course, but the tracks are **separate** sound recordings nevertheless. The parallel sound tracks are then played back simultaneously through **two** separate electronic amplifiers and separate loudspeakers, which reproduce the stereo illusion. In practice, usually **four** sepa-

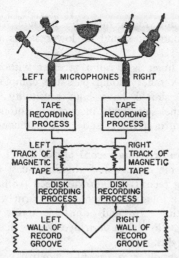

Fig. 182. *Recording Stereo Sound on Tape and Disk*

rate sound tracks are used on a single tape, two stereo channels in one direction and two in the other. Thus, four-track stereo tape is **reversible**, just as the two-track monophonic tape used to be. This is still **the** method used today to make "master" recordings at a live recording session. However, it is relatively difficult to make quantity tape recordings from masters for commercial distribution because of speed differences between commercial and home machines. But, for those who can afford them, commercial stereo tapes are considered the ultimate obtainable in hi-fi reproduction.

Stereo really became popular when it was finally adapted, in 1957, to the phonograph record by the Westrex system of disk recording. The Westrex system permits recording the two separate stereo channels on the two sidewalls of a **single** V-shaped record groove (see Fig. 182). In practice, the two tracks of a stereo master tape are re-recorded through the Westrex process on the left and right walls of the V-grooved master disk, which is then pressed in quantity for commercial distribution.

The pickup cartridges that have been developed to reproduce the complex information recorded in a stereo record groove are veritable marvels of audio engineering. Conventional monophonic cartridges were required only to respond to the lateral (sidewise) vibrations recorded on a mono disk. Stereo cartridges must respond to both lateral and vertical motions (shifted by 45° to fit the walls) recorded in a single groove of a stereo disk. The stereo cartridge must not only translate these complex mechanical vibrations in two planes into the corresponding electrical vibrations, but it must keep the vibrations from each wall completely separate. As if this weren't enough, the same pickup cartridge must also be able to play single-channel or monophonic records with equal fidelity. Essentially, the stylus of

such a cartridge performs motions which are the **vector** resultants of the vertical and lateral sidewall excursions. The magnetic pickup coils of the cartridge then resolve the resultant into the equivalent electrical vectors of the separate stylus excursions and feeds these separate signals into the left and right channels of the stereophonic reproduction system. As we said, the process is quite complex and is really a minor miracle that it works as well as it does.

How Stereo Is Broadcast. Radio was the last medium to take advantage of stereophonic reproduction. This is only natural, since by definition two transmitters and two radio receivers are needed for stereo broadcasting. You will recall that a number of stations (such as WQXR in New York) experimented with two-transmitter stereo broadcasting in the late

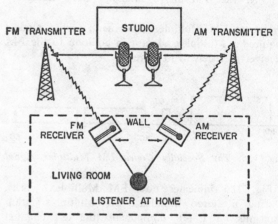

Fig. 183. *Setup for AM-FM Stereo Broadcasting*

1950s. As depicted in Fig. 183, the process involved the AM and FM transmitters of the station, and therefore became known as AM-FM stereo. One channel (the left, for example) was transmitted over the regular FM transmitter and was picked up by an FM receiver at home; the other channel (the right) was transmitted by the AM transmitter and picked up by an AM receiver. Although this works fine in theory, it is not very good in practice since the AM channel is inevitably more noisy and distorted than the FM channel. This leads to imbalances that tend to destroy the stereo illusion.

The solution consists of pure FM stereo (FM-FM) broadcasting, which became a reality in 1961. The ingenious system known as **FM Stereo Multiplex** actually uses only **one** FM transmitter to broadcast both the left and right channels. As shown in the illustration (Fig. 184), the two channels are combined in a predetermined manner in the Multiplex transmitter before transmission and are then separated and restored to their original patterns by the FM Multiplex receiver. This is done automatically

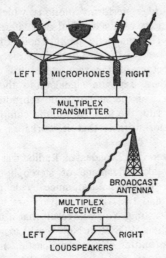

Fig. 184. *FM Stereo Multiplex Broadcasting*

by recent FM Multiplex equipment; earlier models required a **Multiplex adaptor** in addition to the FM receiver to receive FM stereo.

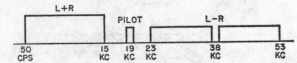

Fig. 185. *The Specially Coded FM Multiplex Signal*

Fig. 185 indicates how FM Multiplex works. During a stereo broadcast the Multiplex station transmits an $L + R$ component (the left channel added to the right), a specially coded $L - R$ component (the left channel minus the right), and a 19-kc **pilot signal** to synchronize the receiver. The FM Multiplex receiver picks up and detects all three components. (A conventional FM receiver detects only the $L + R$ component, which is monophonic.) Special filters then separate the $L + R$ and $L - R$ components. To obtain the **left** channel, the two components are **added** together (since $(L + R) + (L - R) = 2L$), while the **right** channel is recovered by **subtracting** the $(L - R)$ component from the $(L + R)$ component ($= 2R$). The left and right components are then separately amplified and reproduced.

SUMMARY

The response of the human ear to sound is **logarithmic:** to obtain equal increases in **loudness,** the **intensity** of the sound must go up by a factor of 10 each time. Sound intensity and loudness are measured in **decibels,** a logarithmic unit.

Loudness contours show that the response of the ear falls off at both low and high frequencies for sound reproduction at levels *lower* than the original.

The lows and highs must be **boosted,** therefore, when reproducing sound at lower levels than the original.

A **radio tuner** is a radio receiver without audio amplifier. A tuner should pass the complete audio bandwidth, should have high sensitivity and adjacent-channel selectivity, incorporate automatic frequency control (AFC), and have cathode-follower output.

A hi-fi **transcription record player** usually consists of a drive mechanism and aluminum-cast turntable for rotating records at **uniform** speed (without wow or flutter), a long, non-resonant tone arm with minimum tracking error and adjustable stylus pressure, and a **magnetic** phono cartridge with a diamond or sapphire stylus.

The **preamplifier** is the hi-fi control center that ties up the various components. It must provide voltage amplification and **tone equalization** for different recording characteristics, as well as loudness, bass, and treble controls.

The main or **basic amplifier** provides additional voltage amplification and sufficient power amplification to drive the speaker system. It should provide from 10 to 20 watts output with a total **harmonic** and **intermodulation distortion** of no more than 2%.

The dynamic loudspeaker translates an audio signal into sound waves through the interaction (repulsion and attraction) of the magnetic field of a **voice coil,** carrying the audio currents, with a strong magnetic field produced by a permanent magnet.

Since it is very difficult to design a single speaker to reproduce the entire audio band, **coaxial** or **triaxial** speakers with separate **diaphragms** or separate drivers, or completely separate **multiple-speaker systems** are employed. These consist generally of a **woofer** to reproduce the low frequencies, a **tweeter** to reproduce the highs, and sometimes also a **midrange** speaker for medium frequencies. **Crossover networks** must be provided to separate the frequency bands to be fed to each speaker.

Baffles and **loudspeaker** enclosures prevent the sound radiated from the front of the speaker from reaching the sound radiated from the back, which would result in cancellation of **low frequencies.**

The **flat, open baffle** works well, but is too large in size; the **infinite baffle** is an enclosed box that completely prevents the front-radiated sound from reaching the rear; it also raises the resonant frequency of the speaker.

The **bass-reflex enclosure** provides a **port** to release rear-radiated sound that has been inverted in phase. This augments the bass response of the speaker and makes it more uniform.

A **tape recorder** employs a ring electromagnet (recording head) to magnetize an iron-oxide

coated tape longitudinally in proportion to the amplitude and polarity of the audio currents flowing
through the winding of the electromagnet.

Stereophonic (binaural) reproduction requires two
completely separate channels to simulate the effect
of hearing with two ears.

Chapter Seventeen

TELEVISION

Television is one of those miracles which we have
come to accept as part of our everyday world without the slightest sense of mystification. It is taken
for granted that electronics can achieve seeing at a
distance (the literal translation of "television"), just
as we have long ago accepted hearing at a distance,
and are perfectly willing to let electronic "brains"
do our thinking for us. In other words we have become conditioned to the expectation that electronics
can do anything and we've stopped wondering. If
the following "explanations" can bring into focus the
magnitude of the marvel accomplished by television,
it may help to restore your sense of wonderment.

PHYSICAL BASIS

Long before the advent of television, the movies
had taken advantage of the **persistence of vision** of
the human eye to "deceive" us into seeing motion,
when there was none. As every schoolboy knows,
the movies display a series of **still** pictures in rapid
sequence, each picture or **frame** showing a slightly
more advanced phase of the "continuous" action.
When this is done more often than 16 times per
second (in amateur movies; 24 times in professional
movies) the eye is no longer capable of separating
the individual pictures because of its persistence of
vision, and we obtain the impression of a smoothly
blended, continuously progressing motion. Television
uses this same "deception" of conveying moving pictures by sending a rapid series of changing still pictures. Although the motion of an actual scene adds
to the complications, the basic problem of television
really is the transmission and reception of a **still
picture.** Let us see how this is done.

When we look at an actual scene we see a contin-
uum of light and shade, and colors of various wavelengths. This is no longer true when we look at a
(black-and-white) photograph of the same scene.
The photographic print has a limited (though huge)
number of fine silver grains, each being "developed"
to a brightness corresponding to that of the same
spot in the scene. By distributing a tremendous number of these silver grains of varying brightness over
the picture area, the correct proportions of light and
shade in the actual scene are reproduced in the

image. You cannot see the little grains or dots in
the picture, because there are so many of them, but
when the picture is greatly enlarged they become
visible. Moreover, when a photographic print is
"screened" for reproduction in books or newspapers
(a **photoengraving),** the image is broken down into
a much smaller number of **picture elements** of varying light and shade than the fine grain of the original
print, and then these picture elements become
clearly visible. By looking at a newspaper picture,
which employs a fairly rough, clearly visible screen,
you will discover that the picture is actually composed of many black dots, the dark areas containing
large, closely spaced dots, while the light areas consist of smaller, more widely separated dots. Photographic reproductions in books have a finer screen
and you may have to look at them with a magnify-
ing glass to discover the picture dots. The dot structure of an enlarged portion of a picture that has
been screened is shown in Fig. 186.

Fig. 186. *Dot Structure of Enlarged Portion of
Screened Picture*

Another example further illustrates that images
may be composed by assembling a large number
of individual picture elements, or dots. Fig. 187a
shows the outline of a cross composed of relatively

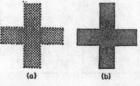

(a)   (b)

Fig. 187. *Cross Composed of Few Dots (a) Blends into
a Solid Figure, When Many More Dots Are Added (b)*

few black dots with white spaces between them. You are not fooled by this, since the dots are clearly evident. However, if you walk about ten feet away from this crude picture, the dots will appear to blend into a solid grayish figure. In (b) of Fig. 187 we have heightened the illusion by providing many more black dots with fewer white spaces between them. The cross now appears to be a solid, gray figure, even at an ordinary reading distance. You must look closely to discover the separate dots.

The basic problem of television now becomes evident: it must break down a distant scene at the transmitter into many small picture elements of varying brightness, send these out **in sequence** via radio waves, an then reassemble all the elements at the receiver in their proper sequence to create a replica of the original picture. There must be a sufficient number of elements and they must be transmitted so fast, that the eye can neither detect their presence nor the process of reassembly. Moreover, a sufficient number of complete images must be sent each second so that the persistence of vision of the eye will blend them into continuous motion. When you think of the tremendous number of picture dots required to make an image and the large number of images to be sent each second (30 per second in television), you will realize that the time alloted to form each picture element is in the order of *millionths of a second*. Only electrons can carry out a task that speedily.

COMPLETE TELEVISION SYSTEM

Fig. 188 is a simplified presentation of a complete television system for the transmission and reception of picture and sound signals. The television station sends out **two separate r-f carriers** over a single antenna, one carrier being **frequency-modulated** by the sound (audio) signal, while the other is **amplitude-modulated** by the picture information or **video signal.** The two carriers are spaced 4.5 mc apart.

At the television transmitter the picture and sound signals are handled separately. The television camera focuses an optical picture of the scene upon an electronic camera tube, which **scans** or breaks down the image into its picture elements and converts the varying brightness of the individual elements into a corresponding electrical, or **video** signal. It also adds several **synchronizing** signals to the video information, which are designed to keep the reassembly of the picture at the receiver in step with the scanning at the transmitter. This **composite video** signal is then strengthened by a number of **video amplifiers** (see Chapter 9) to a level sufficient to amplitude-modulate a radio-frequency transmitter. The carrier with its video modulation is sent out over the TV transmitting antenna.

The sound portion is a conventional FM transmitter. The sound picked up by the microphone is strengthened by an audio amplifier, which frequency-modulates an r-f transmitter with a carrier (center) frequency 4.5 mc *above* the video carrier. The frequency-modulated sound carrier is sent out over the same transmitting antenna used for the video carrier.

The television receiver, too, is a combination of the old and the new. The sound and video r-f signals picked up by the receiving antenna are handled at first *together* by conventional superheterodyne receiving circuits. The desired television channel is selected by tuned circuits and the sound and video signals are strengthened together by a radio-frequency amplifier with sufficient bandwidth to pass both carriers and their modulation sidebands. The r-f signal then heterodynes in the mixer with a locally generated frequency to produce a lower intermediate frequency equal to the difference between the two signals (usually 45.75 mc for the pic-

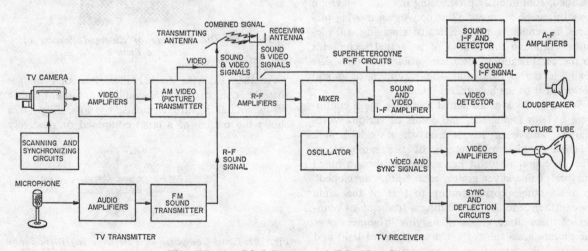

Fig. 188. *Simplified Diagram of Television System*

ture). The sound and video i-f signals are amplified by several stages of i-f amplification and then applied to a **video detector.**

The video detector has two functions: 1. it **demodulates** the composite video signal by means of a diode detector, just as is done in an AM broadcast receiver; 2. it **separates** the sound and video i-f signals. The separation of sound and video is accomplished by beating together (heterodyning) the frequency-modulated sound i-f signal and the amplitude-modulated video i-f signal, which are spaced 4.5 mc apart, you will recall. Because of the detector's partially non-linear characteristic it performs this mixing function automatically. The heterodyning produces a 4.5 mc **frequency-modulated** difference frequency, which is the **sound i-f signal.** Filter circuits in the output of the detector separate this 4.5 mc sound i-f signal from the demodulated composite video signal.

The sound i-f signal is applied to the separate sound portion of the receiver, which is identical to the corresponding circuits in an FM broadcast receiver. The sound signal passes in succession through an i-f amplifier, a limiter and discriminator, or a ratio detector, one or two stages of audio amplification and a loudspeaker.

The demodulated composite video signal from the output of the video detector is applied to the video portion of the receiver. The video signal is amplified by a video amplifier and then reassembled by an electronic beam into a visible image on the face of the picture tube or **kinescope.** The composite video signal is also fed to a "sync" separator where the synchronizing signals are separated from the remainder of the video signal. The sync signals are then applied to the beam deflection circuits to keep the electronic beam that reassembles the image on the picture tube in step with the scanner at the transmitter.

The TV receiver discussed here and indicated in Fig. 188 is known as the **intercarrier** type because of the way the sound i-f signal is obtained by heterodyning the video and sound carriers. There is also an older type of TV receiver, called the **split-sound** receiver, where the sound signal is split off at the mixer and then handled completely separately. This receiver is no longer in use.

It is evident from Fig. 188 that we are already familiar with the sound FM transmitting and receiving circuits (see Chapter 15), video amplifiers (Chapter 9), and the superheterodyne receiving circuits for sound and video (Chapter 15). Let us concentrate, therefore, for the remainder of the chapter on the new and unfamiliar portions, such as the scanning, synchronizing, and deflection circuits, the TV camera and the picture tube.

TELEVISION CAMERAS

The video signal begins its long journey from the TV transmitter to the picture tube in the receiver at the TV camera. The camera must "see" the actual scene to be televised and convert the optical image of the scene into an equivalent electrical image. The picture elements of this electrical image must then be "scanned" to provide a video signal whose instantaneous magnitude corresponds to the brightness of the individual elements. A typical television camera, called the **iconoscope,** is shown in schematic form in Fig. 189.

In brief, the action of an iconoscope is as follows. Light from the illuminated scene (an arrow,

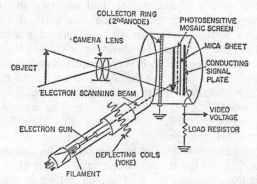

Fig. 189. *Elements of an Iconoscope Camera Tube*

in this case) is focused by means of optical lenses onto a **photosensitive** screen, called the **mosaic.** The mosaic is a coating of millions of light-sensitive silver cesium-oxide globules deposited on one side of a thin sheet of mica. Each photosensitive globule is about one-thousandth of an inch in size and is insulated from all neighboring globules by the mica. The other side of the mica sheet, the **signal plate,** is coated with a conducting film of graphite. The globules insulated by the mica from the metallic coating form myriads of tiny electric **capacitors,** all having the mica dielectric and the metallic signal plate in common. Each light-sensitive globule, therefore, emits electrons and charges up its individual capacitor in accordance with the intensity of the light striking it. (Since electrons are emitted or lost, each capacitor is charged positively.) The entire mosaic plate, thus, has a charge distribution corresponding to the variations in light and shade of the original picture. The upshot is that the mosaic plate stores in its charged globules an electrical image of the optical picture focused upon it.

Obtaining a Video Signal. The electrical image stored on the mosaic screen cannot be transmitted as a whole, but the individual picture elements must be scanned one at a time by discharging the globule-capacitors in an orderly sequence. This is accomplished by an electron scanning beam formed by the electron gun in the narrow elbow of the tube. The action of this electron gun is identical to that of a conventional cathode-ray tube, which we discussed in Chapter 6. The gun contains an electronic lens system of charged electrodes, which produce a sharply focused electron beam. This beam is aimed at the mosaic through the attraction of the highly positive (about 1000 volts) second anode, which consists of a metallic coating on the inside of the glass tube, known as **collector ring.** Horizontal and vertical deflecting coils, mounted at right angles in a yoke around the neck of the tube, provide magnetic deflection of the electron beam to scan the electrical image on the mosaic. As we shall see later, this is done in an orderly fashion from left to right and top to bottom of the mosaic, one line at a time.

The actual process of obtaining a video voltage by scanning the mosaic is very complex, but you can visualize it in simplified form as follows: When the scanning electron beam strikes each globule, the electrons fill in the "holes" left by the previous photoelectric emission of electrons. The beam thus neutralizes the previous positive charge due to photoemission and, in effect, discharges the globule-capacitor. At the instant of discharge a rush of current flows through the load resistor, which is equal to the positive charge stored on the globule and, hence, is proportional to the light illumination of the picture element represented by the globule. This discharge current flowing through the load resistor builds up the **video voltage,** which is fed to the succeeding video amplifier. As the entire mosaic is scanned, the electrical image stored on it is converted successively into a video voltage of varying instantaneous magnitude, which corresponds to the illumination on the individual globules.

Image Orthicon. The video output of the iconoscope is rather low and it requires a brightly illuminated picture to be useful. The iconoscope has been largely replaced, therefore, by another camera tube, the **image orthicon,** which is far more sensitive and can televise anything that is visible with the naked eye. The image orthicon owes its exceptional sensitivity to the **electron multiplier** action of a series of secondary-emission electrodes, or **dynodes.** We have already discussed the electron multiplier in Chapter 6, and other features of the orthicon are similar to those of the iconoscope. There is a third camera tube, known as **vidicon,** which is simpler than the other cameras, but it provides less fineness of detail (resolution) and, hence, is used primarily for televising movie film.

SCANNING AND SYNCHRONIZING

We appreciate by now the need for scanning and understand in a general way how it is done, but we have not yet explored the manner of deflecting an electron beam to obtain the desired **scanning pattern,** and how "sync" signals are used to keep the transmitter and receiver scanning exactly in step. Evidently, if an image is to be assembled on the picture tube of the TV receiver at the same time as the picture elements of the actual scene are being televised, the scanning at the transmitter and receiver must be done in exactly the *same manner* and in perfect synchronism.

Progressive Scanning. Scanning may be carried out in the same way as reading a page in a book. You start at the top, read all the words in the first line from left to right, then return rapidly to the left to read the next line, and so on, until you arrive at the bottom of the page. **Progressive horizontal scanning,** illustrated in Fig. 190, is done in this fashion. As shown in (a), the electron beam at the transmitter and receiver is made to sweep across a horizontal line, covering all the picture elements, and is then quickly returned to the left to scan the next line. Both camera and picture tubes are **blanked out** during this **horizontal retrace period** (shown dashed in Fig. 190) to make the retrace lines invisible. The retraces must be very rapid, of course, in order not to lose valuable picture information. As each horizontal line is scanned, the position of the beam must be progressively **lowered** so that the same line will not be repeated. This is accomplished by a **vertical scanning motion** of the beam from top to bottom, which is superimposed upon the horizontal scanning motion. Moreover, after the scanning beam completes the last, bottom line, it must be rapidly returned to the upper left-hand corner, as shown in (b). This motion is called the **vertical retrace** to distinguish it from the **horizontal retrace** between each line. The vertical retrace is also blanked out. To obtain the maximum amount of picture detail, called **resolution** or **definition,** there should be as many horizontal lines as possible for each image. Because of various practical considerations, the number of horizontal lines has been standarized at a total of 525 per image or **frame.**

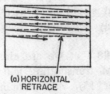

Fig. 190. *Progressive Horizontal Scanning* (*a*) *and* (*b*)

Interlaced Scanning. To produce the illusion of motion the individual still pictures or frames must be displayed sufficiently rapid so that the persistence of vision of the eye will blend them smoothly. In televison, the **frame repetition rate** has been standardized at 30 per second. Despite this high repetition rate, a certain amount of annoying **flicker** is still present. To eliminate this residual flicker, a trick familiar from the movies is employed: **each frame is shown twice** on the face of the picture tube. To accomplish this, a scanning pattern slightly different from progressive scanning must be used. Fig. 191

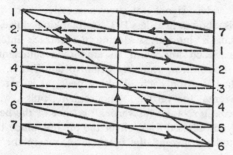

Fig. 191. *Interlaced Horizontal Scanning*

illustrates the **interlaced horizontal scanning pattern,** which divides the total number of lines into two groups of lines, called **fields.**

As you can see, during the presentation of the first field only the odd-numbered lines are scanned, while during the second field all even-numbered lines are scanned. Halfway along the bottom line of the first field (line 7 in the figure), the **vertical retrace** returns the scanning beam to the top of the image and completes the unfinished lines. The remainder of the even-numbered lines are then scanned during field 2. For clarity only seven lines are shown in Fig. 191. Actually, of course, each field consists of exactly one-half of the 525-line total per frame, or 262.5 lines. Since two fields are shown for each frame, the repetition rate of the **fields** is 60 per second, or twice that of the frame repetition rate. Being equal to the a-c power-line frequency in the United States, the 60-cycle scanning frequency simplifies the design of the receiver and transmitter power supply filters.

Note that the scanning patterns illustrated in Figs. 190 and 191 always have the same ratio of picture width to picture height. This ratio of width to height is called the **aspect ratio** and it has been standardized at 4:3; that is, the width is 1.33 times height of the picture frame. This ratio remains the same, whether the actual picture is 2 ft. × 1½ ft. or possibly 20 ft. × 15 ft.

How Scanning Is Obtained. With 525 horizontal lines included in each frame and 30 frames being scanned per second, the total number of horizontal lines scanned per second is 525 × 30, or 15,750 lines per second. The horizontal repetition rate, or **line scanning frequency,** is thus 15,750 cps. Similarly, the **frame scanning frequency** is 30 cps, and since each frame is scanned in two fields, the **field scanning frequency** is 30 × 2, or 60 cps.

Scanning is accomplished by deflecting electron beams in synchronism at the transmitter and receiver. To do this, a **deflection voltage or current** must pull the electron beam horizontally across the tube and vertically up and down at the line and field scanning frequencies, respectively. (A deflection **voltage** is used for electrostatic deflection while a *current* must be used for magnetic deflection.) The amplitude of this voltage or current must rise **linearly** to provide **equal increments in amplitude for equal intervals of time.** At the end of each scanning line or field, the deflection voltage (or current) must snap back to zero in minimum time; that is, the **retrace** must be as short as possible. A waveform that provides this relatively slow linear rise in amplitude and rapid retrace to zero has the appearance of the teeth of a saw and, hence, is called a **saw-tooth waveform.** The saw-tooth scanning waveforms for horizontal and vertical scanning (called **sweep**) are shown in Fig. 192.

Note that the appearance of the saw-tooth scanning waveforms is the same for horizontal and vertical sweep and that only the time relationships have been changed. The time "H" allowable for sweeping a horizontal line (Fig. 192a) is 1/15,750 second, or about 63.5 **microseconds.** This *includes* the retrace, which takes about .15 *H* of the total time, or about 9.5 microseconds. In contrast, the time "V" permis-

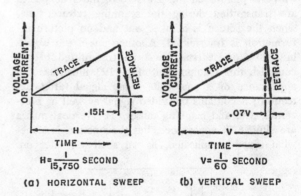

(a) HORIZONTAL SWEEP   (b) VERTICAL SWEEP

Fig. 192. *Saw-Tooth Scanning Waveforms for Horizontal (a) and Vertical Sweep (b)*

sible for sweeping a vertical field (Fig. 192b) is considerably longer, being equal to 1/60 second. The retrace for vertical sweep is only about .07 *V*, or 7% of the trace time.

The required linear saw-tooth scanning waveforms may be obtained either through the approximately linear rise in the **current through an inductance** or the linear rise in **voltage across a capacitor** that is being charged. Actually, neither of these waveforms is quite linear, but is curved in an "exponential" manner. One can get around this by using only a small portion of the exponential waveform, which is approximately linear. Thus, by only *partially* charging a capacitor through a very high resistance, the voltage across the capacitor will build up approximately linear in the manner of a saw-tooth waveform. Then, without waiting for the capacitor to become fully charged and acquire its exponential charging curvature, the capacitor is suddenly discharged through a relatively low resistance and its voltage drops quickly to zero, as required by the retrace portion of the saw-tooth waveform. In practice, the rapid switching from charge to discharge is accomplished by an electron tube oscillator with the aid of the synchronizing signals, as we shall see presently.

SYNCHRONIZATION

To maintain the correct timing of the vertical and horizontal sweep motion and keep the receiver and transmitter locked in step, synchronizing signals must be sent out along with the picture information. These **sync** signals, which have the form of rectangular pulses, are sent out periodically at the line and field scanning frequencies. To keep the horizontal line scanning synchronized, a horizontal sync pulse is transmitted for each horizontal line; hence, the **horizontal sync frequency is 15,750 cps.** Similarly, a vertical sync pulse is sent out for each field to keep the vertical scanning in step; the **vertical sync frequency, thus, is 60 cps.** To prevent interference with the picture on the TV screen, the sync pulses are transmitted during the scanning **retrace time,** when the screen is **blanked out** and no picture information is transmitted. Another pulse, the **blanking pulse** or **pedestal,** must be transmitted to accomplish this blanking out. Fig. 193 illustrates the appearance of a **composite video signal** (after detection); it contains the video signal as well as horizontal sync and blanking pulses. The vertical pulses are similar in appearance, but their action is somewhat more complicated, as we shall see later on.

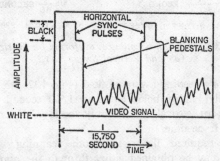

Fig. 193. *Appearance of Composite Video Signal*

Note, first of all, that the *lowest* values of the signal amplitude correspond to the brightest parts of the picture (white), while the highest amplitudes correspond to the darkest parts, or black. This is known as inverse or **negative modulation,** since the instantaneous video signal amplitude is **inversely propor-**

tional to the light intensity. The chief reason why this system has been adopted is its relative freedom from noise interference. With negative modulation, a strong noise pulse will simply blank out the screen momentarily (and probably unnoticeably), while with positive modulation a similar pulse will cause annoying light streaks across the receiver screen.

The video signal for two successive horizontal lines is shown in Fig. 193. A blanking pulse, or **pedestal,** is interposed during the retrace interval between each line, with an amplitude sufficient to drive the electron scanning beam to cutoff (i.e., into the black). Atop each blanking pedestal is perched a horizontal (or vertical) synchronizing pulse, which **starts the horizontal (or vertical) retrace of the beam.** The sync pulses do not affect the picture, since their amplitude is **beyond the black level,** where the picture is blanked out.

Sync Separation. To utilize the sync pulses for controlling the timing of the beam deflection (sweep) oscillator, they must first be **separated** from the composite video signal. This is done by a **sync separator** circuit, which shaves or clips off all signals **below the blanking level,** so that only the sync signals are permitted to pass. A typical triode **base clipper** for accomplishing this, is shown in Fig. 194.

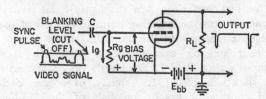

Fig. 194. *Action of Triode Base Clipper*

The triode derives its bias from a grid-leak resistor, Rg, placed in its grid circuit. When the composite video signal is applied through capacitor C to the input of the tube, the grid is driven positive and a grid current flows through Rg, which makes the grid-connected end **negative** with respect to the cathode. The value of the grid resistor is chosen so that the bias voltage developed across it drives the tube to cutoff at the **blanking level,** as shown. As a result the tube is cut off at all times, except during the positive sync pulses, whose amplitudes are larger than the blanking level. The sync pulses permit a momentary flow of plate current, indicated by the **negative** output pulses across R_L in the plate circuit. The polarity of the sync output pulses doesn't matter, since pulses of either polarity may be used to control the timing of the sweep oscillator. If necessary, the polarity can be reversed by means of a **phase inverter** circuit.

Synchronizing Pulses. The appearance of the sync pulses after separation form the composite video signal is illustrated in Fig. 195. The pulses have been

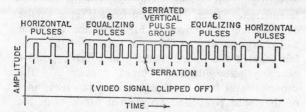

Fig. 195. *Vertical and Horizontal Synchronizing Pulses*

made positive in polarity by means of a phase inverter.

The illustration shows from left to right: 1. three **horizontal** sync pulses for the **last three** horizontal lines of a field; 2. six **equalizing pulses,** somewhat smaller in width than the horizontal pulses; 3. a prolonged **vertical pulse group,** consisting of six individual pulses separated by five **serrations;** 4. another group of six equalizing pulses; and 5. three more horizontal pulses for the **first three** horizontal lines of the next field (second field of a frame). To complete the illustration you would have to add another 522 horizontal pulses, after which the sequence of equalizing and vertical pulses would be repeated. Note that all pulses in Fig. 195 have the same amplitude, but that they differ in width, or the length of time each lasts. Moreover, each vertical pulse extends over a period of three horizontal lines, the serrations being inserted at half-line intervals.

Though not immediately apparent, there are good reasons for this peculiar arrangement of horizontal, vertical and equalizing pulses. The vertical pulse is

made long so that it can easily be separated from the multitude of horizontal pulses. The serrations are inserted into the vertical pulse, so that horizontal synchronization can be maintained over the three-line interval, during which the vertical beam retraces. If horizontal sync is not maintained during the vertical retrace, the interlacing of the scanning pattern may become distorted. Finally, the equalizing pulses are provided to smooth out the differences between vertical sync signals for alternate fields. The scanning for the even and odd fields of each frame must be out of phase by one half-cycle to provide the required interlacing feature. Nevertheless, the vertical sync signal must be the same for both fields. The spacing of the equalizing pulses at half-line intervals assures that horizontal synchronization can be maintained, alternate pulses serving to synchronize the horizontal scanning of even- and odd-numbered fields.

Separation of Vertical and Horizontal Sync Signals. Let us see now how the vertical sync signals are separated from the horizontal ones. Fig. 196 shows what happens.

Part (a) of the figure is a repetition of Fig. 195, showing the synchronizing pulses after separation from the composite video waveform. In (b) the sync pulses have been applied to an electrical circuit, which simulates the mathematical process of **differentiation.** As we shall see shortly, a **differentiating circuit** produces an output waveform that responds only to the **rate of change** of the input waveform. As a result, output pulses from a differentiating circuit occur only at the beginning and end

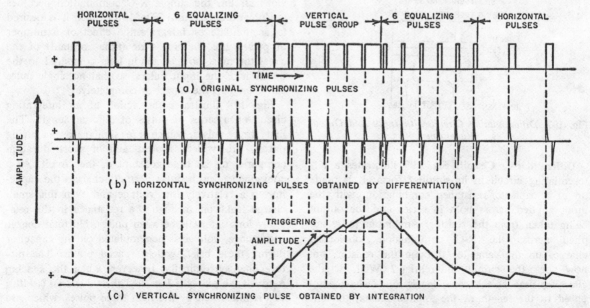

Fig. 196. *Separation of Horizontal and Vertical Sync Pulses: Original Pulses* (*a*); *Horizontal Pulses Obtained by Differentiation* (*b*); *Vertical Pulse Obtained by Integration* (*c*)

of a rectangular waveform, where the waveform changes abruptly. At the beginning of any horizontal, vertical or equalizing pulse, therefore, there will be a short **positive** pulse, signifying that the amplitude of the waveform changes in the positive direction. Similarly, at the end of each sync pulse, regardless of length, there will be a short **negative** output pulse from the differentiating circuit, signifying that the pulse amplitude is abruptly **decreasing** or **going negative.** The upshot is that the differentiating circuit will produce a series of positive and negative output pulses, at the beginning and end of each sync pulse. The **positive** output pulses from the differentiating circuit are used to maintain horizontal synchronization at all times.

In (c) of Fig. 196 the sync pulses have been applied to an **integrating circuit,** which performs the **inverse** of differentiation. During integration the total area (height times width) occupied by the pulses is added up, regardless of their individual shape or rate of change. However, as we shall see, the integrating circuit can perform this addition only for fairly wide, closely spaced pulses, such as the vertical pulse group. For brief, widely spaced pulses, such as the horizontal and equalizing pulses, no addition takes place. As a result, only the vertical sync pulses produce an integrated sync pulse of sufficiently large amplitude for "triggering" the vertical sweep oscillator to initiate vertical scanning.

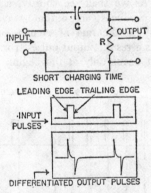

Fig. 197. *Differentiating Circuit with Input and Output Waveforms*

Differentiating Circuit. Fig. 197 illustrates a differentiating circuit in its simplest form. It is made up of a capacitor-resistance combination, with the input applied across both in series and the output being taken from the resistor. Any input pulse applied across the R-C combination, obviously, charges up the capacitor through the resistor, but how does the circuit differentiate? Well, if the **charging time of the capacitor is made short compared to the length of the applied pulses,** the circuit will respond only to **changes** in the pulse

waveform, thus fulfilling the requirement for differentiation. Thus, if a horizontal sync pulse comes along, as shown in Fig. 197, the capacitor will charge rapidly to the peak voltage of the **leading edge** of the pulse, resulting in a brief, **positive charging pulse** through output resistor R. The capacitor will hold its charge during the remainder of the input pulse and no current flows, until the **trailing edge** comes along. With the input voltage suddenly removed, the capacitor discharges rapidly through $R,$ in the **opposite** direction. Thus, a **negative discharge pulse** flows through R during the trailing edge of the input pulse.

The result of the differentiating action is a series of brief positive and negative output pulses, respectively, for the leading and trailing edges of the input sync pulses, *regardless of their length.* The negative output pulses are rejected or clipped off, and the positive pulses are used to synchronize the horizontal sweep oscillator. It thus becomes evident that even the small breaks (serrations) in the vertical pulse group are sufficient to produce output pulses, which will permit the horizontal oscillator to stay in step. The only condition which must be fulfilled is to keep the charging time of the capacitor short compared to the pulse length. This can be attained by choosing the values of the resistor and capacitor so that their product, known as **time constant,** is small compared to the pulse length in microseconds. (The time constant, obtained by multiplying the resistance in ohms by the capacitance in *micro*farads, expresses the approximate charging time in **microseconds.)**

Integrating Circuit. By taking the output from the capacitor rather than the resistor of the previous circuit, the simple R-C combination becomes an **integrating circuit.** (Fig. 198.) Since it is desired to accumulate or **integrate** the effect of a number of pulses, the **charging time** (time constant) of the capacitor must now be made **long** compared to the duration of the input pulses, so that no single pulse can charge up the capacitor completely.

Fig. 198 illustrates the action of an integrating circuit on a series of pulses of different length. The first three pulses, which are similar to equalizing pulses, are widely spaced and of short duration compared to the time constant of the circuit. As a consequence, each pulse *partially* charges the capacitor to a relatively low voltage and even this small charge leaks off through the resistance in the relatively long intervals between pulses. No total charge and, hence, voltage is accumulated on the capacitor during these brief, widely spaced pulses. The picture changes radically, however, when the **spacing between the pulses is less than their duration (width);** this is illustrated by the next three pulses, which are similar to the vertical sync group. Because of its

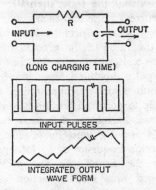

Fig. 198. *Integrating Circuit with Input and Output Waveforms*

longer duration each input pulse now charges the capacitor to a considerably higher voltage; moreover, very little of the accumulated charge can leak off through the resistor during the brief time intervals between pulses. The result is a steady accumulation of charge, or voltage, until the next series of brief, widely spaced pulses arrives; these initiate a progressive discharge of the capacitor. Before this happens, however, the output voltage across *C* has reached a sufficient amplitude to **trigger** the vertical sweep oscillator into operation. (See Fig. 196c.)

We are now able to understand the effect of the equalizing pulses in smoothing out the differences in the vertical sync signals between alternate fields. You will recall that the horizontal sync pulses of alternate (even- and odd-numbered) fields must be spaced **one half-cycle out of phase** with each other to obtain the required **interlaced** scanning pattern. (See Fig. 191.) Because of this phase difference the spacing between the last horizontal sync pulse and the first equalizing pulse also differs by one half-cycle. Thus, the capacitor of the integrating circuit starts to charge up a little earlier during the equalizing pulses of one field than for

the other. If it were not for the equalizing pulses, the triggering voltage that initiates the vertical sweep would be reached a little sooner for one field than for the other, which would result in a poorly interlaced pattern. The equalizing pulses act as a "buffer" between the last horizontal and first vertical sync pulse to compensate for this difference in timing. Though the capacitor accumulates a charge a little earlier during one field, this does not matter, since the equalizing pulses **lose more charge than they contribute,** so that any excess charge initially present is dissipated before the first vertical pulse comes along. The integrating action for the vertical pulses, hence, starts with a clean slate and the triggering amplitude is reached at the same time during alternate fields.

SWEEP OSCILLATOR

Let us now look briefly at a popular oscillator circuit that is used to sweep the electron scanning beam of a picture tube, either horizontally or vertically. The **blocking oscillator** illustrated in Fig. 199 is used for vertical deflection, but by changing the component values it becomes equally suitable for horizontal sweep.

The blocking oscillator is essentially a simple electron tube oscillator of the type discussed in Chapter 11, but it has a few gimmicks added to it, to make it suitable as a linear deflection (sweep) oscillator. Let us first see why it oscillates. As you can see the plate circuit of the triode is coupled back to the grid circuit through the transformer, whose windings are common to both circuits. Because of the 180-degree phase reversal occurring within the tube, the plate voltage is out of phase with the grid input voltage. However, as in any transformer, another 180-degree phase inversion takes place between the primary and secondary of the transformer. As a result of this double phase

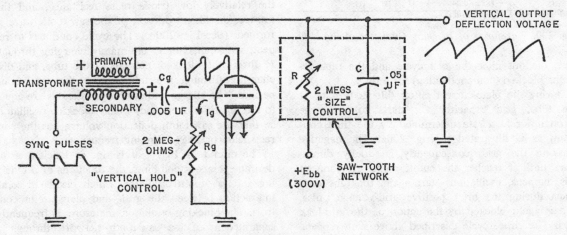

Fig. 199. *Typical Blocking Oscillator for Vertical Deflection*

reversal, the plate voltage coupled back to the grid circuit is **in phase** with the grid voltage and produces **positive** or **regenerative feedback.** With the large amount of positive feedback coupled in by the transformer, the circuit will immediately break into oscillations, as we have explained in Chapter 11. These oscillations would ordinarily continue at a frequency determined by the inductance of the transformer and the stray capacitances. Actually, however, the circuit oscillates only for a single cycle and then "blocks" for a time, as its name implies.

To understand why the circuit of Fig. 199 blocks, the effect of the grid-leak bias has to be taken into account. As you can see no fixed bias is provided, the grid-leak combination Rg-Cg taking its place. The triode is, thus, initially without bias. As soon as oscillations begin, the first positive half-cycle drives the grid **highly positive,** resulting in considerable grid-current flow through grid resistor Rg. The direction of this grid current (i_g) is such that a large **negative bias voltage** is developed across Rg, which drives the tube far into the plate-current cutoff region. (See Fig. 200a.) The negative bias voltage also charges up grid capacitor Cg, which holds its charge for a period depending on the **time constant,** Rg × Cg. The values of Rg and Cg are chosen so that the bias is sufficient to prevent succeeding positive half-cycles of the oscillation from driving the grid above cutoff. The oscillations, there-

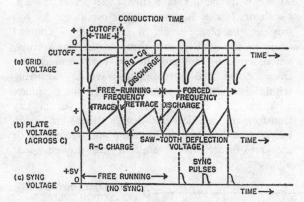

Fig. 200. *Waveforms of Blocking Oscillator of Fig. 199*

fore, die out after the first cycle and the tube remains blocked (non-conducting) for a time.

During the plate-current cutoff (blocked) state of the tube, grid capacitor Cg starts to discharge through Rg at a rate determined by the time constant, as is illustrated in Fig. 200a. The negative bias on the grid, consequently, gradually diminishes until it reaches the cutoff point. As soon as this happens, oscillations start again, the tube conducts during the brief positive grid-voltage pulse, and is again blocked by the action of the grid-leak bias. The entire cycle described above then repeats.

The upshot is that the tube alternately conducts a short plate-current pulse and then cuts off or blocks, as shown by the waveform of Fig. 200a. As you can verify, the value of the product of Rg and Cg (i.e., the time constant) in Fig. 199 has been chosen to obtain a **pulse repetition frequency** in the 60-cps range, as required for vertical scanning. Grid resistor Rg is variable to permit adjusting the repetition frequency. For a **horizontal deflection oscillator,** the values of Rg and Cg must be selected to produce a pulse repetition frequency of 15,750 cps.

Producing the Saw-Tooth Sweep Voltage. We have described thus far the action of the basic blocking oscillator. To adapt this circuit (Fig. 199) for deflecting an electron beam, an R-C network must be added in the plate circuit, which produces the required saw-tooth voltage. Note that the R-C network in the plate circuit of Fig. 199, consisting of a 2-megohm variable resistor and 0.05-microfarad capacitor, has a time constant that is *ten times* that of the grid circuit network Rg-Cg. The large time constant (corresponding to about 10 cps) is chosen, so that plate capacitor C will charge only to a fraction of its full charge during each grid-voltage cycle. As a result, the capacitor operates only over a small, approximately **linear** portion of its exponential charging curve and, thus, produces a **linear trace.**

The action of the R-C saw-tooth network is simple. Whenever the blocking oscillator is cut off, the capacitor is charged through "Size" control R by the positive plate-supply voltage, Ebb. This forms the linear deflection voltage or **trace,** illustrated in Fig. 200b. The amplitude of the deflection voltage, and hence the **vertical picture size,** is adjustable by the variable "Size" control, which controls the amount of charging current into C. (A similar "Width" control for adjusting the horizontal picture width is present in a horizontal deflection oscillator.) As soon as the tube conducts during the positive grid-voltage pulse, capacitor C discharges rapidly through the relatively low triode plate resistance, and the retrace (discharge) portion of the saw-tooth wave is formed. (See Fig. 200b.) The cycle continues to repeat, with capacitor C alternately charging through R during the blocked intervals of the tube, and discharging through the tube during the conducting periods. The charge and discharge intervals are equal to the repetition frequency of the blocking oscillator, so that the saw-tooth deflection voltage has the correct field (or line) scanning frequency.

The circuit of Fig. 199 is not as efficient as is desirable, since it combines the functions of a blocking and saw-tooth oscillator, which results in some interaction between the grid- and plate-circuit controls. The blocking oscillator, therefore, is frequently separated from the saw-tooth network through a separate **discharge tube.** The discharge tube is another triode, whose grid and cathode is connected in parallel with the blocking oscillator tube, and the saw-tooth network is inserted into the plate circuit of the discharge tube. Since the grid of the discharge tube is connected to that of the blocking oscillator, it has the same waveform (Fig. 200a), and hence is switched on and off at the frequency of the blocking oscillator. The plate-circuit action of the discharge tube is identical with that described for the single-tube oscillator. The blocking oscillator and discharge tubes are usually combined in the single envelope of a dual-triode, such as a type 6SN7.

Obtaining Synchronization. Although the **free-running frequency** of the blocking oscillator is controlled by its grid-circuit network (Rg-Cg), this frequency is neither exact nor stable, and hence some means must be provided to **lock the oscillator exactly in step** with the scanning frequency of the TV transmitter. This job is done by the sync signals, as you will recall. In practice, the **positive** sync pulses are applied in series with the grid winding of the transformer (Fig. 199), so that their amplitude **subtracts** from the **negative** bias voltage produced by the oscillator. The triggering action is illustrated in Fig. 200c. Here the first vertical sync pulse comes along at a time when the negative bias voltage has declined considerably, but has not as yet reached the cutoff value. The slight additional positive voltage (about 5 volts) of the sync pulse is sufficient to raise the tube above cut-

off and **trigger the next cycle** of the oscillator. The oscillator then goes through its complete charge and discharge cycle, until the next sync pulse again triggers it into action, just before reaching the cut-off bias value. Note that the sync triggers always occur slightly ahead of the time when the oscillator would start up by itself and, hence, the **forced frequency** due to sync action is **somewhat higher than the free-running frequency** of the oscillator. To attain the proper timing of the sync pulses, therefore, the free-running frequency of the oscillator must be **set** at a value **lower** than **the sync frequency.** The "Vertical Hold" control *Rg* in the grid circuit permits adjusting the free-running oscillator frequency so that it will **lock in** with the arriving sync pulses.

Note also, in Fig. 200, that only **positive** sync pulses can raise the grid of the tube above cutoff. The **negative** horizontal sync pulses produced by differentiation (see Fig. 196b) are, therefore, **automatically rejected** by the horizontal blocking oscillator. As a matter of fact, since the horizontal line scanning frequency of 15,750 cps must be held extremely close, the horizontal sync circuits are somewhat more complicated than those for the vertical deflection oscillator. We shall go into this matter shortly.

PICTURE TUBE AND ASSOCIATED CIRCUITS

The crux of any television system is the **picture tube,** which displays the video information sent out

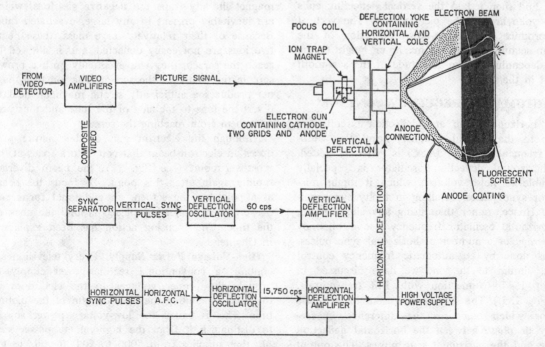

Fig. 201. *TV Picture Tube and Associated Circuits*

by the TV transmitter. The picture tube in practically all modern television receivers is a **magnetic-deflection cathode-ray tube,** which we have discussed at length in Chapter 6. (See Fig. 35.) We need not go into details, therefore, of how the horizontal and vertical deflecting coils in the **yoke** around the neck of the tube produce the proper scanning of the beam produced by the **electron gun.** Instead let us pick up some loose ends connected with the picture tube and its associated circuits. Fig. 201 shows in block-diagram form the video and deflection signals applied to the tube and the circuits generating them.

Scanning Raster. Before any picture information can be displayed on the face of the picture, a picture frame or **scanning raster** must be provided, which corresponds to the interlaced scanning pattern illustrated in Fig. 191. This is accomplished by the synchronizing and deflection circuits we have described in detail. The demodulated **composite video signal** from the video detector is amplified by one or more video amplifiers and then applied to the **sync separator.** Here a **base clipper** circuit separates the sync pulses from the video information, an **integrator** obtains vertical pulses and a **differentiator** circuit provides the horizontal sync pulses. The vertical sync pulses are then applied to the **vertical deflection oscillator** to keep its pulse repetition (sweep) frequency in step with the 60-cps field scanning frequency at the transmitter. The synchronized vertical deflection (saw-tooth) voltage from the output of the oscillator is strengthened by a power amplifier stage, the **vertical deflection amplifier,** and then fed to the **vertical deflecting coils** in the yoke around the picture tube. The vertical coils produce the up-and-down deflection of the electron scanning beam, the maximum **height** of the raster depending on the amplitude of the saw-tooth current in the coils.

AUTOMATIC FREQUENCY CONTROL

The **horizontal** sync and deflection circuits are similar to the vertical circuits, except that **automatic frequency control (AFC)** is generally added. The horizontal deflection oscillator is especially vulnerable to noise voltages, which it might mistake for sync pulses, resulting in faulty synchronization. Hence, rather than using any single pulse, the horizontal oscillator frequency is controlled by the *average* of a number of horizontal sync pulses. This is done by an automatic frequency control circuit, similar to the one we have discussed in Chapter 15 in connection with FM transmitters. (See Fig. 135.) The AFC circuit contains a **phase detector which compares the difference in frequency or phase between the horizontal deflection voltage and the horizontal sync pulses.** The output

of the phase detector is a **d-c control voltage** that is proportional to this frequency or phase difference. The d-c control voltage is applied to the grid of the horizontal blocking oscillator and thus **changes its bias** by the amount necessary to bring the oscillator back into synchronism with the **average frequency** of the horizontal sync pulses. The 15,750-cps horizontal sweep frequency is then amplified by the **horizontal deflection amplifier** and finally applied to the horizontal **deflection coils** in the yoke. This produces the width dimension of the scanning raster in accordance with the amplitude of the horizontal saw-tooth deflection current.

Picture Signal. Now that we have the proper scanning raster on the face of the picture tube, we are ready to display the picture. This is accomplished by modifying the brightness of the electron scanning beam in accordance with the instantaneous amplitude of the picture signal. As illustrated in Fig. 201, the output from the video amplifiers is applied to the grid-cathode circuit in the electron gun of the picture tube. The grid will thus control the instantaneous beam current impinging on the fluorescent screen in exact relation to the amplitude of the picture signal. As a result, a replica of the individual picture elements scanned at the transmitter is produced by the changes in brightness of the scanning spot on the fluorescent screen of the picture tube.

Picture Tube Accessories. The magnetic-deflection picture tube illustrated in Fig. 201 contains a few accessories which are worth noting. A magnet, known as **ion trap,** is placed around the neck of the tube in the vicinity of the electron gun. This magnet literally traps the **negative gas ions,** which are inevitably present in any large, evacuated tube. Because of their relatively large mass, these negative ions are not easily deflected, and if allowed to reach the screen, they will eventually burn a brown spot in the center of the screen. The ion trap magnet produces a sufficiently strong magnetic field to divert the ions to the side of the tube and thus prevent them from reaching the screen.

Although the electron gun of the picture produces an electron beam that comes to a focus at the crossover point (see Fig. 31), the beam diverges rapidly again after this point. To focus the beam at a point on the screen, an additional **focus coil** or permanent magnet is placed around the neck of the tube. The focusing action has been explained in Chapter 6.

High-Voltage Power Supply. Every television set contains a conventional rectifier-power supply to provide the proper voltages to the electrodes of the tubes, including the electron gun of the picture tube. This is called the **low-voltage power supply** to distinguish it from the **high-voltage power supply** that supplies from 5000 to 10,000 volts to the anode of the picture tube. A high voltage of this order is required to produce an intense electron beam that strikes the screen with sufficient force to attain a bright picture over the entire area.

Note that this final accelerating anode takes the form of a **conductive coating** (called **Aquadag**) on the inside wall of the tube near the flared portion, almost up to the screen. Since the anode coating is on the sidewall of the tube, the electron beam—though accelerated by it—does not strike it. What actually happens is this: the electrons strike the screen (in accordance with the deflection voltages), where they produce considerable **secondary electron emission** due to the force of the impact. These secondary electrons are immediately attracted to the highly positive anode coating from where they are returned to the cathode via the high-voltage power supply. Thus, the electron beam has a complete circuit for the roundtrip from cathode to screen, to the anode coating, and back to the cathode through the power supply.

In practice, a trick is used to obtain the required high voltage from the horizontal output stage. With the horizontal deflection voltage applied to the highly inductive deflection coils, a sharp inductive "kickback" (counter emf) of several kilovolts is produced during the retrace or "flyback" portion of the saw-tooth current. The reason for this kickback energy is that the combined inductance of the yoke coils and horizontal output transformer **resist the sudden change in current** during the retrace. By inserting an additional winding in the horizontal output transformer the kickback voltage may be further stepped up to the required value of the anode voltage. The output from the high-voltage secondary winding is then rectified and filtered, and finally applied to the external anode connection on the picture tube. Because of its origin, the circuit is frequently called **flyback** or **kickback power supply.**

TELEVISION BANDWIDTH AND CHANNELS

We have seen that a television system must transmit and reproduce a great deal of information. Each second 15,750 horizontal lines (525 each 1/30th second) must be scanned and each line contains a great many individual picture elements. Because of the huge amount of information to be transmitted in an extremely short period of time, very high video signal frequencies, up to 4 megacycles, are produced. In addition to the video information, a complete FM sound channel must be added to the overall TV bandwidth. The result is that a channel **6 mc wide** must be provided to transmit the complete TV (picture and sound) signal. Obviously, this tremendous frequency range cannot be saddled upon a carrier in the broadcast frequency range from 535 to 1600 kilocycles. The Federal Communications Commission (FCC) has therefore provided for 13 television channels, each 6 mc wide, in the VHF bands from 54 to 88 mc (channels 2 to 6) and from 174 to 216 mc (channels 7 to 13). In addition, a band accommodating channels 14 to 83 have been provided in the UHF range from 470 to 890 mc. These are the frequencies you actually select when you turn the channel knob on your set to the desired channel.

Vestigial Sideband Transmission. If television produces video frequencies up to 4 mc, the **two sidebands** of the amplitude-modulated video carrier alone would embrace 8 megacycles, or more than the entire video-plus-sound channel. To conserve space, therefore, the video carrier is transmitted on a **single sideband,** a process which we have touched upon in Chapter 15. To avoid distortion and simplify the receiver circuits, however, a small **vestigial** portion of the lower video sideband is also transmitted. This is known as **vestigial sideband transmission,** and it is the process used in U.S. television.

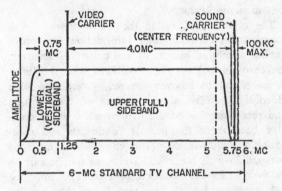

Fig. 202. *Standard U.S. Television Channel*

Fig. 202 illustrates the distribution of the standard 6-mc television channel. If the limit of the lower sideband is arbitrarily labeled zero frequency, the vestigial sideband extends up to 1.25 mc, where the video carrier is located. The upper video sideband extends at full amplitude from 1.25 to 5.25 mc, a range of 4 mc, and at reduced amplitude up to 5.75 mc. The FM sound carrier is positioned 4.5 mc above the video carrier at a center frequency of 5.75 mc, and has two sidebands extending over a maximum of 100 kc. To attain this small bandwidth, **narrow-band** frequency modulation is used, with a frequency deviation of ±25 kc maximum. (You will recall that the sidebands exceed the deviation from the center frequency.) The small frequency band to the right of the sound channel is left open to avoid interference with an adjacent TV channel.

In practice, the 6-mc TV channel range is placed, of course, at a much higher frequency. Thus, channel 2 extends from 54 to 60 mc, channel 4 from 66 to 72 mc, channel 5 from 76 to 82 mc, and so on. The relative distribution of each channel, however, is as shown in Fig. 202.

COLOR TELEVISION

The essential principles of color television are the same as for black-and-white, but the actual circuits are so much more complicated that we can only hint how it is done. At the TV transmitter the illuminated scene is televised with three TV cameras, each provided with an optical filter to transmit a particular color. The **colors red, green, and blue** are used, since all other colors can be reproduced by their proper combination. The color outputs of each TV camera are then combined into two basic signals for transmission over the standard 6-mc channel. One of these signals is called the **luminance signal,** it contains only the **brightness variations** of the picture, just as the ordinary video signal. A conventional black-and-white receiver can receive this luminance signal and, thus, reproduce a color telecast in **monochrome.**

The other signal, called the **chrominance signal,** contains the essential color information for reproducing a colored image. At the color TV receiver, the chrominance signal is **combined** with the luminance signal to recover the red, green, and blue video signals. The actual process of transmission and recombination of the two signals is quite complex because of the narrow bandwidth and **compatibility** requirements. (The color signal must be **compatible** with black and white, so that it can be received in monochrome on any set.) The amplified red, green, and blue video signals are then applied to a **tricolor** picture tube for reproduction of the colored image. At present, the tricolor tube has three separate electron guns, one for each color. The three guns are so oriented that their beams pass at slightly different angles through the holes of a **shadow mask** and each strikes a phosphor dot that glows in the color represented by the beam. A single-gun tube, called **chromatron,** is in development, which will probably replace this arrangement.

SUMMARY

Television is achieved by the consecutive transmission of **still pictures,** each broken up into thousands of tiny **picture elements.** The **persistence of vision** accounts for the illusion of **motion.**

A scanning electron beam in the TV camera tube breaks up the electrical image of the scene stored on the **mosaic** into the individual picture elements and generates a video signal whose instantaneous amplitude corresponds to the brightness of the elements.

The **scanning pattern** (or **raster** at the picture tube) consists of 525 **interlaced** horizontal lines for each picture or **frame,** which is broken up into two alternately even- and odd-numbered **fields.** Thirty frames, or 60 fields, are transmitted each second. The **horizontal line scanning frequency,** thus, is 525 × 30, or 15,750 per second. The **vertical field scanning** frequency is 60 cycles per second.

To keep the transmitter and receiver scanning in step, **horizontal and vertical sync pulses** are transmitted for each line and field, respectively, at 15,750 cps and 60 cps. The sync pulses together with the picture information make up the **composite video signal.**

The TV picture signal is transmitted over an **amplitude-modulated** video carrier, using a 4-mc wide upper sideband and a **vestigial** lower sideband. The sound signal is transmitted over a **frequency-modulated** carrier, using ±25-kc **narrowband deviation.** An entire television channel is 6 megacycles wide.

A **superheterodyne receiver** is used to pick up the TV sound and video signals. In the **intercarrier-type** receiver the signals are separated by heterodyning them in the **video detector.** Separate video and sound circuits are used from this point on.

Chapter Eighteen

RADAR AND NAVIGATIONAL AIDS

Radar (Radio Detection and Ranging) detects objects or "targets" and determines their distance (range) and direction (azimuth). It accomplishes this by sending out short bursts of radio energy that are bounced back to the sender by a distant target. The time it takes the radiowave to complete its two-way journey indicates the distance between sender and target.

Radar's basic principles are as old as radio itself. Both Heinrich Hertz and Marconi experimented with reflected ultrashort waves, the "radio echoes" used in radar. Hertz demonstrated that electric waves could be reflected from plane or curved metal surfaces in accordance with the same laws that apply to light waves. By measuring the wavelengths and the frequency of the impulses Hertz calculated their velocity and found it to be the same as that of light—i.e., an electromagnetic radiation.

Nothing much was done after those early experiments and it remained for the genius Nikola Tesla to recognize and point out the highly practical application of the radio echo in the magazine *Century,* in June 1890. Thus Tesla foresaw Radar in 1890 and he explained it by using the simple analogy of the sound echo. According to this analogy anyone who has ever shouted at the top of his lungs from a cliff and heard his echo bounce back, has engaged in an activity akin to radar: the sound projected forth in a certain direction has bounced off some cliff or mountain top and been returned to the sender, or "transmitter."

If we happen to know that sound travels about 1100 feet each second we can determine the approximate distance to the reflecting object from the length of time it took for the echo to come back to us. For example, if it takes two seconds for an echo to return after the initiation of the shouting, we will know immediately that the sound waves must have traveled a distance of 2 × 1100 or 2200 feet and, since it takes just as long for the sound to reach the reflecting object as for the echo to come back, the target distance must be 1100 feet.

We also know, of course, the approximate direction in which the target (the cliff or mountain top) is located from the direction in which we were shouting. We could even make a crude map of the nearest hills or mountains by shouting in various directions and timing each particular echo coming back to us. This would be a rather rough way of scanning the surrounding area. If we were to shout continuously, however, while making the map, we would probably never hear the weak echoes returning to us and thus lose most of the information. A more efficient way would be to send a short burst of sound in a particular direction, wait for the echo to come back and time it, and then repeat the same process in various other directions, until we had completed the map.

SIMPLE RADAR SYSTEM

The process of sending out wave energy in short bursts is called pulsing and it is used in radar. The basic operation of radar is almost exactly the same as we have just outlined for the example of making a crude map by means of sound waves and their echoes. The essentials of a simple radar system are shown in Fig. 203. The only part that is different from the sound analogy is the indicator unit, which consists of a viewing tube similar to that used in

a television set. This tube displays the original transmitted pulse as well as the returned echo pulse along a horizontal base line that is marked in terms of distance. Other indicators exist that are capable of drawing a map of the area searched by the radar antenna, but the simple system shown here suffices for giving information of the range and direction of the target.

Operation. The block diagram (Fig. 203) shows the operation of a radar set at a glance. The **timer**

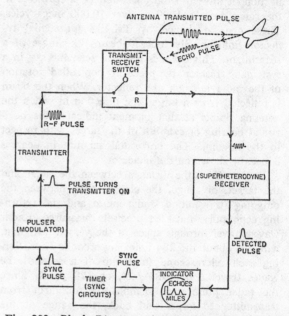

Fig. 203. *Block Diagram of a Simple Radar System*

assures that a single pulse from the **pulser** turns on the transmitter at exactly the same time as the indicator begins to draw the pulses along its scale. The transmitter, a high-power **magnetron,** then sends out short bursts of VHF radio energy through a fast-acting Transmit-Receive, or T-R switch (shown in the "Transmit" position) to the rotating antenna which, in turn, sends it out into space. At the same time the transmitted pulse is recorded visually on the left-hand edge of the indicator scale. If the transmitted pulse happens to hit a target (a ship or plane or stationary object) somewhere in space, a weak echo will be returned to the antenna. This echo signal becomes a maximum when the target is exactly in the center of the transmitted beam— that is, when the antenna axis is lined up with the target. In practice, this is achieved by turning the antenna until a maximum echo pulse is obtained.

In the meantime the fast-acting T-R switch has automatically turned to the "Receive" position and guides the weak echo pulse along to the receiver, where it is amplified and demodulated from its r-f carrier by a superheterodyne circuit. The output of the receiver is applied to the cathode-ray indicator tube, where it is displayed as a small "pip" to the right of the large originally transmitted pulse, called "main bang." The distance between the original pulse and the echo pulse at the right can be read off directly in miles or yards, thus giving the "range" of the target. The direction the antenna is pointing at gives the "azimuth" of the target.

AZIMUTH AND RANGE MEASUREMENT

Since radar relies upon the principle of wave reflection, it must use the **quasi-optical microwaves** in the range from a 1000 to over 10,000 megacycles. These can be shaped by dish-like (parabolic) reflectors into sharply focused beams like those of a searchlight. This beam is made to **scan** the sky in a systematic fashion by precisely controlled rotation of the radar antenna and reflector. When the beam is reflected from a target, the direction in which the antenna points at that moment indicates the horizontal bearing or **azimuth** of the target with respect to the antenna. The horizontal azimuth in degrees is usually shown on the indicator.

To determine the distance between the radar and the target, or **range**, the elapsed time between the outgoing (transmitted) radar pulse and the returning echo pulse must be precisely measured. Radar waves travel through space at the velocity of light, which is about 186,284 miles per second. Thus, during each **microsecond** (millionth of a second) the waves travel about **984 feet, or 328 yards.** Since the radar pulse must complete a roundtrip from transmitter to target and back, the range of the target is **one-half** the total distance traveled by the pulse, or about **164 yards** for each microsecond of elapsed time. This elapsed time is conveniently measured by the linear (saw-tooth) sweep of an electron beam across the screen of a cathode-ray tube. By deflecting the beam horizontally at a *known* rate of motion (say, .01 inch per microsecond), the resulting trace forms a time scale or base, which may be calibrated in inches, or directly either in miles or yards of **range.**

As an example, assume that a radar antenna sends out a 1-microsecond long pulse at the same time as the cathode-ray tube indicator starts sweeping a beam horizontally across the face of the tube at a rate of 1 inch per 100 microseconds. Say the radar pulse strikes and is reflected by an airplane (target) at a range of 32,800 yards from the radar antenna. The situation is depicted in Fig. 204.

The 1-μsec. pulse is seen leaving the radar antenna simultaneously with the indication of a .01-inch long "main bang" at the left edge of the radar indicator (called "scope"). The pulse travels the distance of 32,800 yards to the airplane in 100 microseconds, and after reflection from the target requires another 100 microseconds to return to the

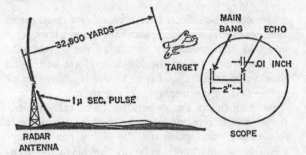

Fig. 204. *Timing of Reflected Radar Pulse on Screen of "Scope"*

antenna and receiver. Accordingly, a weak echo pulse, or **pip,** is recorded on the face of the scope at a distance of 2 inches, or 200 microseconds, from the main bang. This distance must be measured between the leading edges of the main bang and echo pulse, since the pulse itself lasts 1 μsec. and takes up 0.01 inch of the time base. Of course, the single indication of the pip after 200 microseconds is far too brief to be visible. The indication must be repeated many times each second to be perceived, by sweeping the beam across the scope at the scanning rate of the antenna (usually about 15 to 20 r.p.m.).

Fig. 205 illustrates how the ranges of several targets may be indicated on the face of the same scope. The three time bases, in inches, microseconds, and thousands of yards, are equivalent to each other, the sweep velocity being 1 inch per 100 microseconds. Usually, only the "yard" scale is shown.

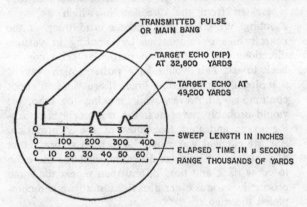

Fig. 205. *Range Indication of Several Targets on Radar Scope*

Plan Position Indicator (PPI). The radar scope of Fig. 205 does not indicate the direction (azimuth) of the target at all. There are several methods for displaying the azimuth on the same indicator, the most useful being the **plan-position indicator (ppi),** which draws a rough **map** of the area scanned by the radar beam.

As shown in Fig. 206, the electron beam on the

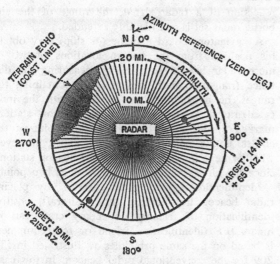

Fig. 206. *Typical Plan Position Indication (PPI)*

face of the ppi-scope is deflected in successive **ra-dial lines** through an entire circular area. The center of this circle indicates the position of the radar. The **distance** from the center along each radial line is the time base of the sweep and, hence, represents the **range** of a possible target. The **direction** of each radial line with respect to North or **zero degree reference** represents the **azimuth** or **horizontal bearing** of the target from the radar. To make the azimuth presentation accurate, the rotation of the scanning radar antenna must be synchronized, of course, with the direction of the radial beam deflection, so that each radial line points—at the moment it is swept—in the same direction as the radar antenna at this same instant. The motor that rotates the radar antenna, therefore, must be electrically coupled to the ppi-scope.

TYPES OF RADAR

Radar has been used in war and peace for a variety of applications. Its original purpose in World War II was to give **early warning** of approaching enemy planes. Early warning radars, such as used in the DEW (Distant Early Warning) line, are large ground-based installations employing huge rotating parabolic reflectors or arrays of stationary dipole antennas, which look somewhat like bedsprings. (See Fig. 207.)

The war-time applications of radar were soon expanded to include ground control of interceptor aircraft (GCI) and the direction of anti-aircraft artillery and searchlights by means of radar. These so-called **fire control radars** not only detect the enemy planes, but also automatically direct the anti-aircraft guns that fire upon them. This job is done by **servomechanisms.** Warning and fire control radars eventually became airborne and are now included in most military airplanes.

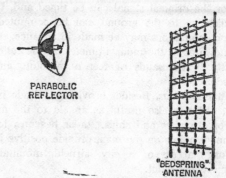

Fig. 207. *Parabolic Reflector and "Bedspring" Antennas Used for Early Warning Radars*

Radar Altimeters. Since radar can measure the distance to targets many miles away, it obviously can do the much simpler job of measuring the distance from a plane to the ground directly beneath it. Conventional pulse-type radars may be used to measure the absolute altitude of an airplane above the surrounding terrain, provided the plane flies at a considerable height. At very low altitudes (below about 1000 ft.) however, the pulse-type radar becomes unsatisfactory, since for the short distances involved the echo pulse tends to merge with the transmitted main pulse. (Remember that a 1-microsecond pulse takes up a range of 984 feet on the screen of the indicator.) Most radar altimeters, for this reason, do not use pulses but **continuous, frequency-modulated waves.** A typical setup is shown in Fig. 208.

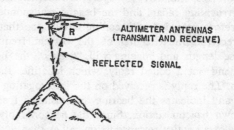

Fig. 208. *Principle of Radar Altimeter*

The transmitting antenna of the radar altimeter sends out a vertical radio beam to the ground with a continuously changing frequency. At the instant the signal leaves the antenna, its frequency will have a certain value. The signal is then reflected from the ground and returns to the receiving antenna of the altimeter. The receiver contains a **phase discriminator which compares the frequency (or phase) of the returning signal with that of the signal now being transmitted.** Since some time has elapsed for the echo of the original signal to return, the frequency of the signal now being transmitted has changed by some amount, of course. With the frequency deviation in cycles per second being known, the roundtrip of the original signal can be timed and, hence, the distance to the ground can be computed. All these computations may be made in advance, so that the indicator of the radar altimeter can be **calibrated directly** in thousands of feet of absolute airplane altitude.

Radar Beacons. Besides providing altitude indications, radar is also useful as an aid to the navigation of airplanes and ships. **Radar beacons,** located at known points on the map, provide positive identification markers of a city, airfield, mountain, or other specific point.

As illustrated in Fig. 209, a radar beacon consists of a ground-based radar transmitter-receiver,

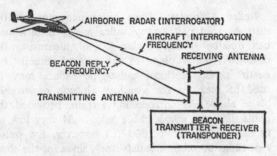

Fig. 209. *Operation of Radar Beacon*

called a **transponder,** with separate receiving and transmitting antennas. Instead of sending out pulses and receiving echoes, a beacon receives a pulse from an airborne or shipborne radar set and sends back a reply. But it is not desirable that the beacon should respond to every spurious radar signal. The **interrogating** radars and the beacon transponder are therefore provided with coding circuits, so that only a received radar signal of the correct frequency, pulse length and spacing will "trigger" the beacon to send out a coded reply, which identifies the station. The reply is received on the interrogating radar set and indicates the bearing and distance from the known beacon station. Since the beacon reply consists of an active retransmission rather than a passive

reflection of a radar signal, the range of the airborne radar set is considerably extended.

A navigator aboard a plane (or ship) may obtain his **position** from the known locations of two or more radar beacon stations by measuring the distance (range) to each station. After obtaining the ranges, he need only draw a circle with the indicated range around each of the beacon stations marked on his map; the intersection point of the circles is his position. He must know, however, whether he is in front or behind the beacon stations, since the circles will usually intersect at two points.

Identification Friend or Foe (IFF). By placing radar beacon transponders inside aircraft, positive identification of friendly or enemy craft can be made. **IFF** (identification friend or foe) equipment is based on the same principles as illustrated in Fig. 209 for the conventional radar beacon. In this case, a ground-based radar set sends out an **interrogation pulse** at a certain frequency, whose pulse length and spacing has been coded in a specific manner. The beacon transponder of a friendly aircraft will be triggered by this interrogation pulse and a coded reply automatically will be sent back to the radar station, identifying the aircraft. An enemy aircraft, presumably, does not know the code and, hence, its beacon transponder (if any) will *not* be triggered into sending out the properly coded reply.

Ground-Controlled Approach (GCA). Radar, finally, comes to the rescue of the airplane pilot who must make a blind landing in bad weather, but does not have the special equipment and instruments to do it with. A skilled crew of ground operators can "talk him down" by means of **ground-controlled approach** radar equipment. Basically the ground equipment at the airport consists of two microwave radar sets, which are usually installed in a single trailer placed adjacent to the runway. One of the radars, known as the **search system,** locates all aircraft within 30 miles or so of the airport and thus provides a radar map of the vicinity on the ppi indicator. The other radar, called the **precision system,** pro-

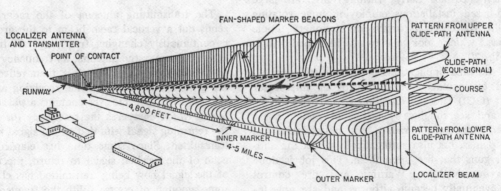

Fig. 210. *ILS Landing Approach Pattern*

vides continuous information regarding the position of the incoming aircraft with respect to the runway. The plane may thus be safely talked down along the sloping **glide path** along which it must approach the runway.

INSTRUMENT LANDING SYSTEM (ILS)

A blind-landing system that existed already before the advent of the GCA radar equipment and is still used by the majority of airports, is known as the **instrument landing system** or **ILS.** In this landing approach system some of the complexity of the ground-based GCA radar equipment is shifted to special radio receiving and indicating equipment, carried aboard the airplane. Fig. 210 illustrates the complicated ILS landing pattern radiated by highly directional radio (*not* radar) transmitters on the ground.

As is evident from Fig. 210, the ILS scheme obtains a sloping glide path to the runway by the intersection of sharply defined **vertical localizer** and **horizontal glide-path** beams, generated by separate transmitters. In addition, low-power beacon transmitters produce **fan-shaped vertical marker beacons** that indicate the aircraft's position at various distances from the runway.

Localizer Transmitter and Receiver. The localizer transmitter, located at the far end of the runway, provides the horizontal (left-right) guidance which enables the pilot to steer the required course to the runway. Operating at a frequency of about 110 megacycles, the transmitter sends out two patterns, amplitude-modulated at 150 and 90 cps, respectively. These two signal patterns overlap along a line formed by the extension of the runway. The 150-cycle signal area to the right of the approaching aircraft is called the **blue area,** while the 90-cycle pattern to the left of the approaching aircraft is called the **yellow area.** The localizer transmitter has a range between 40 and 80 miles, depending on the aircraft's altitude. A horizontal section of the three-dimensional localizer field pattern is shown in Fig. 211.

The ILS receiver aboard the airplane is equipped with an **approach indicator,** which enables the pilot to follow the glide-path and localizer beams. As illustrated in Fig. 211, the approach indicator has two **crossed** indicating needles, one vertical and the other horizontal. The **horizontal** needle indicates whether the aircraft is **above** or **below** the proper glide path, while the **vertical** needle shows whether the plane is to the left or the right of the on-course localizer beam.

The **localizer receiver** in the plane filters out the 90- and 150-cycle modulation components of the localizer signal and applies them to opposite sides of the zero-center localizer needle. If the aircraft

is flying off-course in the blue area of the localizer pattern, the 150-cycle modulation signal prevails and the localizer needle is deflected into the blue sector of the approach indicator. If the plane is flying in the yellow area, however, the 90-cycle signal prevails, and the needle is deflected into the yellow

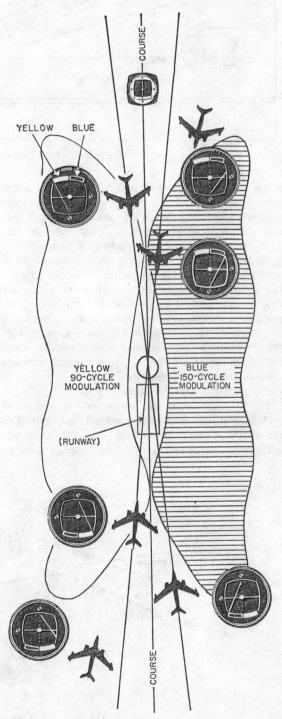

Fig. 211. *Localizer Field Pattern and Indications*

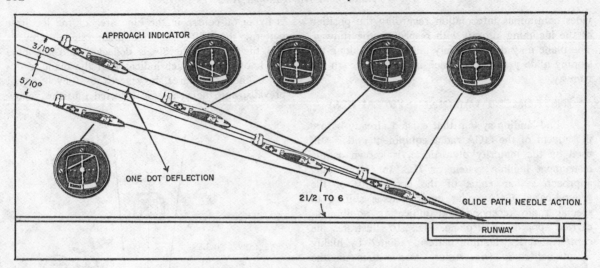

Fig. 212. *Action of Glide-Path Indicator Needle*

sector. Finally, if the craft is flying **on-course** in the overlapping area of the yellow and blue patterns, the two modulation signals are of **equal strength** and the vertical localizer needle will *not* be deflected. The pilot attempts to fly this **equi-signal course.**

Glide Path. The glide-path transmitter, located approximately 750 feet from the approach end of the runway, transmits upper and lower glide-path beams, modulated also at 150 and 90 cycles, respectively. The beams are about 0.8 degrees wide and form an angle between 2.5 and 6 degrees with the runway. The location of the glide path is determined by the area of equal signal strength between the two beams, as shown in Fig. 210. The frequency of the glide-path signal is about 333 mc.

Fig. 212 illustrates the action of the **horizontal glide-path needle** in the ILS approach indicator. The glide-path receiver demodulates the 150- and 90-cycle components of the signal and applies them to

opposite sides of the zero-center glide-path needle, just as for the localizer. When the airplane is above the proper glide path, the 150-cycle signal prevails, and the glide-path needle of the indicator is deflected **downward**, signifying that the pilot must **fly upward.** When the plane is **below** the glide path, the 90-cycle signal is stronger, and the glide-path needle is deflected **upward.** Clearly, the needle always points in the direction toward which the aircraft must be flown to approach the glide path. When on the proper glide path, the two modulating signals will be of equal strength and the glide-path needle remains **horizontal.** As a consequence, the pilot always attempts to fly the plane so that the two needles of the approach indicator will **cross at the center.**

Marker Beacons. In addition to the localizer and glide-path transmitters, there are usually two to three VHF marker beacon transmitters, operating

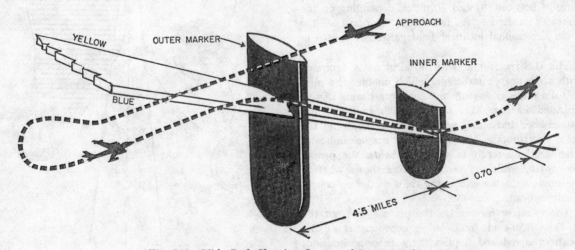

Fig. 213. *Glide Path Showing Outer and Inner Markers*

at 75 mc, which transmit vertical, fan-shaped markers to indicate the position of the aircraft with respect to the runway. As shown in Fig. 213, the outer marker is generally at a distance of 4.5 to 5 miles from the runway, while the inner marker is about 0.7 to 1 mile distant from the runway. Sometimes an additional "boundary marker" is placed at about 250 feet distance from the approach end of the runway and marks its boundary. Each of these marker beacon transmitters sends out a differently coded signal, which is received on the marker beacon receiver aboard the airplane and identifies it. Differently colored flashing lights on the aircraft instrument panel often give additional visual identification of the marker over which the plane is passing at the time.

The ILS system can bring a plane down by the use of instruments alone, but to attain a smooth landing the pilot usually relies on his vision rather than instruments during the last few seconds before touchdown. The runway is generally visible for a few hundred feet even in heavy overcast. The big disadvantage of the system is, of course, that it requires special equipment aboard the aircraft.

RADIO RANGES

The chief purpose of **radio range stations** is to provide a number of courses or beams on which aircraft may fly. Air paths between cities throughout the world are charted by means of these ranges.

We shall consider two types of radio ranges in use in the United States. They are the low frequency range and the VHF omnirange. The low frequency range is one of the earliest electronic aids to air navigation, while the VHF omnirange is a more recent development, dating back to about 1947.

LOW FREQUENCY RANGE

There are over 400 low frequency radio range stations throughout the United States. Each of these stations sends out "on-course" signals which interlock with those of surrounding stations to form easily followed airways. Radio range stations operate on frequencies between 200 kc and 400 kc.

The low frequency radio range depends on the superposition of two figure-8 directional antenna patterns at right angles to provide on-course signals. When two vertical antenna towers are properly fed with radio-frequency energy they produce a figure-8 radiation pattern, as seen from above. This pattern clearly indicates the directional nature of the radiation, as illustrated in Fig. 214.

In Fig. 214, the relative strength of the signal radiated in any direction is represented by the length of the line drawn in that direction from the center (O) to the edge of the pattern. Thus, maximum signals are radiated in directions *OA* and *OB*, while

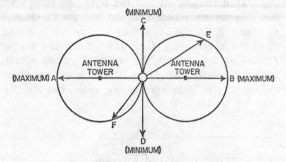

Fig. 214. *Figure-8 Radiation Pattern of Two Vertical Antenna Towers*

minimum signals are radiated in directions *OC* and *OD*. (These minimum signals are shown exaggerated for clarity, but in practice they are practically zero.) In intermediate directions, the signal radiated is of intermediate strength, as shown by lines *OE* and *OF*. The signal radiated in the direction *OE,* for example, is much stronger than the signal in the direction *OC* but weaker than that radiated in direction *OB*.

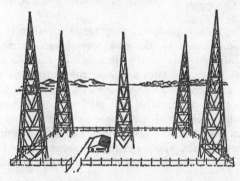

Fig. 215. *Typical Radio Range Station. The Four Outside Towers Are Used for Transmission of "On-Course" Signals; the Center Tower Permits Separate Voice Communications*

If two pairs of vertical towers are placed in a square pattern, as shown in Fig. 215, the result will be two figure-8 radiation patterns at right angles to each other. (The fifth [center] tower shown in the illustration is used for transmitting weather reports.) With two figure-8 patterns available the production of on-course signals becomes a simple matter. The code signal *A* (which looks like this •—) is transmitted over one pair of diagonally opposite towers, and the code signal *N* (which looks like this —•) is transmitted over the other pair of towers. In this way four signal zones, called **sectors,** are produced, two of which carry the *A* signal, while the other two carry the *N* signal. Between the four sectors there are four regions of overlapping, where both signals are present, as shown in Fig. 216.

The *A* and *N* code signals consist of a medium-pitched tone (about 1000 cycles) and are therefore audible in the airborne **radio range receiver.** From

Fig. 216 it is apparent that the following signals are heard in various zones: 1. in the *N* sectors a clear *N* is heard; 2. in the *A* sectors a clear *A* is heard; 3. in the overlapping *AN* regions there are narrow (3° wide) lanes where the *A* and *N* sound equally strong. However, the *A* and *N* signals are not heard individually in these overlapping zones, but the key-

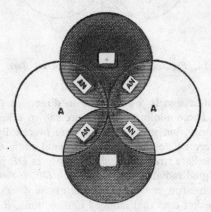

Fig. 216. *"A" and "N" Sectors of Radio Range*

ing is so arranged that they blend together to give a continuous tone, the **on-course** signal. As long as the aircraft flies in the direction of the on-course signal lanes, the pilot will hear the continuous tone, but as soon as the craft deviates from the course either the *A* or the *N* will predominate, depending on the direction of the course deviation. As is seen from Fig. 217, the double figure-8 defines **four on-course signals,** where a continuous tone is heard. The *A* and *N* transmissions are interrupted frequently by

a three-letter call signal, identifying the station from which the transmission originates.

What is a "Cone of Silence"? Directly above a radio range station there exists an area where practically no signal is heard. This area has the shape of an inverted cone, as shown in Fig. 217, and is known as the **cone of silence.** A pilot passing through a cone of silence would know that he was directly over the range station whose signals he had been receiving.

How Radio Range Signals Are Received. No special equipment is needed to receive radio range signals. Any receiver that can be tuned to the frequency of radio range stations (200 to 400 kc) can receive the **aural** radio range signals. To provide visual indication, however, special indicators are required. These generally consist of a **radio compass** that indicates the course and a **left-right indicator** which shows whether the station is to the left or right of the aircraft.

Limitations of Low Frequency Range. The low frequency radio range suffers from a number of flaws which may cause the on-course signals to be misleading. The major defects are the **night effect, multiple courses,** and **bent courses.** Night effect, caused by irregular reflections of the ionosphere, results in fading of the on-course signals in an erratic pattern. Multiple and bent courses are caused by multiple reflections from mountains and bending of radio waves when they pass from one type of terrain to another. All of these effects may combine to bring about the pretty picture of an aircraft bumbling along between spurious and bent *A*'s and *N*'s in a wide area of confusion.

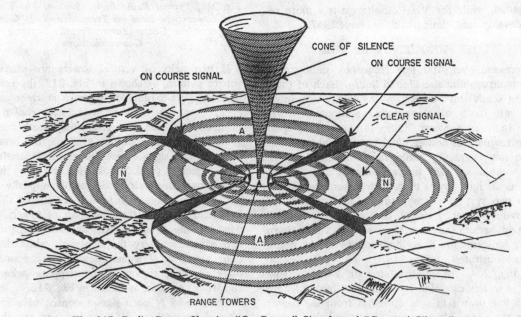

Fig. 217. *Radio Range Showing "On-Course" Signals and "Cone of Silence"*

In addition to these defects, the low frequency radio range has some inherent limitations. As we have seen the double figure-8 defines only *four* course directions, which is insufficient to chart the increasingly heavy air traffic in all possible directions. Furthermore, by providing the pilot only with **direction,** the range gives him insufficient information for complete air navigation.

VISUAL VHF OMNIRANGE (VOR)

The adoption of the visual **VHF omnirange** (abbreviated **VOR**) marked a big step forward in radio navigation. Operating in the VHF region from 112 to 118 megacycles, VOR overcomes most of the disadvantages caused by low frequency radio propagation, such as night effect and multiple beams. Furthermore, VOR has the big advantage over the low frequency range that it provides an **unlimited number of visual course legs** for the pilot to fly. The visual indicator relieves the pilot from listening continuously to on-course signals.

How the Rotating VOR Pattern Is Obtained. The transmitter of the VOR system utilizes five special loop antennas, four at the corners of a square and a fifth in the center, to radiate two types of signals. One of these is a 30-cycle signal which modulates the transmitter output and is fed to the center loop of the antenna system. This loop provides a **circular, non-directional** field—that is, the signal strength is the same in all directions around the antenna. The VOR receiver uses the signal as a **reference** voltage. The remaining transmitter output is fed to the four loops at the corner of the square. Each pair of diagonally opposite loops provides the familiar figure-8 radiation pattern. However, by varying the magnitude of the signal voltage to each pair of loops at the rate of 30 cycles per second, the direction of the pattern is constantly changing so that the figure-8 effectively **rotates** through a full circle (360 degrees) at 30 revolutions per second. This rotating figure-8 pattern combines with the stationary circular pattern to produce a rotating **heart-shaped (cardioid)** field pattern, as shown in Fig. 218.

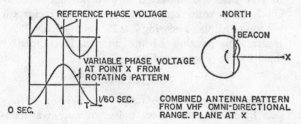

Fig. 218. *VOR Phase Relations and Rotating Cardioid Pattern (Plane at X)*

VOR Receiver. The VOR receiver is equipped to receive both the stationary 30-cps, non-directional

signal and the rotating directional cardioid pattern. By comparing the phase of the variable voltage due to the rotating field with that of the reference voltage, the relative direction of the aircraft with respect to the range station can be determined. For example, if an aircraft is located at point X in the illustration above, the peak of the heart-shaped pattern will be facing the craft, which is due East of the station. At this point maximum signal strength is received. This corresponds to a certain phase (or time) difference between the peak of the rotating pattern (variable voltage) and the reference voltage. For any other location of the aircraft with respect to the station the phase difference between the variable signal and the reference voltage will be different because of the rotation of the pattern. Thus every direction of the aircraft with respect to the station will correspond to a definite phase relationship between the two signals. This is translated by the indicators in the receiver in terms of **compass headings** of the aircraft with respect to the range station.

Limitations of VOR. While superior to the low frequency range VOR still suffers from a number of inherent disadvantages and limitations. Although providing an unlimited number of course legs, VOR —like the low frequency range—can only provide **direction** to the pilot. As we shall see shortly, however, both direction (bearing) and distance (range) are needed for rapid position finding. Furthermore, the frequencies on which VOR operates—although superior to low frequencies—are not free from terrain effects which may cause occasionally large inaccuracies in bearing. Even at best, VOR is accurate only within three degrees of bearing—which is not very precise.

VOR-DME NAVIGATION SYSTEM

If an electronic system could furnish the pilot with *both* direction and distance to a fixed known point, he would know his exact map position at all times without the need for additional computations. The VOR-DME system does just that. It was adopted by the Civil Aeronautics Administration in 1948, but is already becoming obsolete and will probably be replaced by the superior **Tacan** system. As its name implies, the VOR-DME system is a combination of the VHF Omnirange (VOR), we have just discussed, and UHF Distance Measuring Equipment (DME), operating at about 1000 megacycles. The VOR portion of the system provides the aircraft with the horizontal bearing or course to the station in the manner with which we have become familiar. Let us now briefly touch on the operation of the DME portion of the system, which provides the aircraft with **continuous range information.**

How the Distance Measuring Equipment (DME) Operates. Essentially a radar beacon, DME utilizes the familiar echo principle to provide distance indication. As shown in Fig. 209, an airborne "interrogator" set sends out **pairs** of regularly-spaced radio pulses. When these interrogating pulses are received by any DME ground beacon, they "trigger" it, causing it to send out pairs of "reply" pulses, which are coded to identify the ground beacon station. The airborne DME equipment continuously measures the time interval between its own interrogation pulses and the reply pulses. Since radio waves travel at a definite speed (about 186,284 miles per second) the time interval between interrogation and reply is directly proportional to the distance of the aircraft from the ground station. The airborne equipment converts the time delay into nautical miles and indicates the range directly on a dial on the instrument board. DME uses **spaced pairs** of pulses to permit distinguishing them from various noises and DME pulses sent out by other aircraft. To supply range service to a number of aircraft at the same operating frequency, ten different possible spacings between the pulses are provided for. Special pulse "decoders" in the airborne and ground receivers pass only pulse pairs of the prescribed spacing and reject all others, thus avoiding confusion.

TACAN NAVIGATION SYSTEM

Like the VOR-DME system the TACAN (Tactical Air Navigation) system provides an aircraft with continuous distance (range) and bearing (azimuth) information from a ground beacon station, located within a line-of-sight range up to 195 nautical miles. Tacan operates completely in the UHF band from 962 to 1213 megacycles and thus overcomes the limitations of VOR caused by relatively low-frequency operation. The system has a total of 126 channels in the UHF band, each being capable of providing full service for over 100 aircraft. Tacan is remarkably accurate. It provides compass direction with less than one degree error and is capable of indicating true distance from the ground station within about 600 feet. Because of its UHF operation Tacan can use a very small ground station antenna which permits its installation on Air Force mobile units and Navy carriers.

The Tacan system consists of an airborne "interrogator-responsor" set and a ground or shipboard surface beacon. The airborne set is generally provided with **two** indicators, one for showing the **range in nautical miles,** the other for indicating the **bearing** (azimuth) of the aircraft on the station.

Fig. 219 shows how the position of an aircraft may be charted by means of Tacan information. The airborne radio set is first tuned in to the channel

of a Tacan ground station within its range, in this case channel No. 47. Distance interrogation pulses will then automatically be sent out by the set and will trigger the ground beacon. The beacon sends out synchronized reply pulses and also identifies itself aurally in International Morse Code characters. When the reply pulses are received by the airborne set, special **range circuits** measure the elapsed time between the interrogation and reply pulses and display the time difference directly as distance from the beacon (in nautical miles) on the range indicator. In this case the Tacan range indicator shows a distance of 85 nautical miles.

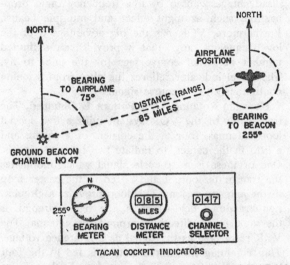

Fig. 219. *Charting Aircraft Position by Use of TACAN*

In addition to the distance reply pulses, the ground station continuously transmits a series of coded and modulated radio-frequency pulses to provide "reference" and "variable" bearing signals. When these signals are received by the airborne set, special **azimuth circuits** compare the phase of the reference and variable bearing signals and display the difference as bearing to the beacon ground station directly in degrees on the azimuth indicator. In this case the indicated airplane-to-beacon bearing is 255° with respect to North. (The reciprocal beacon-to-airplane bearing is 225° − 180°, or 75°.) Hence, the position of the aircraft is completely fixed by the Tacan readings.

SUMMARY

Radar determines the range and bearing to targets by sending out short pulses of microwave energy that are reflected by the distant target. The time it takes the radio wave to complete its two-way journey indicates the distance between transmitter and target. The direction in which the radar antenna points indicates the horizontal bearing (azimuth) to the target.

Radar consists essentially of a **timer** (sync), **pulser** (modulator), **magnetron transmitter, transmit-receive (T-R) switch,** a **scanning antenna, superheterodyne receiver,** and **cathode-ray tube indicator.** The timer assures synchronism between the transmitted pulse and the indicator time base. The pulser turns on the transmitter to send out a radar pulse over the antenna, with the T-R switch in transmit position. The reflected echo pulse from the target is amplified and detected by the superhet receiver and is displayed on the screen of the indicator. The distance between transmitted and echo pulse along the time base indicates the range from the target. In a **PPI indicator** a map-like presentation is given.

A **radar altimeter** uses frequency-modulated, continuous waves to indicate an aircraft's absolute altitude by comparing the frequency of the transmitted waves with that of the reflected (echo) waves.

A **radar beacon** is a transmitter-receiver **(transponder)** that replies to an interrogating aircraft signal and identifies itself. This permits determination of the range and bearing to the known beacon.

In **ground-controlled approach** (GCA) radar, an airplane is talked down to a blind landing by means of ground-based search and precision radars.

ILS (Instrument Landing System) uses **localizer, glide path,** and **marker beacon** transmitters to provide a sloping glide path for instrument (blind) landing approach of an aircraft. Amplitude modulation of the beam pattern by 150 and 90 cps provides left-right and up-or-down indication on a **crossed-pointer approach** indicator. Coded marker beacon signals locate the aircraft's position.

Low-frequency **radio range stations** provide **four** aircraft course legs through the overlapping **(equi-signal** zones) of two figure-8 "A" and "N" patterns. The pattern is obtained by four vertical radio towers.

The **visual VHF omnirange (VOR)** provides an unlimited number of visual course legs through the phase comparison of a **cardioid-shaped rotating radiation pattern** with a **fixed 30-cps reference signal.**

The VOR-DME uses a VHF omnirange to provide course indication and a modified radar beacon to obtain distance indication. The distance-measuring equipment (DME) operates on 1000 megacycles.

The TACAN (Tactical Air Navigation) system is a microwave pulse system that provides both bearing (azimuth) and distance (range) information through modulated pulse signals.

Chapter Nineteen

NEW HORIZONS IN ELECTRONICS

The field of electronics is undergoing an extended revolution, which will make many of the techniques and devices you have studied in this book obsolete within the span of the next decade. The explorations of the physicists into solid-state behavior have brought forth not only the transistor and the tunnel diode we have described, but a whole host of fantastic new devices, among which only the **varistor diode** (a nonlinear resistor), the **cryotron** (a superconductive control device), the **solar cell** or battery, the **maser** and the **laser** (solid-state microwave and light amplifiers) shall be mentioned. This explosion of new devices has been assisted by recent advances in sophisticated manufacturing techniques, such as the evaporation of extremely thin (atomic and molecular) films on a glass or other substrate and the micromachining of extremely tiny parts by means of electron and laser beams. The combination of solid-state concepts and new manufacturing techniques has made it possible to integrate many of the functions of formerly separate electronic components into extremely compact micromodules or integrated circuits consisting of a single solid-state block or crystal. We shall limit our discussion to this major new development and then go on to describe the perhaps most fascinating of the new devices—the laser.

MICROELECTRONICS
AND INTEGRATED CIRCUITS

Microelectronics or microcircuits—the concept of making electronic devices many times smaller and lighter than heretofore possible or envisaged—originated with the rigid weight and space requirements of missile and space research. Man's reach into space has become an incidental boon to the entire electronics industry, which not only profits by the vast reduction in size and weight of electronic equipment, but by the accompanying reduction in power requirements and the increase in overall reliability. (The latter was one of the primary reasons why the microminiaturization was started in the first place.) Present-day electronic systems have become fantastically complex, consisting frequently of hundreds of thousands of separate parts and components, which results in large-system unreliability and possibility of breakdowns. By reducing the number of components and connections to fewer, highly reliable parts and integrated circuits, microelectronics promises to achieve marked increases in system reliability.

Depending upon the manufacturer and techniques used, microelectronics goes under many names, such as integrated microcircuits, molecular electronics, micromodules, functional electronic blocks, Solid Circuits (trade mark of Texas Instruments Corp.), monolithic silicon blocks, and others. Regardless of name, however, microcircuits are here and at work, though primarily in military gear. You can get an idea of what's involved and may be achieved in the not-too-distant future from the announcement that Stanford Research Institute is exploring the feasibility of a computer consisting of some **hundred billion** active components, which are to be mounted on 100 one-inch-square frames, with the entire assembly to be packed into the space of a few cubic inches. This fantastic packing density is to be achieved by using electron beam techniques to micromachine individual components to dimensions of approximately one-millionth inch (1 micron) each.

The general approaches to microelectronics can be divided into the following categories.

1. **Component-oriented Approaches.** These have been going on for some time and consist primarily of miniaturization of discrete conventional electronic parts and hardware, transistorization, printed-circuit techniques, and tighter packaging. To achieve higher packing densities, conventional components are being redesigned to standard forms and shapes, such as wafers, cards, cordwood, welded modules, high-density blocks, etc. They are known variously as **microblocks, microelements,** or **micromodules.** The illustration (Fig. 220) shows an exploded view of a typical micromodule manufactured by the Radio Corporation of America.

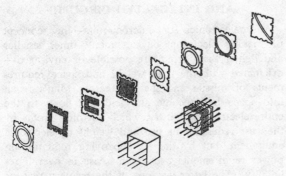

Fig. 220. *Sections of a Micromodule*

2. **Circuit-oriented (Two-dimensional) Approaches.** This scheme endeavors to deposit an entire electronic circuit—resistors, capacitors, etc.—on a flat-base material (usually glass or ceramic) called a "substrate." The components and interconnections are deposited on the substrate in the form of thin layers or films, so that the circuit becomes essentially two-dimensional. The two-dimensional circuits can then be stacked vertically to form a three-dimensional equipment module. Film deposition is accomplished by vacuum evaporation or sputtering. Various thin films have been developed, such as magnetic films, superconducting films, insulator and conductor films, and others. For example, metallic resistors may be made by evaporation of Nichrome on alumina substrate, and capacitors can be made by successive evaporations of metal (such as tantalum) and dielectric. To date only **passive** (non-amplifying) components—resistors, capacitors, and inductors—are being made by thin-film techniques, but investigations are proceeding to make active components

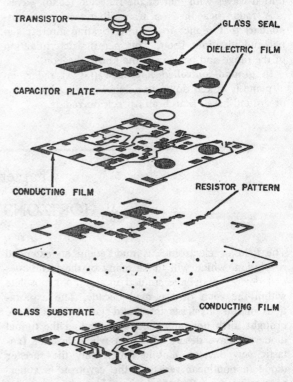

Fig. 221. *The Use of Thin Films to Construct a Two-Dimensional Circuit*

(transistors) in the same way. Figure 221 shows a typical two-dimensional circuit made up of a variety of thin films.

3. **Function-oriented Approaches.** Variously known as **molecular electronics, integrated microcircuits,** or **functional electronic blocks,** this most recent and most advanced approach attempts to integrate entire circuit functions into a solid substrate. The essential idea of the functional-block scheme is to alter and rearrange the internal physical properties of a solid material, such as a germanium or silicon crystal, to

perform a circuit function more complicated than that performed by an individual component. The single crystal of an integrated microcircuit replaces a number of conventional electronic components, such as resistors, capacitors, diodes, and transistors. The microcircuit is usually formed by the controlled diffusion of impurities into small blocks of silicon and under a layer of silicon-oxide. Various techniques, such as vapor growth, planting, etching, and alloying, transform the structure of the solid block into a "multi-zone" crystal, capable of functioning as a complete electronic circuit or subsystem. Special-purpose amplifiers, flip-flops, and various high-speed computer logic elements are already available in the form of functional electronic blocks. The arrangement of an R-C phase-shifting network in a diffused silicon block is shown in Fig. 222.

The various approaches to microelectronics are already beginning to pay off in complete micro-miniaturized electronic systems. For example, microelectronic computers have been built by various manufacturers and are now operating in various military and space applications. Further advances in this fascinating field are rapidly being made.

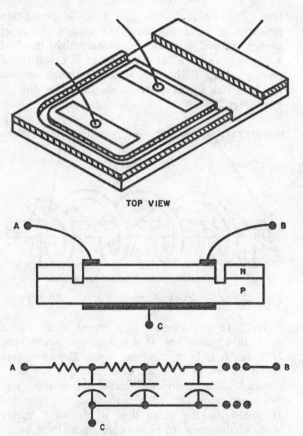

TOP VIEW

Fig. 222. *A Diffused-Silicon Block Containing a Phase-Shifting Network with Distributed Resistance and Capacitance*

THE LASER—
A NEW COMMUNICATION MEDIUM

Solid-state research has recently brought forth the **laser** (acronym for Light Amplification by Stimulated Emission of Radiation), which can produce kilowatts of "coherent" light in a narrow beam that will someday carry fantastic amounts of information. The tremendous bandwidths possible at light frequencies can solve once and for all the problems of channel crowding that must be faced in the portion of the electromagnetic spectrum now used for communications. To fulfill the promise of "visible" communications on back of a 500 million megacycle (5×10^{11} cps) carrier frequency, the problems inherent in detectors and modulators of this extreme bandwidth still need to be solved. Hundreds of firms are betting that it can be done. To get an idea of what this would mean, reflect for a moment that the visible light region from 4000 to 7000 angstroms wavelength (hundreds of millions of megacycles) can accommodate as many as 80 million television channels.

But the laser is not only a powerful new communications medium, capable of carrying messages over gigantic distances, but it is a fundamental discovery in physics, comparable to the invention of the vacuum tube—with all the developments of radio, radar, and television still to come. The laser's almost perfectly parallel and intensely concentrated beams already perform delicate surgery, make radar 10,000 times more precise, weld microscopic wires, and can vaporize any known material. Focused laser beams may make possible the death rays of science fiction, burning off everything that stands in the way of the beam. Such laser death rays are seriously being considered and studied. As a preview of things to come, a group of M.I.T. engineers recently sent a beam of red laser light up to the moon and lit up a spot on the moon's surface that was only two miles in diameter. By contrast, the tightest microwave beam of the type used in radar would have "illuminated" an area over 500 miles wide.

How It Started. The ingenious principle of the laser was first worked out in 1958 by Dr. Charles Townes, then a physicist at Columbia University, and his brother-in-law, Dr. Arthur Schawlow, of Bell Laboratories. Their reasoning was based upon fundamentals of atomic physics which had been known for more than thirty years. Dr. Townes and his colleagues had first made use of these principles a few years earlier to invent a device very similar in its essentials to the laser—the so-called **maser** (for Microwave Amplification by Stimulated Emission of Radiation). The maser produces or amplifies extremely stable narrow-band microwave frequencies, which are used in "atomic clocks" and low-noise, supersensitive microwave receivers. The laser resulted from an extension of the maser principle and, for this reason, is frequently called **optical maser.** Thus in learning about lasers you will also be able to comprehend the principle of the maser, if you should run across it.

The theories of Townes and Schawlow—which we shall explore presently—were proved correct in the summer of 1960, when Dr. Theodore Maiman of Hughes Aircraft built the first working laser, essentially a pencil-thin rod of synthetic ruby. Since then more than 400 firms and laboratories have rushed into the field and about 50 different types of lasers have been built in various places. During 1962-3 approximately 30 million dollars were spent on laser research, much of it by the U. S. Government.

How the Laser Works. Physicists Townes and Schawlow started from the fact that all light results from the activity of electrons, the tiny, negatively charged particles that circle like planets around the nuclei of atoms. As we have found in the first chapter of this book, the electrons travel in certain prescribed orbits (or shells) around a sunlike nucleus. You will recall that only a fixed series of orbits is possible within each atom. However, electrons can **jump** from one orbit to the next, depending upon whether they gain or lose energy. When they acquire energy from an outside source, such as an electric current, they temporarily jump to higher, more energetic (and faster) orbits, farther away from the nucleus of the atom. Within a few microseconds, however, the accelerated electrons fall back to their original orbits, or energy levels, giving up their extra acquired energy in the process. This energy is emitted in the form of electromagnetic radiation—radio waves, X-rays, or if within the visible portion of the spectrum: light. Whether it emanates as light or not depends upon the atom and the amount of energy imparted to the electrons. (See Fig. 223.)

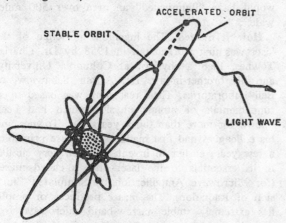

Fig. 223. *Electron Orbits*

To produce light, then, the electrons must first absorb a sufficient amount of energy from an external source. In an electric light bulb, for example, this energy comes from an electric current. The light radiated by the bulb comes in all wavelengths, colors and directions—reinforcing and canceling each other at random in wild disorder, as shown in the illustra-

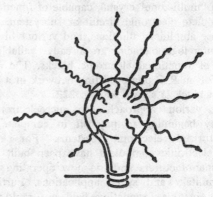

Fig. 224. *Random Light Waves*

tion (Fig. 224). In the first laser—a pencil-thin pink rod of synthetic ruby—the outside source of energy, known as **pump,** is ordinary light itself of a greenish variety. The light source is essentially a photographer's flash tube, which is wound around the ruby like a corkscrew, as shown in the illustration (Fig. 225).

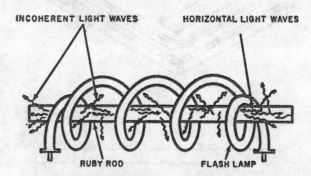

Fig. 225. *Stimulating Natural Random Fluorescence*

Synthetic ruby consists of a crystal of aluminum oxide in which a few of the aluminum atoms have been replaced by chromium atoms. The chromium atoms absorb green light, thus leaving a red color —which is the reason why rubies are naturally red. When the photoflash light surrounding the ruby rod is turned on, the green light of the flash excites some of the electrons in the chromium atoms, pushing them up to higher orbits. As these electrons fall back to their normal levels, the absorbed green light is re-emitted as red light, giving rise to the well-

known red fluorescence of the ruby. The trouble with this natural fluorescence is that it is as incoherent (random) as any other kind of light. Thus far, then, we have only stimulated a ruby into natural red fluorescence, but have not licked the problem of coherent light.

To tame the millions of unruly electrons in the ruby—to make the laser "lase"—Townes and Schawlow devised a complicated system. The excited electrons, which have jumped to higher orbits, are extremely unstable and are ready to give up their excess energy. By exposing these excited electrons to light waves of the same wavelength as those they are about to emit, the electrons can be triggered to emit their excess energy sooner than they ordinarily would and exactly in step (in phase) with the triggering waves, so as to reinforce them. Thus a large pool of excited electrons is needed. The light stimulated spontaneously by the first few electrons would then stimulate other excited electrons in the pool instantly to emit their own identical light waves. In this way one would obtain a vastly amplified light wave, as shown in the illustration (Fig. 226). Unfortunately, however, these amplified waves would soon be lost in the general disorder of the incoherent light.

EXCITED ATOMS

UNEXCITED ATOM

Fig. 226. Stimulated Emission of Radiation from Excited Atoms Reinforces the Stimulating Light Wave

One final step was needed: The initially emitted light waves had to be made to pass right back into the pool of excited electrons, and back and forth through the excited electrons millions of times. Only in this way would the powerful chain reaction be started that would result in a burst of coherent light. This crucial final step was accomplished by placing mirrors at both ends of the ruby rod. One mirror was less heavily silvered than the other, so that it was partially transparent and would permit

light to exit from this end (see Fig. 227). Of course any light waves that were traveling at odd angles to the axis of the ruby rod would soon be lost, but those that were horizontal (i.e., parallel to the axis) would soon build up, stimulating more electrons each trip, until they were amplified billions of times. If the laser rod were a multiple (n) of the wavelength of the emitted light in length, a powerful standing wave of coherent light would be built up. In practice, the entire build-up takes place in a fraction of a second after the pumping light is applied, and at the end of this interval a powerful beam of coherent red light shoots out of the semitransparent end of the laser. The beam emerges in a narrow cone, parallel to the axis of the rod and mirror system. Operation of the laser is, of course, discontinuous, consisting of a single light pulse at a time.

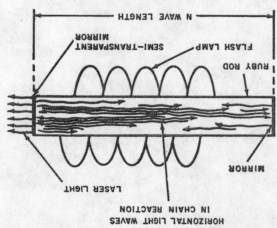

MIRROR
RUBY ROD
FLASH LAMP
SEMI-TRANSPARENT MIRROR
N WAVE LENGTH
HORIZONTAL LIGHT WAVES IN CHAIN REACTION
LASER LIGHT

Fig. 227. The First Ruby Laser

Since the time in 1960 Dr. Maiman demonstrated the first ruby laser, laser action has been observed in a variety of substances, such as neodymium glass and calcium-fluoride doped with rare earth atoms, gallium arsenide, etc. Moreover, within a few months continuous lasers, though of much lower power, were devised, consisting of a mixture of neon and helium gases. The continuous gas laser looks much like an ordinary neon tube. More recently, a solid-state injection laser has been invented, in which laser action can be turned on and off at will.